AAPG TREATISE OF PETROLEUM GEOLOGY

The American Association of Petroleum Geologists
gratefully acknowledges and appreciates the leadership and support
of the AAPG Foundation in the development of the
Treatise of Petroleum Geology.

STRUCTURAL TRAPS I
TECTONIC FOLD TRAPS

COMPILED BY
EDWARD A. BEAUMONT
AND
NORMAN H. FOSTER

TREATISE OF PETROLEUM GEOLOGY
ATLAS OF OIL AND GAS FIELDS

PUBLISHED BY
THE AMERICAN ASSOCIATION OF PETROLEUM GEOLOGISTS
TULSA, OKLAHOMA 74101, U.S.A.

ISBN 0-89181-580-5
ISSN 1043-6103

Available from:
The AAPG Bookstore
P.O. Box 979
Tulsa, OK 74101-0979

Phone: (918) 584-2555
Telex: 49-9432
FAX: (918) 584-0469

Association Editor: Susan Longacre
Science Director: Gary D. Howell
Publications Manager: Cathleen P. Williams
Special Projects Editor: Anne H. Thomas
Science Staff: William G. Brownfield
Project Production: Custom Editorial Productions

Table of Contents

TREATISE OF PETROLEUM GEOLOGY
ADVISORY BOARD

Edward C. Roy, Jr.
Eric A. Rudd
Floyd F. Sabins, Jr.
Nahum Schneidermann
Peter A. Scholle
George L. Scott, Jr.
Robert T. Sellars, Jr.
Faroog A. Sharief
John W. Shelton
Phillip W. Shoemaker
Synthia E. Smith
Robert M. Sneider
Stephen A. Sonnenberg
William E. Speer
Ernest J. Spradlin
Bill St. John
Philip H. Stark
Richard Steinmetz
Per R. Stokke
Denise M. Stone
Donald S. Stone
Doug Strickland
James V. Taranik
Harry Ter Best, Jr.
Bruce K. Thatcher, Jr.
M. Ray Thomasson
Jack C. Threet
Bernard Tissot
Donald F. Todd
M. O. Turner
Peter R. Vail
B. van Hoorn
Arthur M. Van Tyne
Ian R. Vann
Harry K. Veal*
Steven L. Veal
Richard R. Vincelette
Cecil von hagen
Fred J. Wagner, Jr.
William A. Walker, Jr.
Anthony Walton
Douglas W. Waples
Harry W. Wassall, III
W. Lynn Watney
N. L. Watts
Koenradd J. Weber
Robert J. Weimer
Dietrich H. Welte
Alun H. Whittaker
James E. Wilson, Jr.
John R. Wingert
Martha O. Withjack
P. W. J. Wood
Homer O. Woodbury
Walter W. Wornardt
Marcelo R. Yrigoyen
M. A. Yukler
Mehmet A. Yukler
Robert Zinke

* Deceased

American Association of Petroleum Geologists Foundation
Treatise of Petroleum Geology Fund*

Major Corporate Contributors
($25,000 or more)

Chevron Corporation
Mobil Oil Corporation
Oryx Energy Company
Pennzoil Exploration and Production Company
Shell Oil Company
Union Pacific Foundation

Other Corporate Contributors
($5,000 to $25,000)

Cabot Energy Corporation
Canadian Hunter Exploration Ltd.
Conoco Inc.
Marathon Oil Company
The McGee Foundation, Inc.
Phillips Petroleum Company
Texaco Philanthropic Foundation
Transco Energy Company

Major Individual Contributors
($1,000 or more)

C. Hayden Atchison
Richard R. Bloomer
A. S. Bonner, Jr.
David G. Campbell
Herbert G. Davis
Paul H. Dudley, Jr.
Lewis G. Fearing
James A. Gibbs
George R. Gibson
William E. Gipson
Robert D. Gunn
Merrill W. Haas
Cecil V. Hagen
Frank W. Harrison
William A. Heck
Roy M. Huffington
Harrison C. Jamison
Thomas N. Jordan, Jr.
Hugh M. Looney
Jack P. Martin
John W. Mason
George B. McBride
Dean A. McGee
John R. McMillan
Grover E. Murray
Rudolf B. Siegert
Robert M. Sneider
Jack C. Threet
Charles Weiner
Harry Westmoreland
James E. Wilson, Jr.

The Foundation also gratefully acknowledges the many who have supported this endeavor with additional contributions.

*Contributions received as of January 10, 1990.

PREFACE

THE ATLAS OF OIL AND GAS FIELDS AND THE TREATISE OF PETROLEUM GEOLOGY

The *Treatise of Petroleum Geology* was born during a discussion we had at the 1984 Annual AAPG Meeting in San Antonio, Texas. Our discussion led us to the conviction that we should write a state-of-the-art textbook in petroleum geology, aimed not at the student, but at the practicing petroleum geologist. The project to put together one textbook gradually evolved into a series of three different sets of publications: the Reprint Series, the Atlas of Oil and Gas Fields, and the Handbook of Petroleum Geology; collectively these publications are known as the *Treatise of Petroleum Geology*. The Treatise is one of the Diamond Jubilee projects commemorating AAPG's 75th anniversary in 1991.

Together with input from the Advisory Board of the Treatise of Petroleum Geology, we designed this entire effort so that the set of publications will represent the cutting edge in petroleum exploration knowledge and application: the Reprint Series to provide useful and important published literature, the Atlas as a collection of detailed field studies to illustrate the various ways oil and gas are trapped, and the Handbook as a professional explorationist's guide to the latest knowledge in the various areas of petroleum geology and related fields.

The Atlas is part of AAPG's long tradition of publishing field studies. Notable AAPG field study compilations include *Structure of Typical American Fields*, published in 1929 and edited by Sidney Powers; and the more recent *Memoir 30—Giant Fields of 1968-1978*, published in 1981 and edited by Michel T. Harbouty. The Treatise Atlas continues that tradition but in a different way: Papers in this Atlas follow a format meant to make access to particular data easier. We also intend for publication of papers following this format, or a similar one, to become a tradition, with many field studies published as a part of the Atlas in subsequent years.

Hundreds of geologists from all parts of the industry and the world participated in this first compilation of the Atlas. We gratefully acknowledge the generous contribution they have given of their time, resources, and knowledge. Some of the field study authors were directly involved in the discovery of the fields they describe.

PURPOSE OF THE ATLAS

The Atlas is designed to help geologists become more efficient explorers and developers of oil and gas fields by making them aware of the myriad ways oil and gas are trapped. The Atlas will be a primary source for locating much of the information necessary for creating prospects. Over the years the Atlas will augment the *AAPG Bulletin* by serving as a repository for descriptions of oil and gas fields whose petroleum geology can be reinterpreted as technology evolves.

The primary tool of the explorationist is imagination. Wallace E. Pratt remarked that the unfound field must first be sought in the mind. What is imagined is based on what is remembered. In other words, memory is the direct link to what is created in the mind. To create ideas that lead to the discovery of new fields, the mind of the geologist builds from its knowledge of petroleum geology.

Next to the firsthand experience of having prospects tested with the drill bit, studying developed fields is the best way for the geologist to load the mind with information needed to create plays and prospects. In addition, being familiar with the many ways oil and gas are trapped allows the geologist to get beyond the noise inherent in all exploration data and to close gaps in the data. Before one can find new fields, one must understand how oil and gas are generated, how they migrate, and how they are trapped. One must then be able to visualize these processes, using available data in prospective areas.

FORMAT OF THE ATLAS

To make data access easier, all field studies in the Atlas follow the same format. Once users become familiar with the format, they will know where to look in any of the field studies for the information they seek. Different fields from different parts of the world can be compared and contrasted easily.

The following is a general format outline for field studies in the Atlas:

- Location
- History
  - Pre-Discovery
  - Discovery
  - Post-Discovery
- Discovery Method
- Structure
  - Tectonic History
  - Regional Structure
  - Local Structure
- Stratigraphy
- Trap
  - General Description
  - Reservoir(s)
  - Source(s)
- Exploration Concepts

CRITERIA FOR INCLUDING A FIELD

Fields described in the Atlas were selected using two main criteria: (1) trap type, and (2) geographic distribution. Our ultimate goal is to have a field study from each major petroleum-producing province and to have an example of all known trap types. Neither size nor economic importance is, of itself, a criterion. Many fields that are not giants are included because they are geologically unique, or because they are outstanding examples of geological detective work and original thinking, or because they are important historically and led to the discovery of many other fields.

GROUPING OF FIELDS INTO SEPARATE VOLUMES

We considered several ways to group fields in these volumes. One obvious way was by geography. Other ways included by reservoir rock type, basin type, and trap type. We chose trap type because the purpose of the Atlas is to make explorationists more effective oil and gas trap finders, regardless of where they look.

Grouping oil and gas field studies into separate volumes by trap type is a difficult exercise. We decided to group the fields into volumes by designating them as structural or stratigraphic traps. Most traps are a combination of structure and stratigraphy. Some traps are obviously more a consequence of one than the other; however, many are not. The continuum that exists between purely stratigraphic and purely structural traps makes grouping some fields into a particular volume difficult. A further complication is that many fields contain more than one trap type.

Our criterion for choosing the stratigraphic or structural classification for a field is simply this: would the field be there if one of those factors were not? The trap is structural if structure is the primary reason for its existence, stratigraphic if it exists because of stratigraphy. If enough papers were available to subdivide the trap type further into something more specific, such as anticlines, we grouped papers into that subdivision.

PAPERS SELECTED FOR *STRUCTURAL TRAPS I: TECTONIC FOLD TRAPS*

This first book in the *Atlas of Oil and Gas Fields* series contains studies of fields that exist because of the presence of an anticline; without the anticline there would be no trap. The fields described in this volume illustrate the complex nature of a trap type that some explorationists mistakenly regard as simple and therefore not worthy of close scrutiny. However, if one looks closely enough he or she will find that fields with anticlinal traps are like all other fields—each has its own peculiar personality.

Oil and gas traps exist because many factors came together in one place. Considering all of the geologic factors that must occur in both time and space, it is a wonder that traps exist at all. When you read the papers contained in this volume, notice how similar—yet different—these anticlinally trapped fields are. For example, many of the fields in this volume are sourced from continentally derived organic matter. All have sandstone reservoirs. Most have shale seals, but one, Sarir, has an evaporite seal. Many are faulted anticlines. Some of the anticlines formed in a compressional tectonic setting, whereas others formed in a tensional tectonic setting. How do these variables affect the amount and distribution of the oil and gas?

One of the most important aspects in each study is the history of exploration. What seems obvious today usually was not obvious when the fields were found. Sometimes what was expected was not what was encountered. People discovered these fields by creating concepts based on a limited amount of information. Information at their disposal was limited by the technology available at the time. Drilling and discovery show how closely concept matches reality. Knowing the history of discovery may help explorationists realize that problems, seemingly insoluble at one time, eventually were solved. It also is instructive to learn about the sequence of thinking that solved these problems.

Authors have been asked to reflect upon the thought processes and techniques used in each discovery and to suggest how one might solve the problem in a better way if faced with the same circumstances today. In other words, what lessons does this particular hydrocarbon accumulation teach us, so that we can all become better explorationists?

Study these fields carefully to become a better creator of prospect concepts, and enjoy learning about the fascinating science of petroleum geology.

Good hunting!

Edward A. Beaumont and Norman H. Foster, Editors

Maui Field

W. O. ABBOTT
Occidental International Exploration and Production Company
Bakersfield, California

FIELD CLASSIFICATION

BASIN: Taranaki
BASIN TYPE: Interarc Basin
RESERVOIR ROCK TYPE: Sandstone
RESERVOIR ENVIRONMENT OF DEPOSITION: Coastal and Paralic Coal Basin
RESERVOIR AGE: Upper Eocene
PETROLEUM TYPE: Gas and Condensate
TRAP TYPE: Anticline

LOCATION

Maui field is of interest because of its isolation from other world-giants and its uniqueness as a hydrocarbon accumulation in New Zealand. The nearest big fields are in Bass Strait, Australia, more than 2000 km to the west. Maui field is located offshore approximately 85 km southwest of New Plymouth, North Island, New Zealand (Figure 1). It is situated near the center of the Taranaki basin on the eastern side of the relatively stable Western platform, adjacent to the South Taranaki graben along the Cape Egmont fault system in an area of extensional tectonics (Figures 2 and 3). Water depths on the Tasman Sea continental shelf vary from 104 to 112 m (340–365 ft) in the field confines. The discovery well coordinates for Maui field are 39°40′15″S; 173°18′35″E.

Maui field is the largest offshore commercial field found to date in New Zealand. Estimated ultimate recovery is 5.3 tcf of gas, 230 million bbl natural gas liquids (NGL) and 300 million bbl of oil. It is ranked as the 254th largest in the compilation by Carmalt and St. John (1986). There are two commercial onshore fields (Kapuni and McKee) and one noncommercial field (Motorua). In addition, the Kupe South area is presently being evaluated as a potential gas condensate field (Figure 2).

HISTORY

Pre-Discovery

Several oil seeps and numerous gas seeps have been found in New Zealand, and they attracted attention to the country's oil and gas potential early in the nineteenth century. Exploration began in New Zealand as early as 1839 with the digging of pits near oil seeps. The first exploratory well was spudded near New Plymouth (Figure 1) in 1865 in the Taranaki basin. This shallow onshore well, located 80 km northeast of Maui field, discovered a minor amount of oil and is credited with the discovery of Motorua field in 1866. The Motorua field discovery is also attributed to wells drilled in 1904 and 1934; it was in the latter year that continuous but small production was established. At abandonment in 1972, Motorua had produced only about 216,000 bbl of oil.

Shell-B.P.-Todd began exploration in New Zealand in 1955. This led to the Kapuni field strike in 1959, a recoverable reserve of 991 BCFG and 41.3 million bbl NGL. Extensive marine surveys were started in 1965 by Shell-B.P.-Todd, with an aeromagnetic survey and both surface and helicopter gravity work. This program revealed evidence of structures and depocenters in the Taranaki bight and was followed by reconnaissance and then detailed reflection and refraction seismic work. An offshore oil seep in Taranaki bight reportedly was the focal point of some of the marine geophysical surveying and contributed to the discovery of Maui field. Three onshore Taranaki basin wells and six Wanganui basin wells were drilled in the interim between the Kapuni and Maui discoveries. All were dry holes.

Discovery

Oil and gas exploration had been underway for 130 years in New Zealand when the 5.3 tcf to 8.4 tcf offshore giant Maui field was discovered in 1969. The discovery well, Maui #1, was drilled 51 km offshore by Discovery II in 111 m of water by Shell, B.P., and Todd (Figures 3 and 4). The Maui structure is a broad, low relief feature with two main culminations set on the eastern edge of the Western platform. Maui #1 tested the southernmost closure. The principal objective of the discovery well was the upper part of the Kapuni Group sequence at 2691 m, which had hydrocarbon indications (Figure 5). Further indications were obtained at 2991.6 m. The well was drilled to 3511 m (T.D.) where it penetrated conglomerates near basement, and operations were suspended on 26 March 1969.

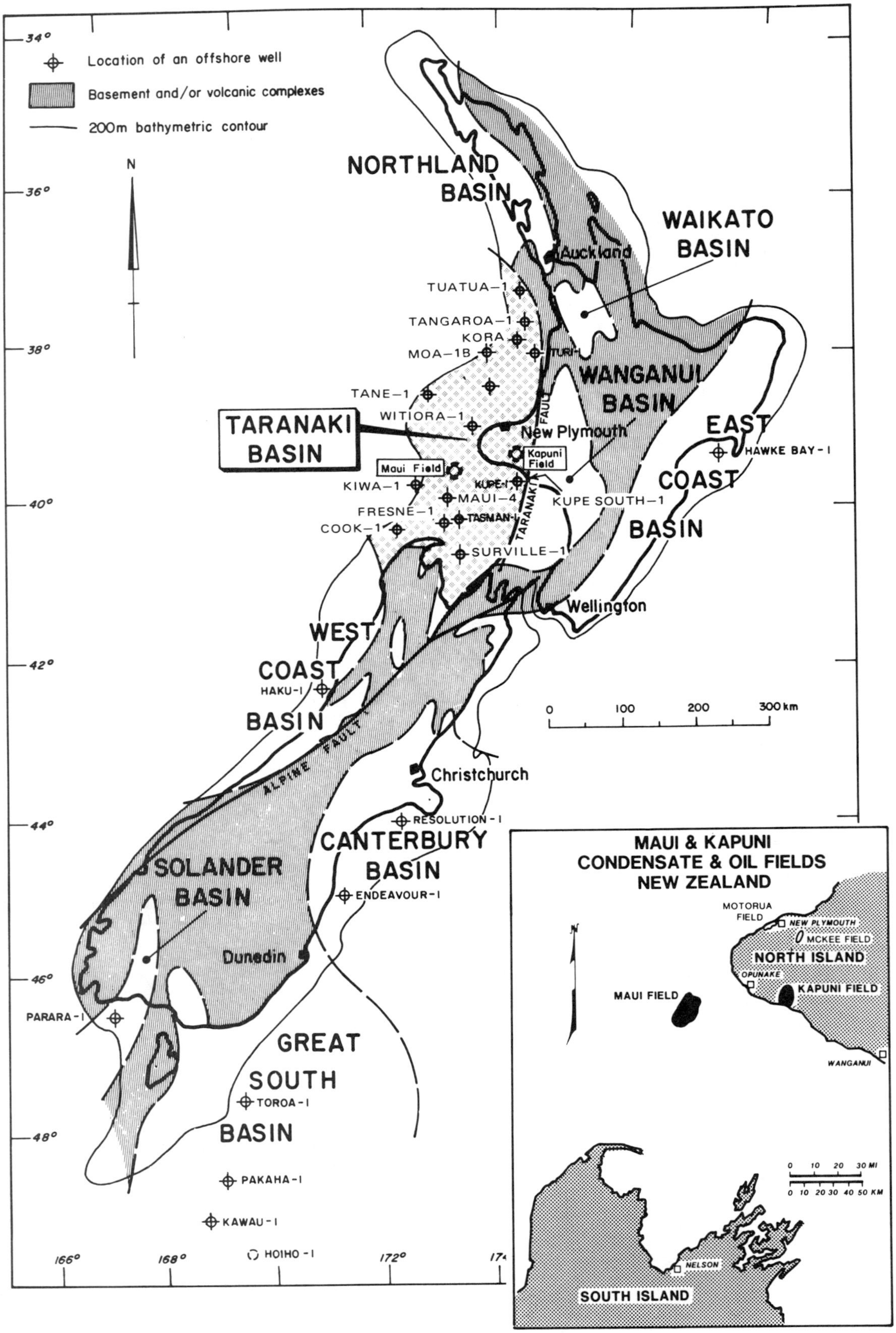

Location of an offshore well
Basement and/or volcanic complexes
200m bathymetric contour
N
NORTHLAND BASIN
WAIKATO BASIN
TUATUA-1
TANGAROA-1
KORA
MOA-1B
TURI-1
TANE-1
WITIORA-1
TARANAKI BASIN
WANGANUI BASIN
New Plymouth
Kapuni Field
Maui Field
KIWA-1
KUPE-1
MAUI-4
FRESNE-1
TASMAN-1
COOK-1
SURVILLE-1
KUPE SOUTH-1
TARANAKI FAULT
EAST COAST BASIN
HAWKE BAY-1
Auckland
Wellington
WEST COAST BASIN
HAKU-1
ALPINE FAULT
Christchurch
RESOLUTION-1
CANTERBURY BASIN
ENDEAVOUR-1
SOLANDER BASIN
Dunedin
PARARA-1
GREAT SOUTH BASIN
TOROA-1
PAKAHA-1
KAWAU-1
HOIHO-1
0 100 200 300 km
34°
36°
38°
40°
42°
44°
46°
48°
166°
168°
172°
MAUI & KAPUNI CONDENSATE & OIL FIELDS NEW ZEALAND
MOTORUA FIELD
NEW PLYMOUTH
MCKEE FIELD
NORTH ISLAND
OPUNAKE
KAPUNI FIELD
MAUI FIELD
WANGANUI
0 10 20 30 MI
0 10 20 30 40 50 KM
NELSON
SOUTH ISLAND

Figure 1. Index map and sedimentary basins in New Zealand (after Pilaar and Wakefield, 1984).

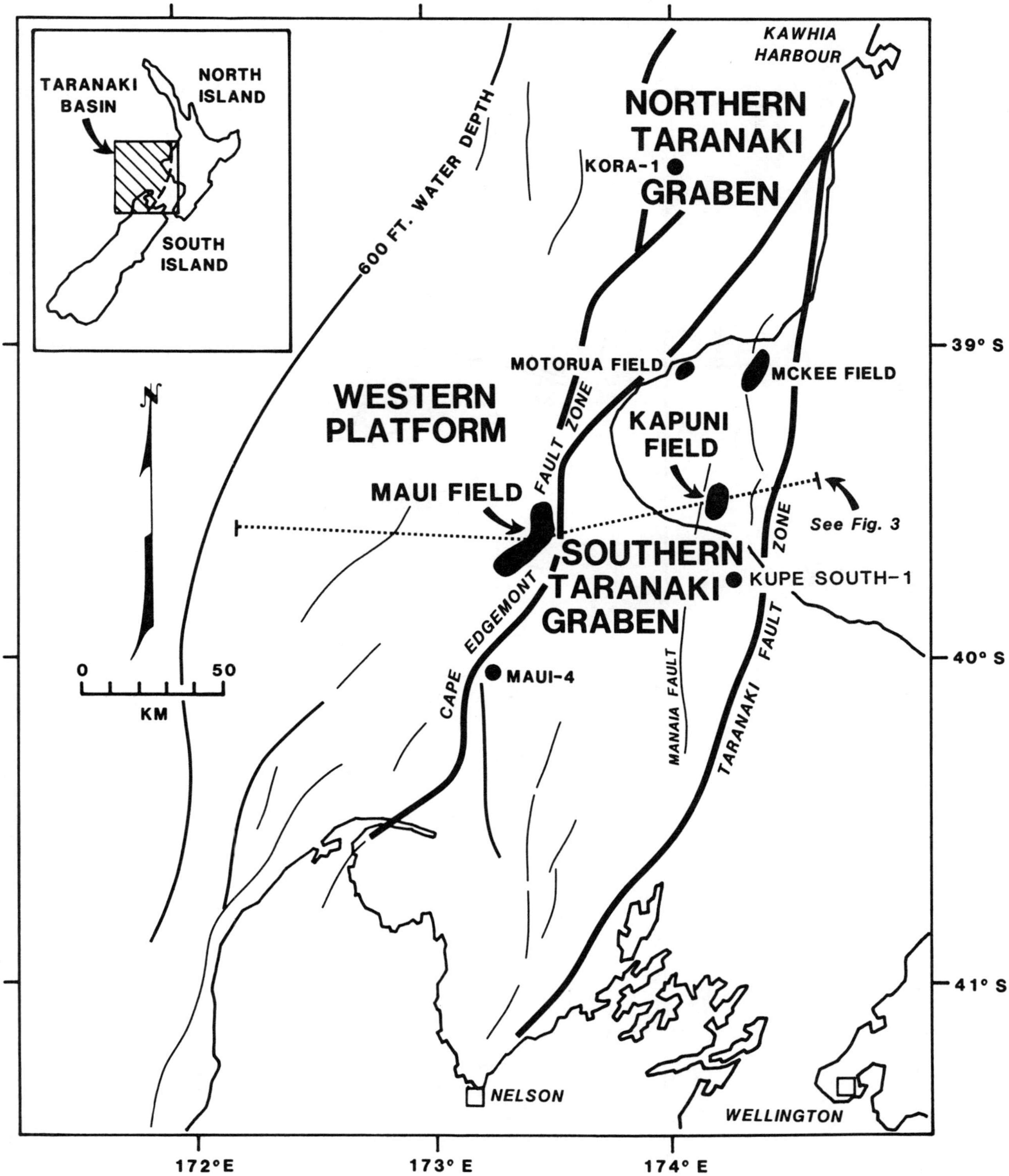
TARANAKI BASIN
NORTH ISLAND
SOUTH ISLAND
KAWHIA HARBOUR
NORTHERN TARANAKI GRABEN
KORA-1
600 FT. WATER DEPTH
MOTORUA FIELD
MCKEE FIELD
WESTERN PLATFORM
KAPUNI FIELD
MAUI FIELD
FAULT ZONE
See Fig. 3
SOUTHERN TARANAKI GRABEN
KUPE SOUTH-1
CAPE EDGEMONT
MAUI-4
MANAIA FAULT
TARANAKI FAULT ZONE
0
50
KM
NELSON
WELLINGTON
39° S
40° S
41° S
172°E
173° E
174° E

Figure 2. Taranaki basin, major structural elements (after King and Robinson, in press).

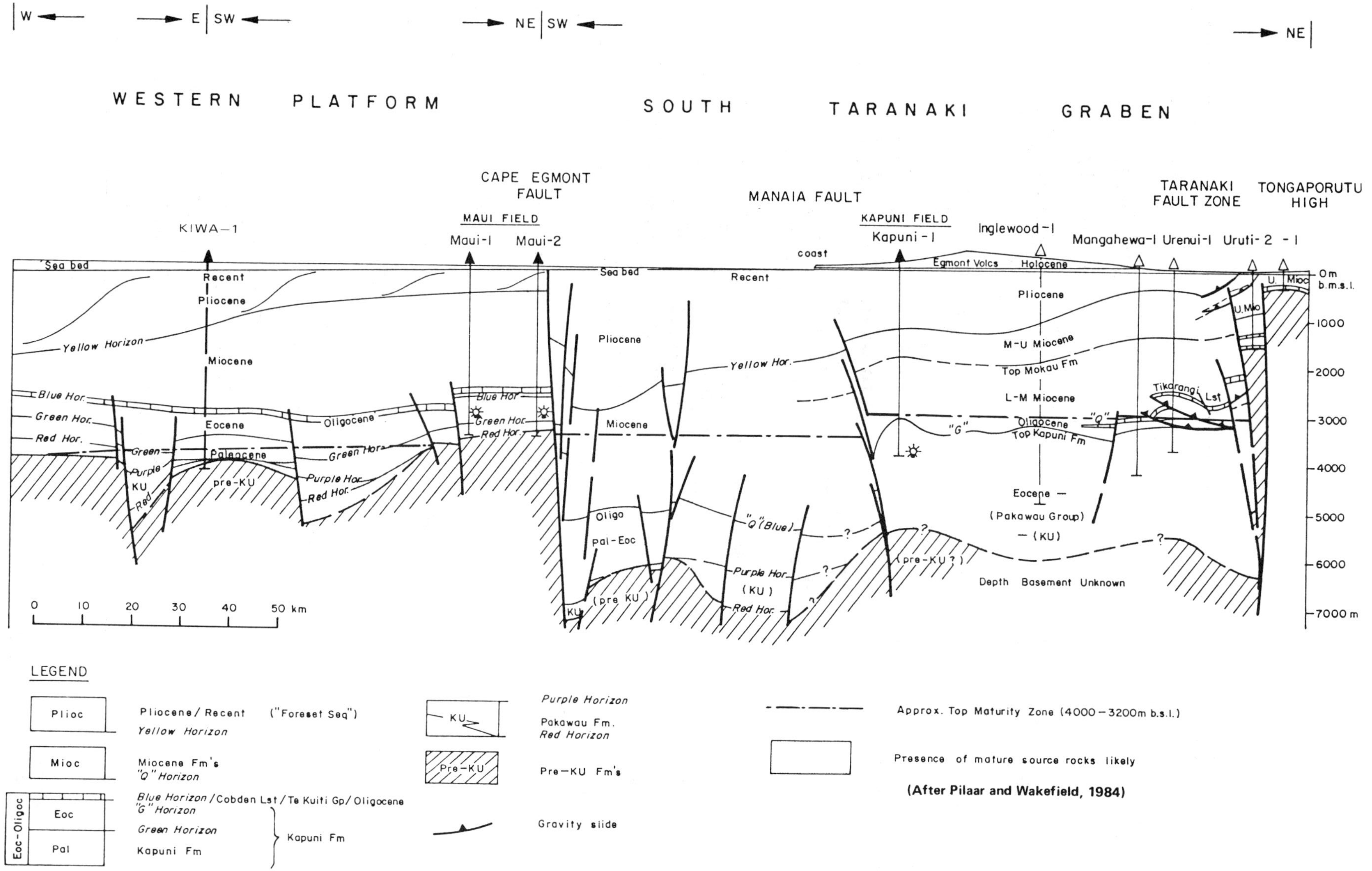
W
E | SW
NE | SW
NE
WESTERN PLATFORM
SOUTH TARANAKI GRABEN
CAPE EGMONT FAULT
MANAIA FAULT
TARANAKI FAULT ZONE
TONGAPORUTU HIGH
KIWA-1
MAUI FIELD
Maui-1 Maui-2
KAPUNI FIELD
Kapuni-1
Inglewood-1
Mangahewa-1 Urenui-1 Uruti-2 -1
coast
Sea bed
Recent
Pliocene
Miocene
Eocene
Paleocene
pre-KU
Oligocene
Yellow Horizon
Blue Hor.
Green Hor.
Red Hor.
Purple Hor.
Egmont Volcs
Holocene
M-U Miocene
Top Mokau Fm
L-M Miocene
Top Kapuni Fm
Tikorangi Lst
Eocene
(Pakawau Group)
(KU)
Depth Basement Unknown
Oligo
Pal-Eoc
"Q"(Blue)
"G"
U. Mioc
U.Mio
0m b.m.s.l.
1000
2000
3000
4000
5000
6000
7000 m
0 10 20 30 40 50 km
LEGEND
Plioc
Pliocene/Recent ("Foreset Seq")
Yellow Horizon
Mioc
Miocene Fm's
"Q" Horizon
Eoc-Oligoc
Eoc
Pal
Blue Horizon/Cobden Lst/Te Kuiti Gp/Oligocene
"G" Horizon
Green Horizon
Kapuni Fm
Kapuni Fm
KU
Purple Horizon
Pakawau Fm.
Red Horizon
Pre-KU
Pre-KU Fm's
Gravity slide
Approx. Top Maturity Zone (4000-3200m b.s.l.)
Presence of mature source rocks likely
(After Pilaar and Wakefield, 1984)

Figure 3. Taranaki basin, structural cross-section. See Figure 2 for location. (After Pilaar and Wakefield, 1984.)

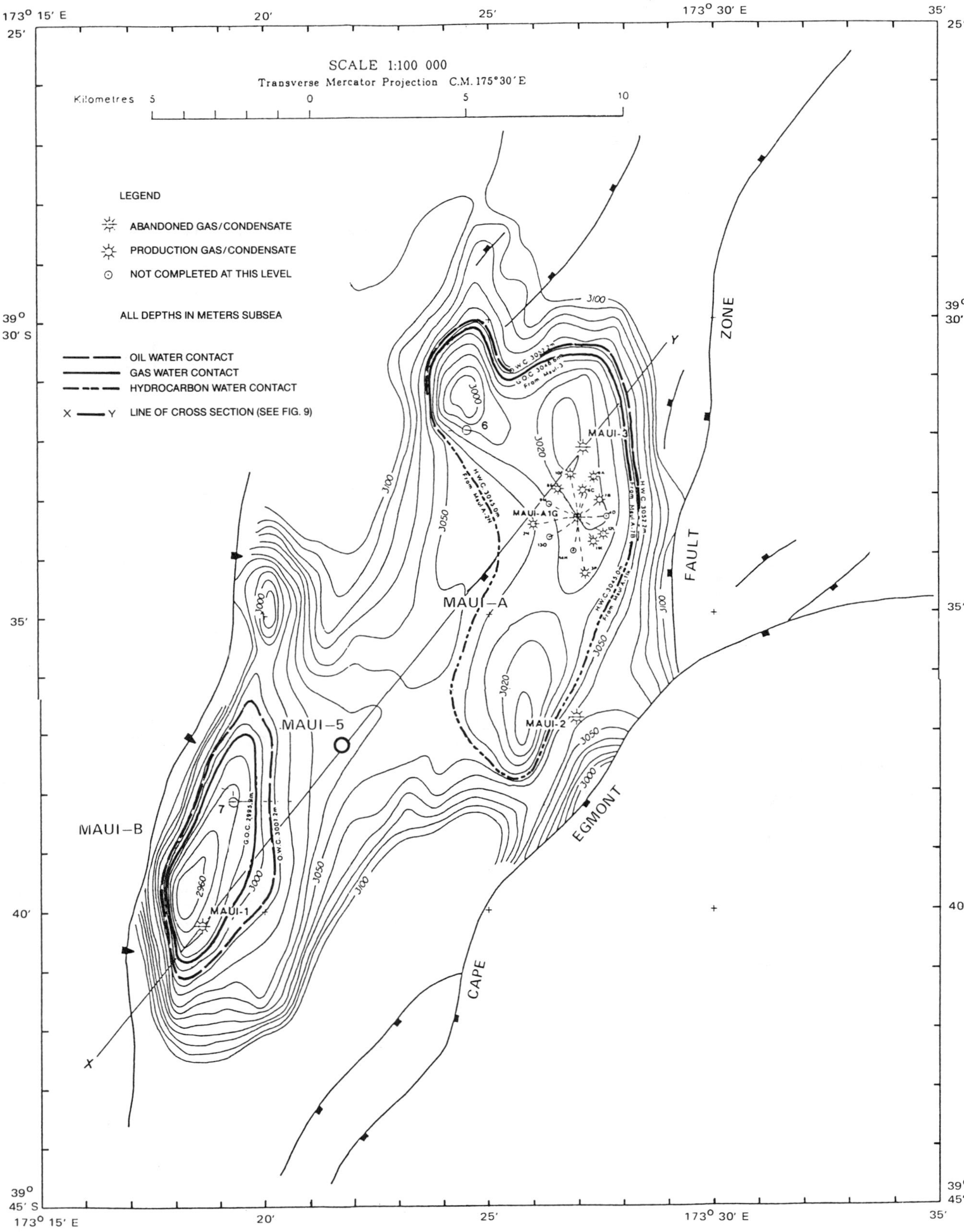
173° 15′ E
20′
25′
173° 30′ E
35′
25′
39° 30′ S
35′
40′
39° 45′ S
SCALE 1:100 000
Transverse Mercator Projection C.M. 175°30′E
Kilometres 5 0 5 10
LEGEND
ABANDONED GAS/CONDENSATE
PRODUCTION GAS/CONDENSATE
NOT COMPLETED AT THIS LEVEL
ALL DEPTHS IN METERS SUBSEA
OIL WATER CONTACT
GAS WATER CONTACT
HYDROCARBON WATER CONTACT
X Y LINE OF CROSS SECTION (SEE FIG. 9)
MAUI-A
MAUI-B
MAUI-1
MAUI-2
MAUI-3
MAUI-5
MAUI-A1G
CAPE
EGMONT
FAULT
ZONE
X
Y

Figure 4. Top Kaimiro Formation (D_1 sand) migrated depth contours. Contour interval, 10 m. (After Haskell, 1986.)

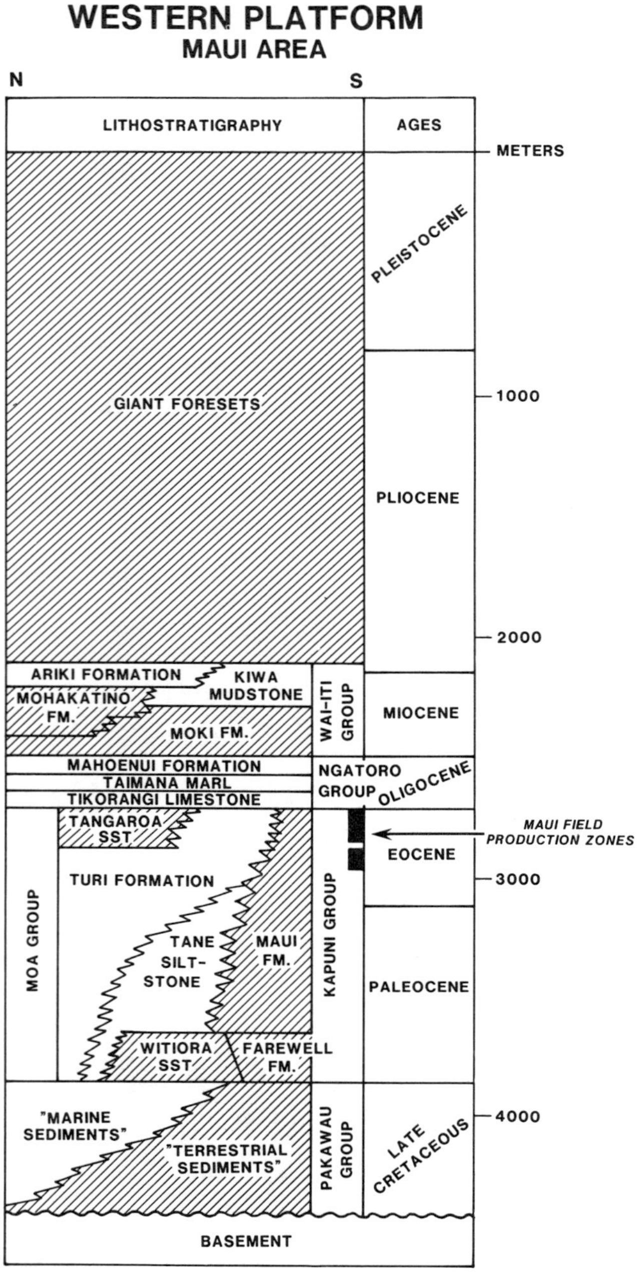
WESTERN PLATFORM
MAUI AREA
N
S
LITHOSTRATIGRAPHY
AGES
METERS
PLEISTOCENE
GIANT FORESETS
1000
PLIOCENE
2000
ARIKI FORMATION
KIWA
MUDSTONE
MOHAKATINO
FM.
MOKI FM.
WAI-ITI GROUP
MIOCENE
MAHOENUI FORMATION
TAIMANA MARL
TIKORANGI LIMESTONE
NGATORO
GROUP
OLIGOCENE
TANGAROA
SST
MAUI FIELD
PRODUCTION ZONES
EOCENE
TURI FORMATION
3000
MOA GROUP
TANE
SILT-
STONE
MAUI
FM.
KAPUNI GROUP
PALEOCENE
WITIORA
SST
FAREWELL
FM.
"MARINE
SEDIMENTS"
"TERRESTRIAL
SEDIMENTS"
PAKAWAU GROUP
LATE CRETACEOUS
4000
BASEMENT

Figure 5. Generalized stratigraphic section, Western platform. Note Maui field production zones. (After Robinson and King, in press.)

Post-Discovery

The confirmation well (Maui #2) was drilled about 15 km to the northeast soon after the Maui #1 discovery. Of the three gas wells drilled in the initial program, Maui #3 is the best producer. The Maui #5, #6, and #7 wells were drilled as part of the field development program (Figure 4).

Production platforms on Maui A and B structures were installed in mid-1976, and the field was put on production in 1978. There are currently 14 production wells on the platform located on the Maui A structure.

Maui #4 well (in the western portion of the South Taranaki graben) was drilled on what later was determined to be a separate feature 40 km south-southwest of the discovery well (Figure 2). It encountered oil in a different pay than the gas-bearing zones of the Maui field, but the 600 BOPD rate from a depth of about 3500 m was considered noncommercial.

Several dry exploratory wells have been drilled in the Taranaki basin since Maui field was discovered. Two recent significant tests in the offshore Taranaki basin have sparked renewed interest in the area. The TCPL Resources, Ltd., 3B Kupe South appraisal well flowed 10,000 bbl/d during three drill-stem tests. It is located 60 km southeast of the Maui field in the South Taranaki graben. The Kora #1 well drilled by Arco Petroleum New Zealand Inc., 115 km north-northeast of the Maui field in the North Taranaki graben, has reported possible commercial hydrocarbons.

DISCOVERY METHOD

After the discovery of the Kapuni field in August 1959, onshore exploration by the Shell-B.P.-Todd consortium continued. During this period, a further 768 km of seismic refraction profiles were shot in the Taranaki and Wanganui basins. An additional 310 km of marine seismic profile was obtained within the 4.8 km offshore limit. The nine exploratory wells drilled in these two basins were all dry. In 1966, the first significant offshore reconnaissance survey obtained 2200 km of four-fold reflection coverage. This showed two major fault blocks divided by the Cape Egmont fault zone. Three structural leads were outlined. In 1967, an aeromagnetic survey and a semidetailed marine seismic survey of 536 km were carried out.

Following the success of Maui #1, preparations were made to drill a number of appraisal wells. Maui #2 was drilled to resolve the previous positioning errors and further delineate the structure. In addition, a seismic survey with 24-fold coverage was carried out over the Maui structure.

Some of the most up-to-date methods of exploration and interpretation were employed in the development of the Maui field. The effective use of gravity and aeromagnetic surveys is worthy of emphasis. These geophysical methods provided the original definitions of the areal extent and depths to basement of the subbasins and grabens. The gravity method effectively located anomalies that were later detailed seismically as prospective traps for hydrocarbons. The quality of the seismic proved to be excellent. This fine-quality data, combined with well control, can be used in the future to define source, seal, and reservoir to locate subtle stratigraphic traps as well as to delineate other structural plays.

STRUCTURE

Tectonic History

During late Paleozoic through Jurassic time, the New Zealand geosyncline was rapidly subsiding and filling with turbidites. The geosyncline essentially was filled, then uplifted and partially stripped during the Rangitata orogeny in Late Jurassic and Early Cretaceous time. New Zealand began to separate from Australia in the Late Cretaceous. At this time, the proto-Taranaki basin existed as the southeastern-most limit of the New Caledonia basin, which extended as far as the North Island of New Zealand (Figure 6). In response to the Tasman Sea rifting, a series of north–south-trending subbasins and half-grabens developed in the Taranaki region. Subsidence in these depocenters was initially rapid but gradually slowed as the rate of crustal cooling diminished. The Eocene and early Oligocene was marked by a period of tectonic stability interrupted in late Oligocene by the onset of rapid subsidence in the Taranaki region (Hayward, 1987). This event is attributed to oblique extension and westerly downthrow on the Taranaki fault, which in turn is probably related to the development of a transform plate boundary through New Zealand and general foundering of the New Zealand platform. Increasing oblique compression on the plate boundary caused several episodes of fault reversal and subbasin inversion.

Emplacement of low-angle overthrusts along the eastern margin of the Taranaki basin occurred from early to late Miocene time. Thereafter, the South Taranaki graben region continued to subside while the area south of the graben experienced pronounced uplift.

Conversely, to the north, extensional tectonics dominated, and movement along an enechelon series of antithetic normal faults produced the North Taranaki graben.

The Western platform remained a relatively stable block throughout most of the Tertiary and was affected only by upper Cretaceous to Eocene normal block faulting.

Increased convergence along the plate boundary during Pleistocene time resulted in the Taranaki Peninsula region being uplifted and tilted. Volcanic extrusions were emplaced during this period.

Regional Structure

New Zealand is a fragmented block of continental crust isolated by the Tasman Sea from its parent continent, Australia, and is in an active seismic-volcanic belt.

The region northwest of the Alpine fault (Figure 1) and west of the Taranaki fault, which includes the Taranaki basin, is a separate tectonic province from southeastern New Zealand. This northwestern region has generally been more stable in the postrift period than its southeastern counterpart, remaining essentially fixed relative to the Challenger plateau and Lord Howe rise. The subsidence and sedimentary history of the Taranaki basin reflects three predominant tectonic regimes: extensional, obliquely extensional, and obliquely compressional. These influences evolved in relation to the changing nature and position of the New Zealand continental block with respect to the Pacific and Indonesian–Australian plate boundary.

The Taranaki basin can be defined by the Taranaki fault and the continental shelf break (approximately 200 m isobath). To the north, the basin merges arbitrarily with the southwestward offshore portions of the Northland basin, while to the south, it overlaps basin and range provinces of the northwestern South Island. Two distinct elements of the Taranaki basin are recognized: the Taranaki graben to the east and the Western platform. The Cape Egmont fault zone separates these elements. The graben is divided into northern and southern segments, differentiated mainly by the greater depths of the northern graben. The two areas are separated by a complex zone of normal faults, downthrown to the northwest (Cook and King, 1987).

Two main fault trends are found within the basin: north-south and northeast-southwest. These tectonic trends control the structural configuration of the Taranaki basin. At their intersections, the north-south trend is usually offset by the younger, northeast-southwest oblique trend. This system of faults is interpreted as a first-order wrench system (Pilaar and Wakefield, 1984).

Local Structure

Maui field is located on a complex, multiple culmination anticline adjoining the major Cape Egmont fault zone on the Western platform. It is a complicated, north-plunging anticlinal fold. The Maui anticline has about 185 m of closure. It is about 33 km long and averages less than 10 km wide. Two main structural highs were tested by Maui #1 and Maui #3 (Figures 4, 9, and 10). Total area under closure is described as about 250 km^2, with about 200 km^2 of potentially productive area. The structure is *not* filled to reservoir capacity with oil and gas condensate. Maui anticline is believed to pre-date the Miocene, and the initial folding may have resulted from shearing caused by movement on the major Alpine–Waimea–Taranaki fault trend and subsidiary faults in the system.

Maui field is faulted on three sides, but only the Cape Egmont fault zone contributes significantly to the trapping mechanism. The trap type can be considered a faulted anticline. None of the literature suggests that stratigraphy, paleogeomorphology, or hydrodynamics play an important role in entrapment.

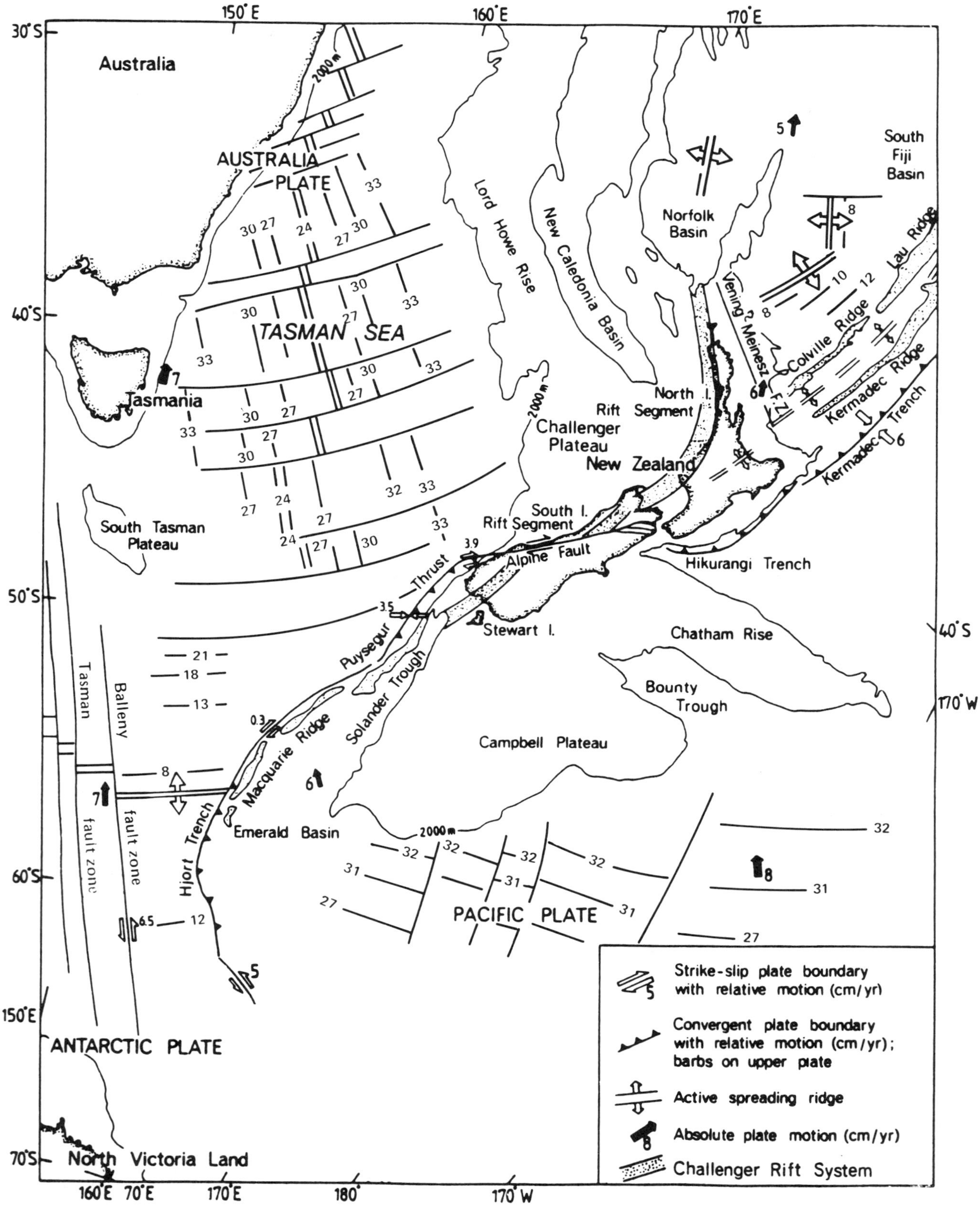

150°E
160°E
170°E
30°S
40°S
50°S
60°S
70°S
150°E
160°E
70°E
170°E
180°
170°W
40°S
170°W
Australia
AUSTRALIA PLATE
TASMAN SEA
Tasmania
South Tasman Plateau
Lord Howe Rise
New Caledonia Basin
Norfolk Basin
South Fiji Basin
Vening Meinesz F.Z.
Colville Ridge
Lau Ridge
Kermadec Ridge
Kermadec Trench
North I.
Rift Segment
Challenger Plateau
New Zealand
South I.
Rift Segment
Alpine Fault
Thrust
Puysegur
Hikurangi Trench
Stewart I.
Chatham Rise
Bounty Trough
Campbell Plateau
Solander Trough
Macquarie Ridge
Hjort Trench
Emerald Basin
Tasman fault zone
Balleny fault zone
PACIFIC PLATE
ANTARCTIC PLATE
North Victoria Land
2000m
Strike-slip plate boundary with relative motion (cm/yr)
Convergent plate boundary with relative motion (cm/yr); barbs on upper plate
Active spreading ridge
Absolute plate motion (cm/yr)
Challenger Rift System

Figure 6. Plate reconstruction and tectonic character of the Australia-Pacific plate boundary showing Alpine fault dislocation of the Challenger rift system through Western New Zealand. (Modified after Kamp, 1986.)

STRATIGRAPHY

A variety of Paleozoic through Jurassic basement rocks underlie the Taranaki basin. These rocks are of igneous, metasedimentary, and low-grade metamorphic rank. It has long been considered a tectonic province in which there are extreme lateral variations and vertical repetition of lithofacies. For clarity, therefore, the best convention used in discussing the stratigraphy is age rather than formation. The stratigraphy of the Western platform is illustrated in Figure 5. The stratigraphy for the entire Taranaki basin is illustrated in Figure 7.

Upper Cretaceous (Pakawau Group)

By Upper Cretaceous time, the topographic effects of an Early Cretaceous mountain-building episode in the proto-New Zealand region (Rangitata orogeny) had been considerably modified by subsequent erosion and, in places, peneplanation. In the Taranaki basin, a block-faulted subdued basin and range topography existed. Initial sediments (Pakawau Formation) were an onlapping sequence of terrestrial fluvio-lacustrine origin deposited in discrete, rapidly subsiding depocenters controlled by normal faulting. In places, uplift and erosion along the normal faults produced thick wedges of conglomerate and coarse sandstone. Where outcropping basement was less controlled by faulting, fringes of terrestrial sediments offlapping high areas occurred. Low-relief areas surrounded the Taranaki basin area, except in the north and northwest where a shallow sea had invaded. As the sea moved southward into fault-bounded embayments, paralic shoreline sands and then shelf muds were deposited (King and Robinson, in press). Lithofacies of the Pakawau Formation range from basal conglomerates and sandstones to alternating kaolinitic quartzose sandstones, carbonaceous mudstones, and humic coals. The coal measures are thought to be a hydrocarbon source for the Maui and Kapuni fields.

Paleocene-Eocene (Kapuni Group)

During the Paleocene, deposition of the Pakawau coal measures continued in the depocenters in the southern and eastern portions of the basin. To the north, the shoreline position fluctuated, apparently advancing and retreating several times in response to local tectonics and/or eustatic controls. Sheet-like sand bodies were deposited over, and in turn overlain by, shelf muds. By late Paleocene time, relative sea-level rise exceeded rate of sediment supply, and a transgression from the northwest took place. The sediments laid down were well-sorted glauconitic sandstones (Island Sandstone) overlain by massive brown-gray mudstones and fine sandy siltstones.

By middle-late Eocene times, renewed deposition of coal measures (Kapuni Formation) was evident. The Kapuni contains a lower unit that is dominated by quartzose sandstones and overlain by carbonaceous shales and coal-rich sequences. The sandstones become dominant toward the top of the Kapuni and finally are overlain by gray-brown calcareous mudstones. The Kapuni provides the reservoirs, seals, and probably the hydrocarbon source for the Maui field.

Oligocene (Ngatoro/Te Kuiti Group)

Gentle submergence continued through the early Oligocene, then subsidence rates increased dramatically and the whole basin deepened (Hayward, 1987). Carbonate deposition was widespread throughout the basin except in the south where proximity of source and shallower water depths produced a higher terrigenous component in the sediments.

Miocene (Wai-iti Group)

Subsidence continued, particularly adjacent to the Taranaki fault. At this time, the transgression reached its maximum extent, both areally and bathymetrically. Already, however, the effects of uplift in the hinterland and to the south and east were being felt in peripheral subbasins. Here, sediment supply exceeded subsidence to a degree that infilling led to terrestrial sedimentation and eventually to the development of the Taranaki graben complex. By contrast, the Western platform was not yet affected by tectonic activity and continued to receive fine-grained sediments rich in planktonic faunas.

Later, sediment supply increase began to have a profound effect on sedimentation patterns in the Taranaki marine basin. Carbonate content diminished and development of large submarine fans became widespread. These formed an actively prograding sedimentary wedge, the precursor of the modern continental shelf. Related to this, a net regressive sedimentary cycle began that has continued to the present day.

This pattern was complicated greatly along the northeast margin of the basin by the onset of oblique compression on the Taranaki fault.

A massive silty and marly mudstone sequence, the Ureni Formation of late Miocene age, marks the last transgressive phase in the Taranaki basin. During the same period, major volcanic activity occurred in the North Taranaki graben, where andesitic plugs intruded into the graben-fill and produced tuffaceous sequences (Pilaar, 1984).

Pliocene-Recent (Giant Foresets)

A widespread Pliocene erosion surface marks the climax of tectonic activity and basin inversion in the eastern and southern areas of the Taranaki basin, while over the Western platform the continental shelf prograded outward to the northwest toward its present limit, forming the Pliocene-Recent "giant foresets" sequence (Pilaar, 1982).

The increased tectonic development in the Taranaki basin and the southern uplift, in particular, coincided with the increased uplift in the hinterland further east and south. Huge quantities of sediment were supplied to the Taranaki basin; the rapidly subsiding North Taranaki graben was overfilled, and shelf progradation to the northwest was accelerated still further. As a result of this increased sediment influx, large channels (up to 5 km across) developed on the continental slope. Originating on the shelf edge, they acted as conduits for sediment redistribution to greater depths (King and Robinson, in press).

TRAP

The trap in the Maui field is a compound anticlinal trap (Figure 8). It owes its origin to movements on the Cape Egmont fault, which underwent compressional and extensional tectonics during Oligocene, Miocene, and Pliocene-Pleistocene time. The mudstones and shales of the Kapuni and Kaiata formations (Eocene) seal the reservoir sandstones. The C_1 shale seals the C_1 sandstone reservoir and the D_1 shale seals the D_1 reservoir. Shows in the overlying Mahoenui Formation suggest that the sealing capacity of the shales was not completely effective and that some gas is migrating up-section. It is noteworthy that the main hydrocarbon accumulations in the Western platform have been found northwest and adjacent to the Cape Egmont fault in the Maui field. This block is structurally high and stable relative to the Taranaki graben, which served as a hydrocarbon kitchen on the opposite side of the Cape Egmont fault. The five different oil-water contacts are indicated on Figure 9. Timing was excellent in that the sealed traps (Oligocene-lower Miocene) pre-date the migration by a considerable period. Time-temperature relationships are not entirely clear, but available data suggest a late expulsion, probably Pliocene, and a favorable thermal history for preservation of the accumulations.

RESERVOIR

Two major hydrocarbon-bearing intervals, characteristic of the entire field, are present in Maui #1. The C sand intervals have 122 m gross and 74 m net of gas pay and 2 m of net oil pay in quartzose sandstones. The D_1 interval has 60 m gross and 11 m net gas and 8 m net oil pay (Figure 9). The original deposits evidently were weathered and leached in nearshore channel and bar environments and thus are cleaner than the typically clay-rich, low-quartz, poorly sorted, low-porosity sandstone found in much of New Zealand. The Maui pay sandstones are medium to fine grained, angular to subangular and friable, with minor kaolinite common to the deeper sandstones. Intergranular porosity ranges up to 25% and permeability is described as good (0.2 to 2400 md). The reservoir drive is believed to be both pressure depletion and water influx.

The C_1 pay zone Includes two main structural culminations (Figures 8 and 9), at an average depth of 2680 m subsea, and the D_1 pay accumulation includes the same culminations at an average depth of about 3000 m (Figures 4 and 9). The C_1 zone contains about 88% of the gas reserves and about 80% of the condensate reserves in Maui field.

Although Maui field reservoir pressures have not been published, one of the problems the operators have had to contend with is high pressures in the lower Tertiary. It is not known whether the higher pressures exist in the reservoirs or shales, or both.

The reservoir properties for the Maui A Drilling Platform are summarized on Table 1 and the reserve estimates for the Maui field are shown on Table 2. A cross-section of the Maui A structure illustrating the reservoir intervals for the C and D sandstones is shown on Figure 11.

Siliciclastic diagenetic studies of the Pakawau and Kapuni Group reservoirs establish them to be chemically and mechanically of stable framework minerals. However, "early" diagenetic features that have had minor effect on the reservoir qualities are plagioclase corrosion and kaolinite neoformation. Kaolinite is the most common, early diagenetic clay mineral observed and is present to a burial depth of about 3 km, while illite and chlorite are formed later (>3 km). Quartz overgrowth has only been observed in samples from deeper than 3 km. Carbonate cemented horizons, although of relatively limited occurrences, have been observed over the entire depth range. Conversely, reservoir quality has been enhanced by secondary porosity development, through dissolution of carbonate cement and framework minerals, as well as by grain fracturing (Van der Lingen et al., in press). Good secondary porosity development, probably due to carbonate-cement dissolution, has been observed in the gas/condensate reservoir of the Maui field.

SOURCE

The Late Cretaceous and Eocene coal measure sequences are the only proven source rocks in the Taranaki basin, and their presence can be established over much of the basin from seismic and well data. Despite their common abundance of leaf cuticles, pollen, spores, and resins relative to woody material, the coals are classified as hydrogen-poor or gas-generative (Pilaar and Wakefield, 1984). However, high volatile perhydrous coals, which could be a source for oil, do occur in the West Coast basin of the South Island and might be found to occur in the Taranaki basin.

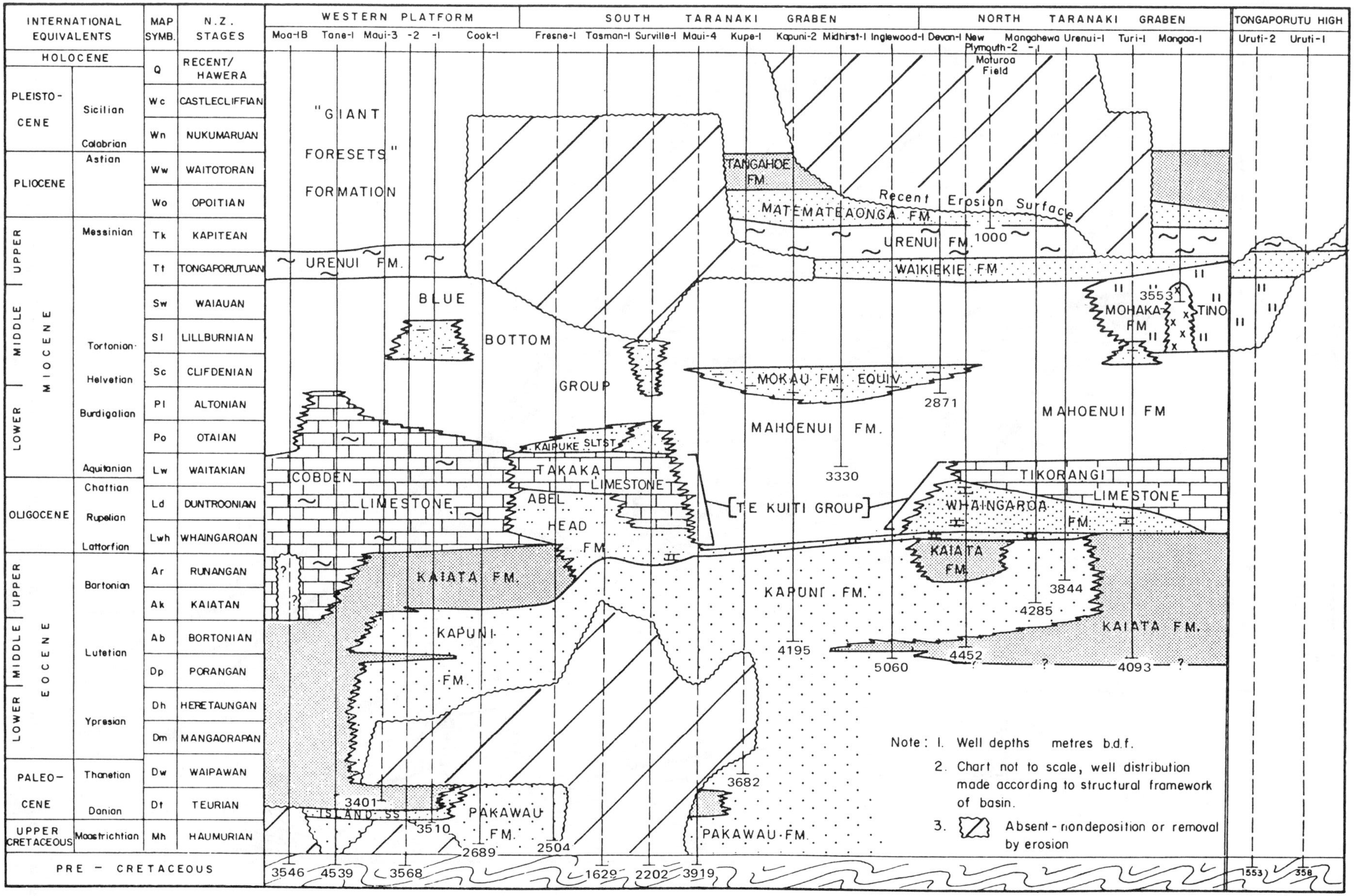
INTERNATIONAL EQUIVALENTS
MAP SYMB.
N.Z. STAGES
WESTERN PLATFORM
SOUTH TARANAKI GRABEN
NORTH TARANAKI GRABEN
TONGAPORUTU HIGH
Moa-1B Tane-1 Maui-3 -2 -1 Cook-1
Fresne-1 Tasman-1 Surville-1 Maui-4 Kupe-1 Kapuni-2 Midhirst-1 Inglewood-1 Devon-1 New Plymouth-2 Mangahewa -1 Urenui-1 Turi-1 Mangoa-1
Maturoa Field
Uruti-2 Uruti-1
HOLOCENE
PLEISTOCENE
PLIOCENE
UPPER MIDDLE LOWER MIOCENE
OLIGOCENE
UPPER MIDDLE LOWER EOCENE
PALEOCENE
UPPER CRETACEOUS
PRE - CRETACEOUS
Sicilian
Calabrian
Astian
Messinian
Tortonian
Helvetian
Burdigalian
Aquitanian
Chattian
Rupelian
Lattorfian
Bartonian
Lutetian
Ypresian
Thanetian
Danian
Maastrichtian
Q RECENT/HAWERA
Wc CASTLECLIFFIAN
Wn NUKUMARUAN
Ww WAITOTORAN
Wo OPOITIAN
Tk KAPITEAN
Tt TONGAPORUTUAN
Sw WAIAUAN
Sl LILLBURNIAN
Sc CLIFDENIAN
Pl ALTONIAN
Po OTAIAN
Lw WAITAKIAN
Ld DUNTROONIAN
Lwh WHAINGAROAN
Ar RUNANGAN
Ak KAIATAN
Ab BORTONIAN
Dp PORANGAN
Dh HERETAUNGAN
Dm MANGAORAPAN
Dw WAIPAWAN
Dt TEURIAN
Mh HAUMURIAN
"GIANT FORESETS" FORMATION
TANGAHOE FM
Recent Erosion Surface
MATEMATEAONGA FM
URENUI FM
URENUI FM
1000
WAIKIEKIE FM
BLUE BOTTOM GROUP
MOHAKA FM
TINO
3553
MOKAU FM EQUIV.
2871
MAHOENUI FM.
MAHOENUI FM
3330
KAIPUKE SLTST
TAKAKA LIMESTONE
COBDEN LIMESTONE
TIKORANGI LIMESTONE
TE KUITI GROUP
ABEL HEAD FM
WHAINGAROA FM
KAIATA FM
KAIATA FM.
KAIATA FM.
KAPUNI FM.
KAPUNI FM.
3844
4285
4195
4452
5060
4093
3401
ISLAND SS
3510
PAKAWAU FM.
PAKAWAU FM.
2689
2504
3682
3546 4539 3568 1629 2202 3919
1553 358
Note: 1. Well depths metres b.d.f.
2. Chart not to scale, well distribution made according to structural framework of basin.
3. Absent - nondeposition or removal by erosion

Figure 7. Taranaki basin, stratigraphic correlation chart (after Pilaar and Wakefield, 1984).

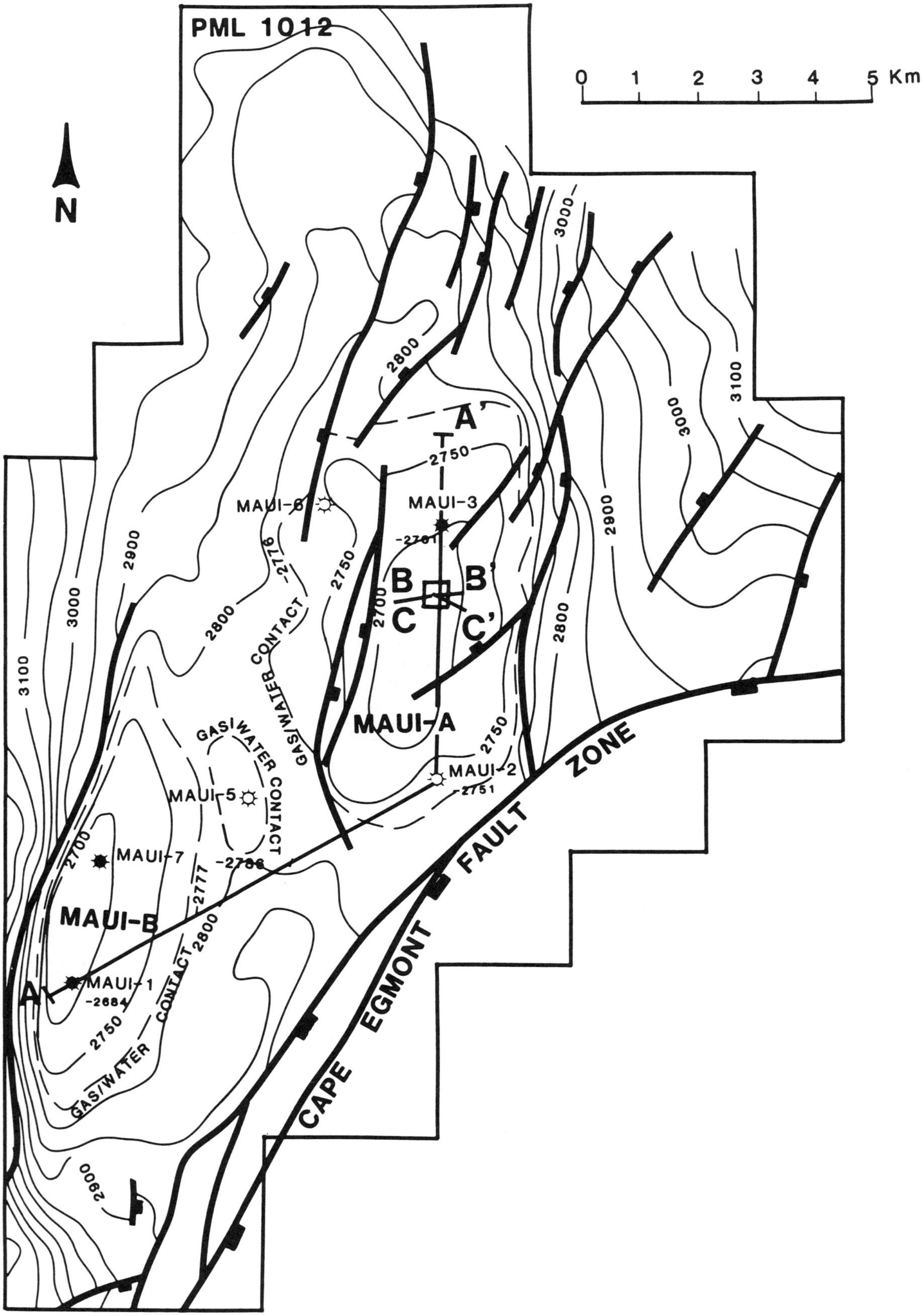
PML 1012
0 1 2 3 4 5 Km
N
MAUI-A
MAUI-B
MAUI-1
-2684
MAUI-2
-2751
MAUI-3
-2761
MAUI-5
MAUI-6
MAUI-7
A
A'
B
B'
C
C'
GAS/WATER CONTACT
-2776
-2777
-2788
2700
2750
2800
2900
3000
3100
CAPE EGMONT FAULT ZONE

Figure 8. Depth structure map top Maui C sands (McKee Formation). Contour interval, 50 ft. See Figure 11 for reservoir properties and cross-sections B-B[1] and C-C[1]. (After Shell-B.P.-Todd.)

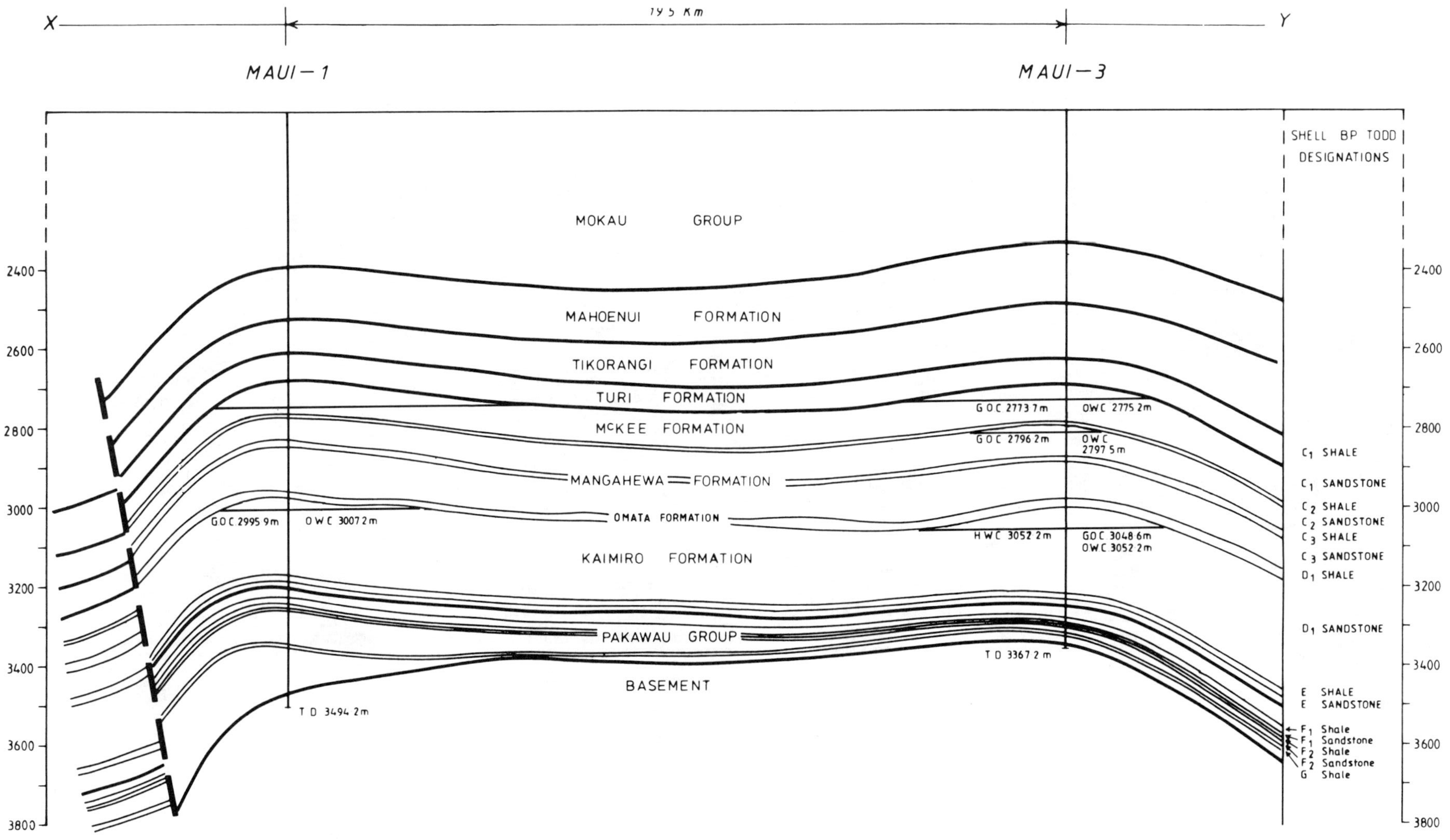
X
Y
19.5 Km
MAUI – 1
MAUI – 3
SHELL BP TODD DESIGNATIONS
MOKAU GROUP
MAHOENUI FORMATION
TIKORANGI FORMATION
TURI FORMATION
McKEE FORMATION
MANGAHEWA FORMATION
OMATA FORMATION
KAIMIRO FORMATION
PAKAWAU GROUP
BASEMENT
GOC 2995.9m
OWC 3007.2m
TD 3494.2m
GOC 2773.7m
OWC 2775.2m
GOC 2796.2m
OWC 2797.5m
HWC 3052.2m
GOC 3048.6m
OWC 3052.2m
TD 3367.2m
C1 SHALE
C1 SANDSTONE
C2 SHALE
C2 SANDSTONE
C3 SHALE
C3 SANDSTONE
D1 SHALE
D1 SANDSTONE
E SHALE
E SANDSTONE
F1 Shale
F1 Sandstone
F2 Shale
F2 Sandstone
G Shale
2400
2600
2800
3000
3200
3400
3600
3800

All depths in metres subsea

VERTICAL SCALE 1:10 000

HORIZONTAL SCALE 1:100 000

Figure 9. Geological cross-section between Maui-1 and Maui-3 (after Haskell, 1986). See Figure 4 for location and Figure 10 for seismic profile.

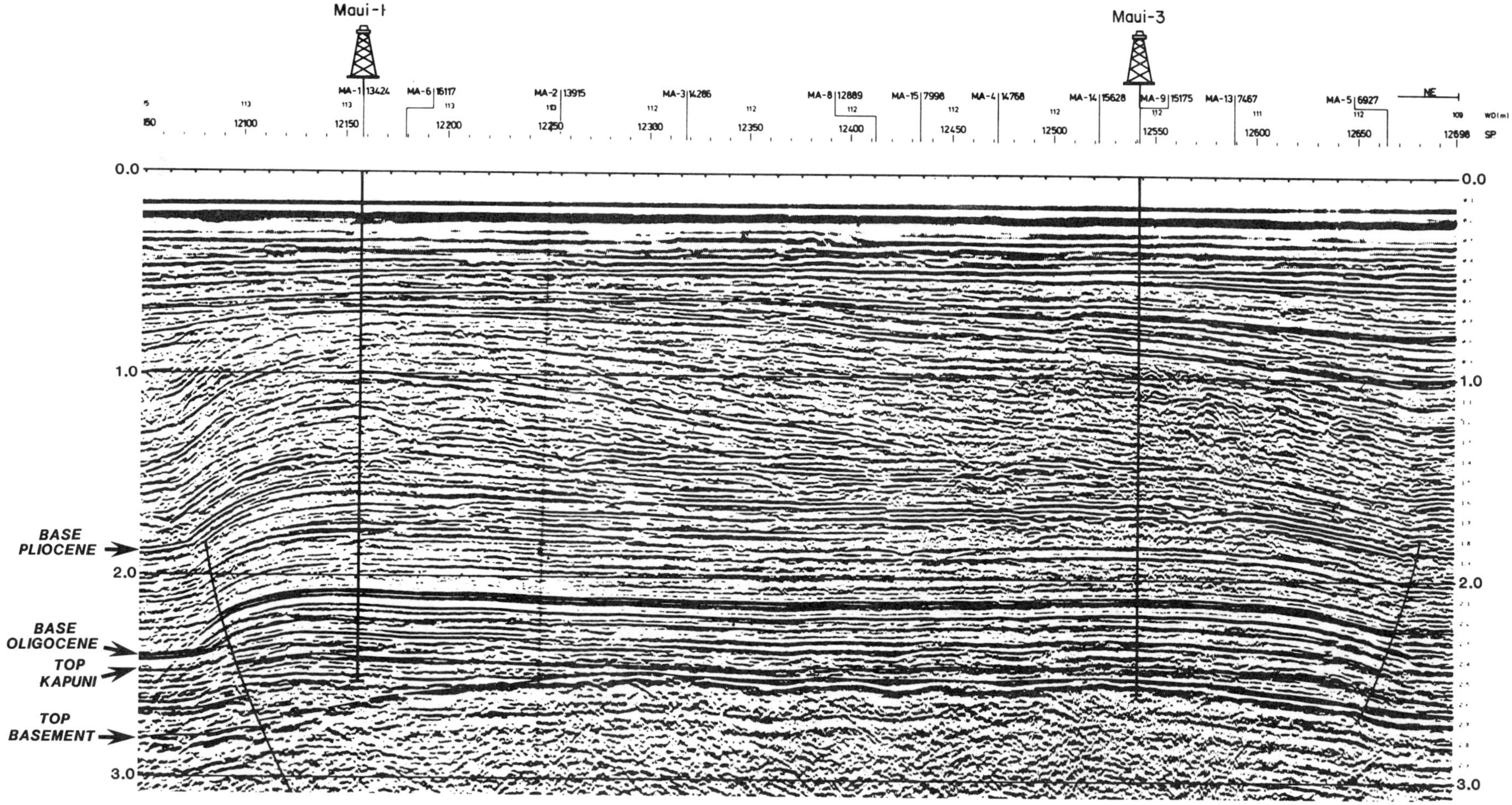
Maui-1
Maui-3
MA-1 13424
MA-6 15117
MA-2 13915
MA-3 14286
MA-8 12889
MA-15 7998
MA-4 14768
MA-14 15628
MA-9 15175
MA-13 7467
MA-5 6927
NE
WD(m)
SP
12100
12150
12200
12250
12300
12350
12400
12450
12500
12550
12600
12650
12698
0.0
1.0
2.0
3.0
BASE PLIOCENE
BASE OLIGOCENE
TOP KAPUNI
TOP BASEMENT

Figure 10. Seismic line between Maui-1 and Maui-3 . (Seismic line from Shell-B.P.-Todd). See Figure 9 for geologic cross-section.

Table 1. Reservoir properties from Maui-A drilling platform.

McKee Formation Reservoir = C_1

Sand Layer	Gross (m)**	Gross (m)**	Porosity (%)**	Interval (m)**	Net/ Permeability (md)**
1	7.7	0.85	17	2750.0-2758.0	14.3
2	3.4	0.98	2	2758.0-2762.2	1654.0
3	5.2	1.00	19	2762.7-2766.7	304.0
4	12.1	0.95	15	2766.7-2778.4	19.0
5	9.3	1.00	23	2778.4-2789.1	8240.0
6	13.7	0.90	17	2789.1-2804.1	39.0
7	2.2	1.00	22	2804.1-2804.9	447.0
8	5.5	0.97	23	2804.9-2808.8	160.0
9	7.4	0.84	15	2808.8-2813.3	146.0

Kaimiro Formation Reservoir = D_1

Sand Layer	Gross (m)	Gross (m)	Porosity (%)	Interval (m)*	Net/ Permeability (md)
1	5.5	1.00	21	3070.0-3073.2	860.0
2	7.7	1.00	19	3073.2-3076.4	533.0
3	11.7	0.84	21	3076.4-3079.4	1540.0
4	6.0	1.00	6	3079.4-3093.0	75.0

* Field average.
** Maui-A1(G) core data.

Table 2

OFFSHORE MAUI FIELD RESERVES ESTIMATES

	HYDROCARBONS INITIALLY IN PLACE			ULTIMATE RECOVERY (*)			CUMULATIVE PRODUCTION TO 31 MARCH 1987	RESERVES AS AT 1 APRIL 1987		
	LOW	MIDDLE	HIGH	LOW	MIDDLE	HIGH		LOW	MIDDLE	HIGH
DRY GAS (PJ)										
MAUI A AREA	3369	4246	5171	2112	2541	3080	626	1486	1915	2454
MAUI B AREA	2074	2649	3239	1287	1625	2032	-	1287	1625	2032
TOTAL	5443	6894	8410	3400	4166	5112	626	2774	3540	4486
CONDENSATE (MMstb)										
MAUI A AREA	121.40	154.70	188.70	56.90	66.70	78.70	24.40	32.50	42.30	54.30
MAUI B AREA	90.60	115.70	140.30	40.80	51.50	63.70	-	40.80	51.50	63.70
TOTAL	212.00	270.40	328.90	97.70	118.20	142.40	24.40	73.30	93.80	118.00
LPG (MILLION TONNES)										
MAUI A AREA	1.51	1.93	2.36	0.88	1.01	1.18	0.406	0.47	0.60	0.77
MAUI B AREA	1.11	1.41	1.72	0.61	0.77	0.96	-	0.61	0.77	0.96
TOTAL	2.62	3.34	4.07	1.49	1.78	2.14	0.406	1.08	1.37	1.73
OIL (MMstb)										
MAUI A AREA	-	-	-	-	-	-	-	-	-	-
MAUI B AREA	136.50	196.20	257.20	8.30	11.80	17.20	-	8.30	11.80	17.20
TOTAL	136.50	196.20	257.20	8.30	11.80	17.20	-	8.30	11.80	17.20

* ASSUMING THE MAUI 'B' DEVELOPMENT IS UNDERTAKEN.
(After Haskell, 1986)

Similarities in the produced liquids from Maui, Kapuni, and Motorua fields suggest a common source for the oil and condensate. The oils of the Taranaki basin are high wax and paraffinic. The chemical and biomarker parameters consistently indicate that the oils are mainly derived from nonmarine source rocks deposited under anoxic and low bacterial freshwater swamp conditions. An exception is oil from the Tangaroa #1 well that has a strong marine influence and represents a different source in the Northern Taranaki graben.

Pilaar and Wakefield (1984) have plotted the vitrinite reflection values measured in the various wells against depth. Maturation curves were drawn: one for wells drilled on the Western platform and another for those drilled in the Taranaki graben (Figure 12). From these data it can be assumed that top maturity level and, consequently, the zones of

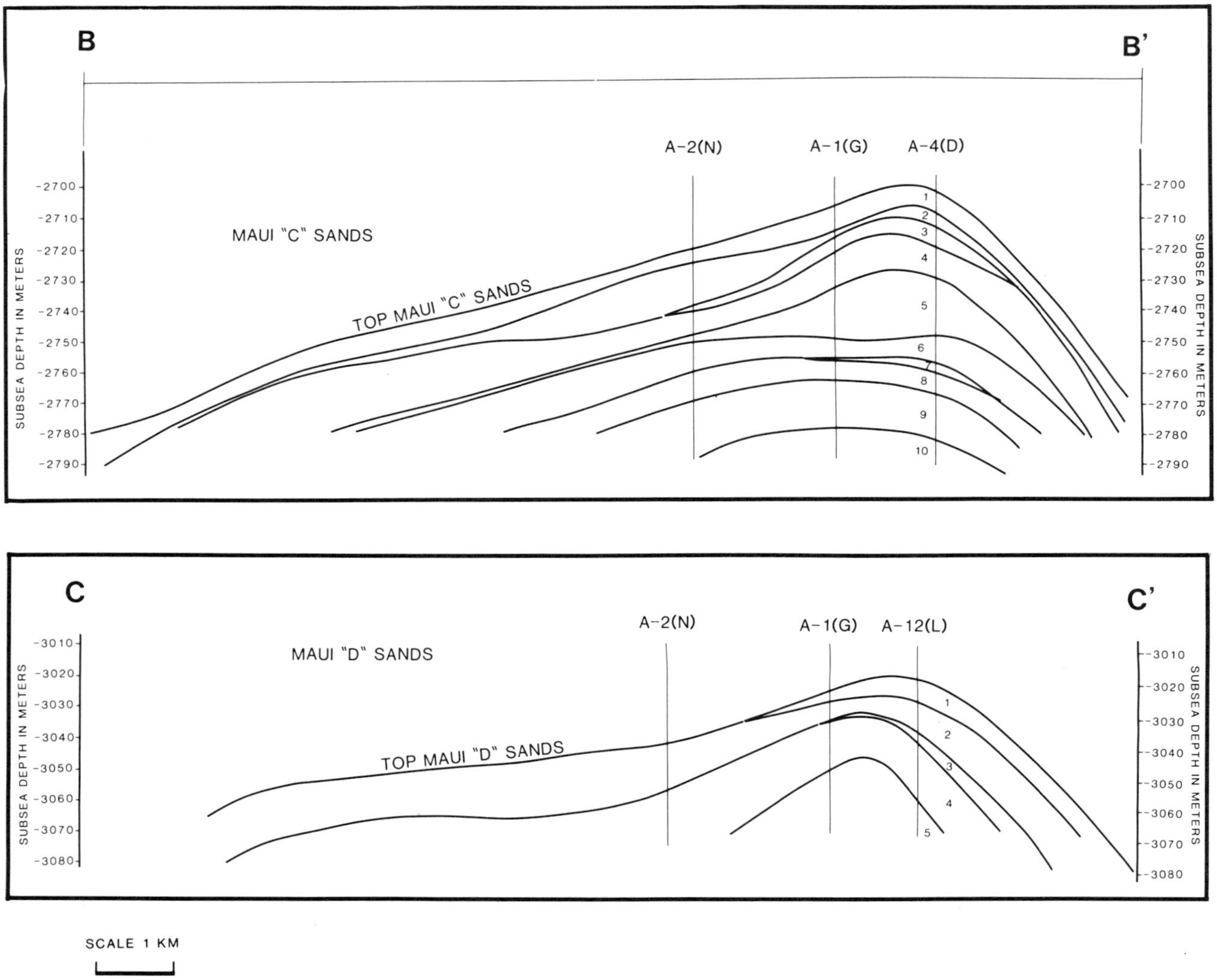
B
B'
A-2(N)
A-1(G)
A-4(D)
MAUI "C" SANDS
TOP MAUI "C" SANDS
SUBSEA DEPTH IN METERS
-2700
-2710
-2720
-2730
-2740
-2750
-2760
-2770
-2780
-2790
1
2
3
4
5
6
7
8
9
10
C
C'
A-2(N)
A-1(G)
A-12(L)
MAUI "D" SANDS
TOP MAUI "D" SANDS
-3010
-3020
-3030
-3040
-3050
-3060
-3070
-3080
SCALE 1 KM

Figure 11. Cross-sections of Maui A structure. Maui A drilling platform area illustrating the reservoir intervals for the C and D sandstones. Reservoir properties shown on Table 1. See Figure 8 for location of cross-sections. (After Haskell, 1986.)

generation and expulsion of hydrocarbons lie at a greater depth on the Western platform than in the graben complex.

Pilaar and Wakefield (1984) indicate that this difference in depth cannot be readily explained by the small difference in geothermal gradients (16-41°C/km) calculated for the two provinces. It seems more likely to be the result of the age difference of the "effective" overburden in the graben being Miocene while it is Pliocene-Pleistocene on the platform.

Pilaar and Wakefield have also constructed burial graphs giving the relationship between the depth of burial and geologic time for each vitrinite reflectance value measured in selected graben complex wells and Western platform wells. The results are given on Figures 13A, B, and C. These curves indicate that present-day depths are close to maximum depths.

The effect of geologic time on the maturation process in the Taranaki basin is illustrated on Figure 14, in which the expulsion areas are outlined as they are thought to be today (Pilaar and Wakefield, 1984).

It is theorized that hydrocarbon charge for the Maui and Kapuni fields is from the deeper portions of the South Taranaki graben. In particular, the Kapuni field appears well placed relative to its underlying expulsion area, whereas Maui relies on deeper areas located in the Taranaki graben for its charge (Figures 12, 13, and 14).

EXPLORATION CONCEPTS

A study of Maui field and its neighbors, Kapuni and Motorua, did not reveal anything surprising. Seeps provide evidence of effective source beds;

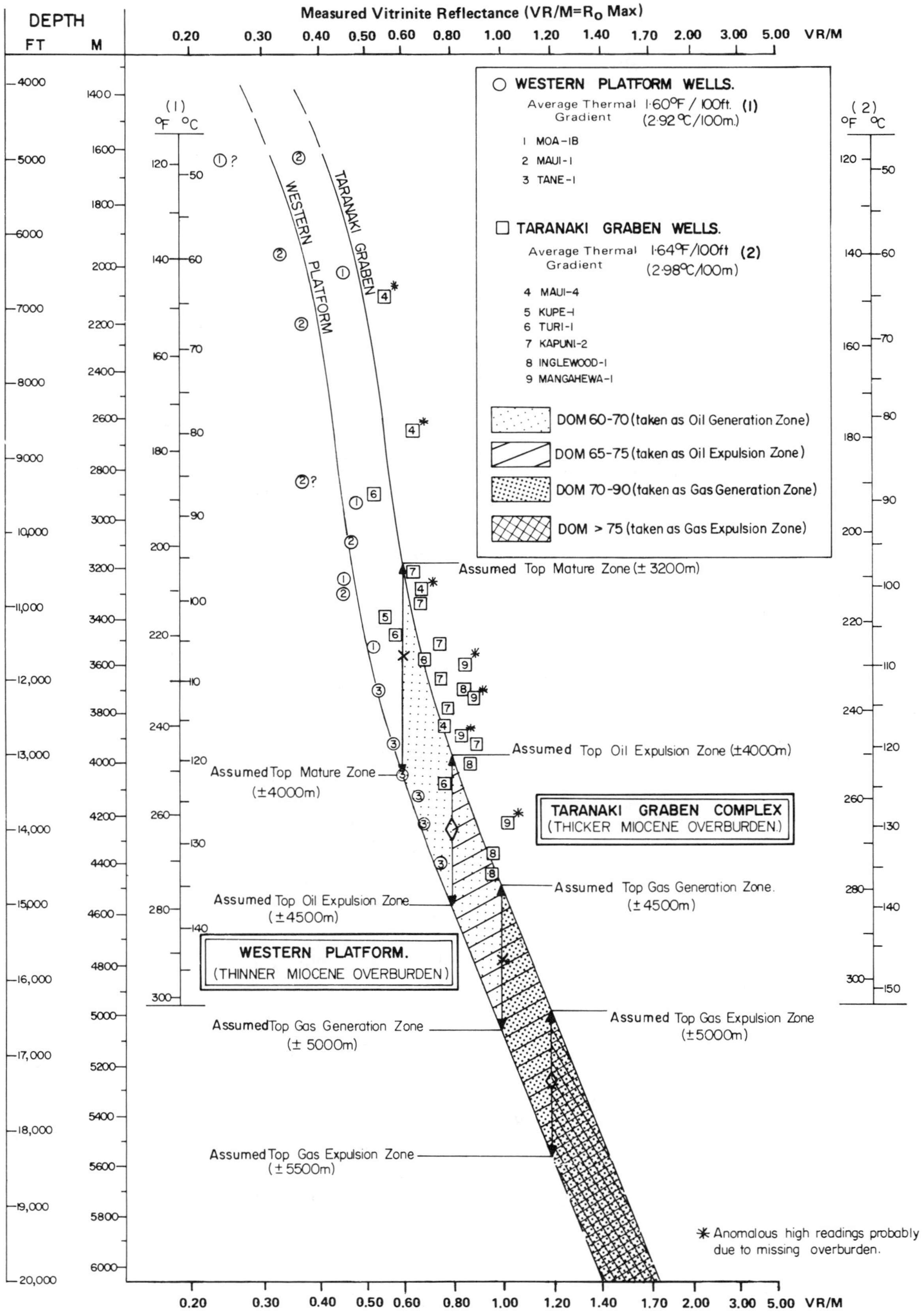

DEPTH
FT
M
Measured Vitrinite Reflectance (VR/M=R_0 Max)
0.20 0.30 0.40 0.50 0.60 0.80 1.00 1.20 1.40 1.70 2.00 3.00 5.00 VR/M
WESTERN PLATFORM WELLS.
Average Thermal Gradient 1·60°F / 100ft. (1) (2·92°C/100m.)
1 MOA-1B
2 MAUI-1
3 TANE-1
TARANAKI GRABEN WELLS.
Average Thermal Gradient 1·64°F/100ft (2) (2·98°C/100m)
4 MAUI-4
5 KUPE-1
6 TURI-1
7 KAPUNI-2
8 INGLEWOOD-1
9 MANGAHEWA-1
DOM 60-70 (taken as Oil Generation Zone)
DOM 65-75 (taken as Oil Expulsion Zone)
DOM 70-90 (taken as Gas Generation Zone)
DOM > 75 (taken as Gas Expulsion Zone)
WESTERN PLATFORM
TARANAKI GRABEN
Assumed Top Mature Zone (± 3200m)
Assumed Top Oil Expulsion Zone (±4000m)
Assumed Top Mature Zone (±4000m)
TARANAKI GRABEN COMPLEX (THICKER MIOCENE OVERBURDEN)
Assumed Top Gas Generation Zone. (± 4500m)
Assumed Top Oil Expulsion Zone (±4500m)
WESTERN PLATFORM. (THINNER MIOCENE OVERBURDEN)
Assumed Top Gas Generation Zone (± 5000m)
Assumed Top Gas Expulsion Zone (±5000m)
Assumed Top Gas Expulsion Zone (± 5500m)
* Anomalous high readings probably due to missing overburden.

Figure 12. Taranaki basin, vitrinite reflectance versus depth (after Pilaar and Wakefield, 1984).

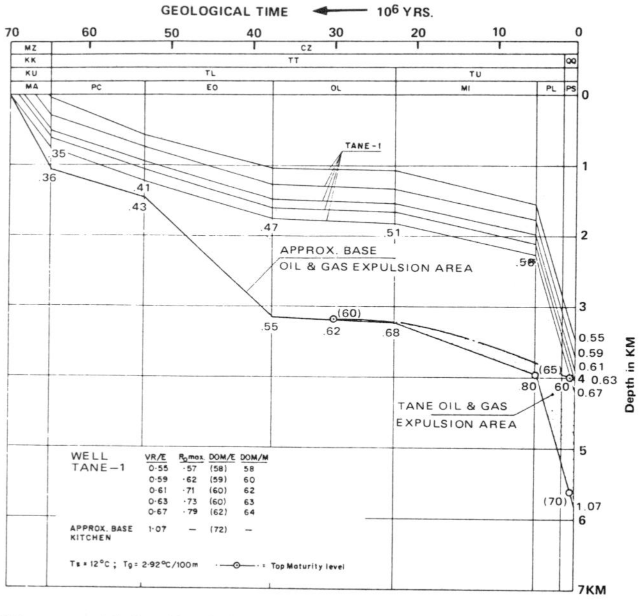
GEOLOGICAL TIME
10⁶ YRS.
TANE-1
APPROX. BASE
OIL & GAS EXPULSION AREA
TANE OIL & GAS
EXPULSION AREA
Depth in KM
7KM

Figure 13A. Burial graph, Western platform (west). DOM, degree of organic metamorphism; DOM/E, estimated; DOM/M, measured; VR/E, vitrinite reflectance equivalent; R.O., vitrinite reflectance of oil.

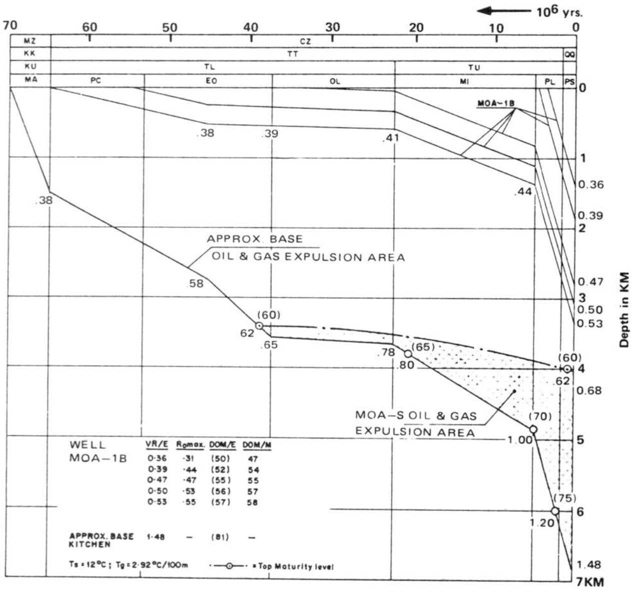
10⁶ yrs.
MOA-1B
APPROX. BASE
OIL & GAS EXPULSION AREA
MOA-S OIL & GAS
EXPULSION AREA
Depth in KM
7KM

Figure 13B. Burial graph, Western platform (east) (after Pilaar and Wakefield, 1984). DOM, degree of organic metamorphism; DOM/E, estimated; DOM/M, measured; VR/E, vitrinite reflectance equivalent; R.O., vitrinite reflectance of oil.

source bed quality deposits were confirmed by drilling through thick shales, mudstones, and coals that are rich in organic matter. The faults, especially the major Cape Egmont system, are logical migration avenues to feed the gas and oil to the traps, which are of a size to hold major reserves, given adequate

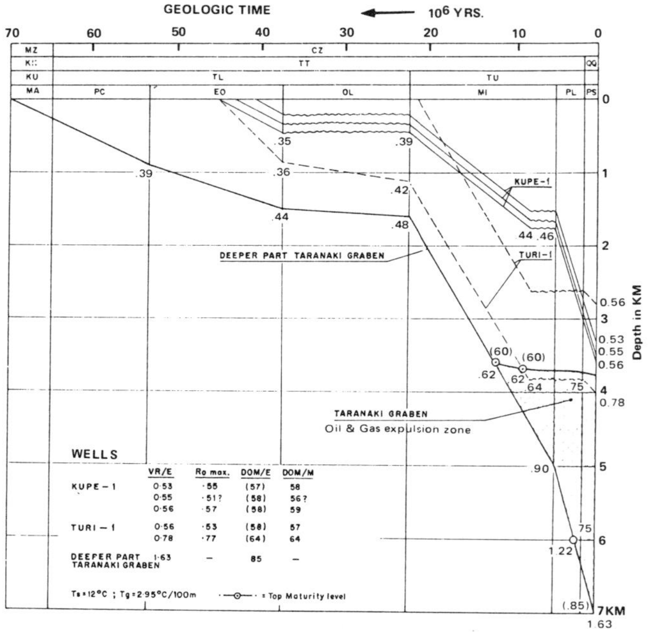
GEOLOGIC TIME
10⁶ YRS.
KUPE-1
TURI-1
DEEPER PART TARANAKI GRABEN
TARANAKI GRABEN
Oil & Gas expulsion zone
WELLS
Depth in KM
7KM

Figure 13C. Burial graph, Taranaki graben complex (after Pilaar and Wakefield, 1984). DOM, degree of organic metamorphism; DOM/E, estimated; DOM/M, measured; VR/E, vitrinite reflectance equivalent; R.O., vitrinite reflectance of oil.

reservoir capacity. Neither Maui nor Kapuni are filled to spill point. This suggests inadequate source, partial isolation from the source, inadequate seals and escape migration, or a combination of these factors.

Using upgraded seismic and well data, the original recoverable reserves were redetermined at 460 million million BTU of dry gas, an increase of 84%. Estimated condensate recovery was 34 billion bbl.

If this field were to be evaluated today using present techniques, 3D seismic might be employed to define the discontinuous fluvial sandstone bodies in the upper Kapuni interval and more precisely map the fault trends and fault plane geometries.

ACKNOWLEDGMENTS

Occidental International Exploration and Production Company offered assistance in preparation of this manuscript. Special assistance was given by L. M. Seibert, typist; H. L. Scott and B. D. De La Cruz, graphics; and F. R. Abbott, research and editing. Conversations with J. M. Winterman, Occidental International Exploration and Production Company, and published data by Shell, B.P., and Todd Oil Services, Ltd., contributed greatly to this publication.

This paper draws heavily on the published work and illustrations of W.F.H. Pilaar and L. L. Wakefield as well as Haskell Exploration Services.

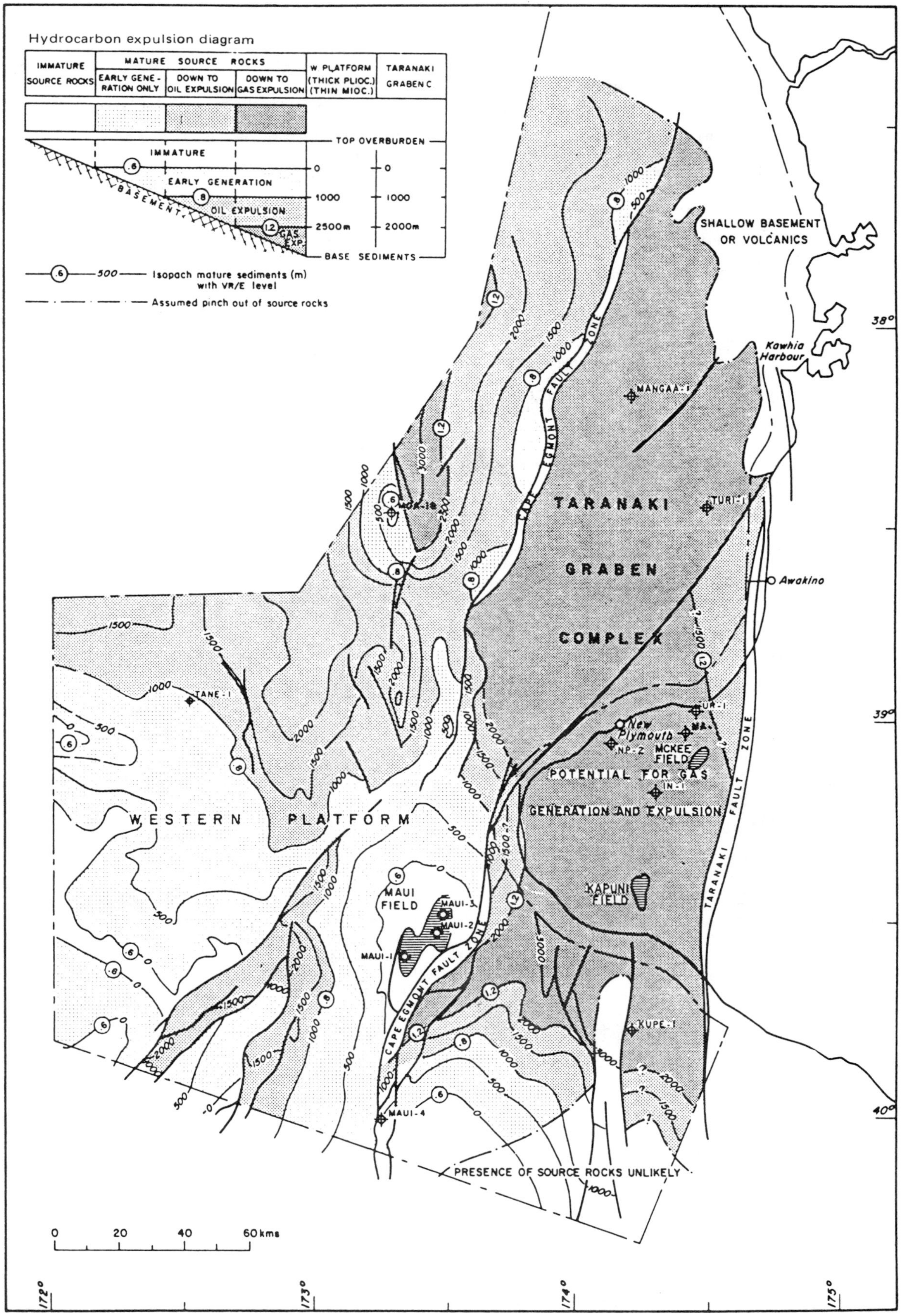
Hydrocarbon expulsion diagram
IMMATURE SOURCE ROCKS
MATURE SOURCE ROCKS
EARLY GENERATION ONLY
DOWN TO OIL EXPULSION
DOWN TO GAS EXPULSION
W PLATFORM (THICK PLIOC.) (THIN MIOC.)
TARANAKI GRABEN C
TOP OVERBURDEN
IMMATURE
EARLY GENERATION
OIL EXPULSION
GAS EXP.
BASEMENT
BASE SEDIMENTS
0
1000
2500m
2000m
Isopach mature sediments (m) with VR/E level
Assumed pinch out of source rocks
SHALLOW BASEMENT OR VOLCANICS
Kawhia Harbour
MANGAA-1
TARANAKI GRABEN COMPLEX
CAPE EGMONT FAULT ZONE
TURI-1
Awakino
TANE-1
MOA-1B
New Plymouth
NP-2
MCKEE FIELD
POTENTIAL FOR GAS GENERATION AND EXPULSION
TARANAKI FAULT ZONE
WESTERN PLATFORM
MAUI FIELD
MAUI-1
MAUI-2
MAUI-3
MAUI-4
KAPUNI FIELD
KUPE-1
PRESENCE OF SOURCE ROCKS UNLIKELY
0 20 40 60kms
38°
39°
40°
172°
173°
174°
175°

Figure 14. Taranaki basin, tentative hydrocarbon generation and expulsion provinces at present (after Pilaar and Wakefield).

REFERENCES

Alexander, R., R. I. Kagi, G. W. Woodhouse, and J. K. Volkman, 1983, The geochemistry of some biodegraded Australian oils: APEA Journal, v. 23, n. 1, p. 53-63.

Analabs (Oil and Gas Division), 1984, Petroleum geochemistry of the Taranaki Basin: New Zealand Geological Survey unpublished open file petroleum report 1013.

Austin, P. M., et al., 1973, Structure and petroleum potential of eastern Chatham Rise, New Zealand: American Association of Petroleum Geologists Bulletin, v. 57, n. 3, p. 477-497.

Barnard, E. J., 1969, Design criteria for Kapuni Field facilities: N.Z. Engineering, v. 24, p. 15-22.

Beddoes, L. R., Jr. (ed.), 1973, Oil and gas fields of Australia, Papua New Guinea and New Zealand: Tracer Petroleum & Mining Publications Pty Ltd., 382 p.

Blake, M. C., et al., 1974, Active continental margins; comparisons between California and New Zealand, *in* C. A. Burke and C. L. Drake, eds., The geology of continental margins: Springer-Verlag New York, Inc., p. 853-872.

Carmalt, S. W., and B. St. John, 1986, Giant oil and gas fields, *in* M. T. Halbouty, ed., Future petroleum provinces of the world: American Association of Petroleum Geologists Memoir 40, p. 11-54.

Carter, R. M., and R. J. Morris, 1976, Cainozoic history of southern New Zealand; an accord between geological observations and plate tectonic predictions: Earth Science and Planetary Letters, v. 3, n. 1, p. 85-94.

Connan, J., 1974, Time-temperature relation in oil genesis: American Association of Petroleum Geologists Bulletin, v. 58, n. 12, p. 2516-2521.

Cook, R. A., 1987, The geology and geochemistry of the crude oils and source rocks of western New Zealand: N.Z. Geological Survey Report PR 1250, 380 p.

Cook, R. A., and P. R. King, 1987, Summary of petroleum geology of onshore Taranaki: Petroleum Exploration News, November, p. 6-11.

Cope, R. N., and I. J. Reed, 1967, The Cretaceous paleogeology of the Taranaki-Cook Strait area: Proceedings Australian Institute of Mining and Metallurgy.

Editorial Staff, 1975, Work starts to develop big Maui Field: Oil and Gas Journal, September 1, p. 52-53.

Editorial Staff, 1975, Maui Field soon to hit full stride: Oil and Gas Journal, October 27, p. 140-155.

Elphick, J. O., and R. P. Suggate, 1964, Depth/rank relations of high volatile bituminous coals: N.Z. Journal of Geology and Geophysics, v. 7, n. 3, p. 594-601.

Fairburn, S. G., 1980, Diagenesis of the Kapuni Formation—offshore South Taranaki: BSc thesis, Victoria University, Wellington, New Zealand, 33 p.

Furzey, D. G., 1970, Kapuni plant treats high CO_2 content natural gas: World Petroleum, September 1970, p. 48-54.

Gibbons, M. J., and S. Fry, 1983, A compilation of the geochemical characteristics of the oils and condensates of New Zealand: New Zealand Geological Survey unpublished open file report 989.

Griffiths, J. R., 1971, Reconstruction of the southwest Pacific margin of Gondwanaland: Nature, v. 234, November 26, p. 203-207.

Harding, T. P., 1974, Petroleum traps associated with wrench faulting: American Association of Petroleum Geologists Bulletin, v. 57, p. 74-96.

Haskell, T. R., 1986, A study of structural development, sandstone depositional systems, and hydrocarbon accumulation: Haskell Exploration Services Ltd., Monograph-1, p. 1-96.

Hayward, B. W., 1987, Paleobathymetry and structural and tectonic history of Cenozoic drillhole sequences in Taranaki Basin: N.Z. Geological Survey Report PAL 122, 63 p.

Hill, P. J., and J. D. Collen, 1978, The Kapuni sandstones from Inglewood-1 well, Taranaki—petrology and the effects of diagenesis on reservoir characteristics: N.Z. Journal of Geology and Geophysics, v. 21, n. 2, p. 215-228.

Hogan, J. A., 1979, Stratigraphy and sedimentology of the Kapuni Formation, Taranaki, New Zealand: MSc thesis, Victoria University, Wellington, New Zealand, 189 p.

Hood, A., C. C. M. Gutjahr, and R. I. Heacock, 1975, Organic metamorphism and the generation of petroleum: American Association of Petroleum Geologists Bulletin, v. 59, p. 986-996.

Kamp, P. J. J., 1986, The mid-Cenozoic Challenger Rift system of western New Zealand and its implications for the age of alpine fault inception: Geological Society of America Bulletin, v. 97, p. 255-281.

Katz, H. R., 1968, Potential oil formations in New Zealand and their stratigraphic position as related to basin evolution: New Zealand Journal of Geology and Geophysics, v. 11, n. 5, p. 1077-1133.

Katz, H. R., 1974, Margins of the southwest Pacific, *in* C. A. Burke and C. L. Drake, eds., The geology of continental margins: Springer-Verlag New York Inc., p. 549-565.

Katz, H. R., 1974, Offshore petroleum potential in New Zealand: APEA Journal, v. 14, n. 1, p. 3-13.

Katz, H. R., 1976, Sedimentary basins and petroleum prospects, onshore and offshore New Zealand: American Association of Petroleum Geologists Memoir 25, Circum-Pacific Energy and Mineral Resources, p. 217-228.

Kear, D. 1965, The Kapuni gas-condensate field, New Zealand—a case study: Third Symposium on the Development of Petroleum Resources of Asia and the Far East: Mineral Resources Development Series No. 29, p. 86-91.

King, P. R., and P. H. Robinson, in press, An overview of the Taranaki region geology, New Zealand: New Zealand Geological Survey DSIR, Lower Hutt, New Zealand.

Kingma, J. T., 1974, The geological structure of New Zealand: New York, John Wiley & Sons, Inc., 407 p.

Knox, G. J., 1982, Taranaki Basin, structural style and tectonic setting: New Zealand Journal of Geology and Geophysics, v. 25, n. 2, p. 125-140.

Knox, G. J, 1982, Taranaki Basin, structural style and tectonic setting: New Zealand Journal of Geology and Geophysics, v. 28, n. 2, p. 197-216.

Krebs, W., 1975, Formation of southwest Pacific island arc-trench and mountain systems; plate or global-vertical tectonics: American Association of Petroleum Geologists Bulletin, v. 59, n. 9, p. 1639-1666.

Lopatin, N. V., 1971, Temperature and geological time as factors in coalification: Akademiya Nauk SSSR, Izviestia, Seriya Geologicheskaya, n. 3 (English translation by N. H. Bosteck, 1972, Illinois State Geological Survey, p. 96-106).

Lowery, J. H., 1986, Vitrinite reflectance and maturation in the 5 km deep Inglewood-1 prospecting well, Taranaki Basin: New Zealand Geological Survey Report M149.

McBeath, D. M., 1976, Kapuni and Maui gas-condensate fields of New Zealand—summary: American Association of Petroleum Geologists Memoir 25, Circum-Pacific Energy and Mineral Resources, p. 211-216.

McBeath, D. M., 1977, Gas-condensate fields of the Taranaki Basin, New Zealand: New Zealand Journal of Geology and Geophysics, v. 20, n. 1, p. 99-127.

Nathan, S., H. J. Anderson, R. A. Cook, R. H. Herzer, R. H. Hoskins, J. I. Raine, and D. Smale, 1986, Cretaceous and Cenozoic sedimentary basins of the west coast region, South Island, New Zealand: Geological Survey of New Zealand and Basin Studies 1.

Nelson, C. S., 1978, Temperate shelf carbonate sediments in the Cenozoic of New Zealand: Sedimentology, v. 15, p. 737-771.

New Zealand Government, 1964, Report on utilisation of Kapuni natural gas: Government Printer, Wellington, 29 p.

New Zealand Government, 1965, Second report on utilisation of Kapuni natural gas: Government Printer, Wellington, 14 p.

New Zealand Government, 1967, Gas purchase contract between Shell (petroleum mining) Company Limited, BP (oil exploration) Company of New Zealand Limited, Todd Petroleum Mining Company Limited, sellers, and the Minister of Mines, buyer: Government Printer, Wellington, 36 p.

New Zealand Government, 1973, Development of the Maui gas field: Government Printer, Wellington, 316 p.

Palmer, J., 1985, Pre-Miocene lithostratigraphy of Taranaki Basin, New Zealand: New Zealand Journal of Geology and Geophysics, v. 28, n. 2, p. 197-216.

Pilaar, W. F. H., and L. L. Wakefield, 1978, Structural and stratigraphic evolution of the Taranaki Basin, offshore North Island, New Zealand: Australian Petroleum Exploration Association Journal, p. 93-101.

Pilaar, W. F. H., and L. L. Wakefield, 1984, Hydrocarbon generation in the Taranaki Basin, New Zealand, *in* G. Demaison and R. J. Murris, eds., Petroleum geochemistry and basin evaluation: American Association of Petroleum Geologists Memoir 35, p. 405-423.

Robinson, P. H., P. R. King, and G. P. Thrasher, in preparation, Lithostratigraphic nomenclature for Taranaki region, New Zealand.
Roux, B., et al., 1980, An improved approach to estimating true reservoir temperature from transient temperature data: Society of Petroleum Engineers of AIME, n. 8888.
Shell, B.P., Todd Oil Services Ltd., 1975, Exploration well proposal structure D (Kupe) PPL 682: New Zealand Geological Survey unpublished open-file petroleum report 643.
Short, K. C., 1962, The stratigraphy of the Taranaki Peninsula structure: Shell, B.P., and Todd Oil Services Limited unpublished report.
Smale, D., and A. C. Morton, in press, Heavy mineral suites of core samples from the McKee Formation (Eocene-Lower Oligocene), Taranaki; implications for provenance and diagenesis: New Zealand Journal of Geology and Geophysics, v. 30, n. 3.
Sprigg, R. C., et al., 1969, New Zealand Basin offers promise: Oil and Gas Journal, June 16, p. 124-129.
Stach, E., et al., 1975, Stach's textbook on coal petrology, 2nd revised edition: Berlin-Stuttgart, Gebr. Borntraeger.
Suggate, R. P., 1950, Quartzose coal measures of west Nelson and north Westland: New Zealand Journal of Science and Technology, Bulletin 31, n. 4, p. 1-14.
Suggate, R. P., 1956, Puponga coal field: New Zealand Journal of Science and Technology, Bulletin 37, n. 5, p. 539-559.
Suggate, R. P., 1973, Coal ranks in relation to depth and temperature in Australia and New Zealand oil and gas wells: New Zealand Journal of Geology and Geophysics, v. 17, n. 1, p. 149-167.
Thompson, J. G., 1982, Hydrocarbon source rock analyses of Pakawau Group and Kapuni Formation sediments, northwest Nelson and offshore South Taranaki, New Zealand: New Zealand Journal of Geology and Geophysics, v. 25, n. 2, p. 141-148.
Van der Lingen, G. J., D. Smale, G. A. Challis, W. A. Watters, and P. H. Robinson, in press, Siliciclastic diagenesis in Paleocene-Eocene reservoir sandstones of the Taranaki Basin: New Zealand Geological Survey DSIR, Lower Hutt, New Zealand.
Van de Watering, W. P. M., 1976, Planning pays in setting Maui structure: Oil and Gas Journal, March 1, p. 107-113.
Walcott, R. I., 1987, Geodetic strain and the deformational history of the North Island of New Zealand during the Late Cenozoic: Philosophical Transactions of the Royal Society of London, A321, p. 163-181.
Waples, D. W., 1980, Time and temperature in petroleum formation; application of Lopatin's method to petroleum exploration, American Association of Petroleum Geologists Bulletin, v. 64, p. 916-926.
Watt, D. S., 1965, Natural gas and oil in western New Zealand: Publications 8th Commonwealth Mining and Metallurgical Congress 5, p. 77-88.
Wilcox, R. E., T. P. Harding, and D. R. Seely, 1973, Basic wrench faulting: American Association of Petroleum Geologists Bulletin, v. 57, p. 74-96.
Williams, H. R., 1968, Production of the Kapuni Field and separation of gas and condensate: New Zealand Engineering 23, p. 458-462.
Wodzicki, A., 1974, Geology of the pre-Cenozoic basement of the Taranaki-Cook Strait-Westland area, New Zealand, based on recent drillhole data: New Zealand Journal of Geology and Geophysics, v. 17, n. 4, p. 747-757.

SUGGESTED READING

Shell Oil Company, 1987, Atlas of seismic stratigraphy: American Association of Petroleum Geologists Studies in Geology 27, v. 1, p. 15-71.

Appendix 1. Field Description

Field name .... *Maui*

Ultimate recoverable reserves .... *Gas, 5.3 tcf (142 billion m³); NGL, 230 million bbl (36.6 million kL); oil, 300 million bbl (est.)*

Field location:

Country .... *New Zealand*
State .... *North Island*
Basin/Province .... *Taranaki basin*

Field discovery:

Year field discovered *1969*
Year second pay discovered *1969*
Third pay .... *NA*

Discovery well name and general location
(i.e., Jones No. 1, Sec. 2T12NR5E; or Smith No. 1, 5 mi west of Sheridan, Wyoming):

First pay .... *Maui 1, 39°40′15″S; 173°18′35″E*
Second pay .... *Maui 1, 39°40′15″S; 173°18′35″E*
Third pay .... *NA*

Discovery well operator

Shell Co. Ltd., B.P. of New Zealand Ltd., and Todd Petroleum Mining Co. Ltd.

(if more than one pay in field, list operators of discovery well in other pays)

Second pay .... *NA*
Third pay .... *NA*

IP in barrels per day and/or cubic feet or cubic meters per day:

First pay .... *Not on production—data not available*
Second pay .... *NA*
Third pay .... *NA*

All other zones with shows of oil and gas in the field:

Age	Formation	Type of Show
None		

Geologic concept leading to discovery and method or methods used to delineate prospect, e.g., surface geology, subsurface geology, seeps, magnetic data, gravity data, seismic data, seismic refraction, nontechnical:

In 1965, seismic reflection and refraction profiles were shot in the offshore Taranaki basin, followed in 1966 by fourfold reflection coverage. In 1967, an aeromagnetic survey and semidetailed seismic survey were carried out.

Structure:

Province/basin type (see St. John, Bally, and Klemme, 1984) .... *Interarc basin*

Tectonic history

After rifting away from Australia in late Paleozoic(?) time, Taranaki basin was part of extensive New Zealand geosyncline until Late Jurassic-Early Cretaceous Rangitata orogeny, during which the geosyncline was uplifted and eroded. Simultaneous movement on nearby transcurrent Alpine fault may have initiated Kapuni structural growth. A major rifting phase began after the Rangitata orogeny and transcurrent fault movement continued. The Taranaki basin began to form in Late Cretaceous time.

Regional structure

The Taranaki basin is composed of two main structural units: a stable western platform area to the west, the Western platform, and a graben system to the east, the Taranaki graben complex. The Cape Egmont fault separates these elements. The graben is divided into northern and southern segments. Subsidence reflects three tectonic regimes: extensional, oblique extensional, and oblique compressional.

Local structure

Faulted north-plunging compound anticline adjacent to the Cape Egmont fault zone in the Western platform.

Trap

Trap type(s)

The Maui structure is a broad, low relief compound anticline with two main culminations. It is bounded on the southeast by major down-to-southeast Cape Egmont fault and on the west and east by normal faults.

Basin stratigraphy (major stratigraphic intervals from surface to deepest penetration in field):

Age	Formation	Depth to Top in ft (m)
Recent	*Superficial deposits*	*Sea floor*
Pleistocene-Pliocene	*Giant foresets-Matemateaonga-Tangahoe*	*1100 (335)*
Miocene	*Urenui-Kiwa mudstone*	*1700 (518)*
Miocene	*Waikiekie*	*2930 (893)*
Miocene	*Mohakatino-Mokau-Moki*	*4870 (1484)*
Oligocene-Late Miocene	*Mahoenui*	*7810 (2380)*
Oligocene	*Te Kuiti-Tikorangi-Kakaka*	*8235 (2510)*
Paleocene/Eocene	*Kapuni*	*8625 (2629)*
Upper Cretaceous	*Pakawau*	*11,400 (3474)*

Location of well in field .... *NA*

Reservoir characteristics:

Number of reservoirs *3*

Formations .... *Kapuni Fm., C1, C2, and D1 sandstone*

Ages .... *Upper Eocene*

Depths to tops of reservoirs .... *9000 ft (2743 m) upper zone (C1 and C2); 10,000 ft (3048 m) lower zone (D1)*

Gross thickness (top to bottom of producing interval) .... *370 m*

Net thickness—total thickness of producing zones

Average .... *76 m in C1 and C2 ss; 19 m in D1 ss*

Maximum ....

Lithology

Quartzose sandstones with some plagioclase, medium to fine grained, angular to subangular and friable. Kaolinite, illite, and chlorite present with some quartz overgrowths.

Porosity type .... *Good intergranular, second porosity due to carbonate cement dissolution*

Average porosity *25%*

Average permeability .... *70-500 md*

Seals:

Upper

Formation, fault, or other feature .... *Oligocene Mahoenui marl and limestone; Eocene shales of Kapuni Fm. especially C1 and D1 shales*

Lithology .... *Marl and shale*

Lateral

Formation, fault, or other feature .... *None*

Lithology ....

Source:

Formation and age .... *Lower Cretaceous Pakawau and Paleocene-Eocene Kapuni*

Lithology .... *Coal*

Average total organic carbon (TOC) .... *NA*

Maximum TOC .... *NA*

Kerogen type (I, II, or III) .... *II and III*
Vitrinite reflectance (maturation) .... *R_o = 0.95–1.0*
Time of hydrocarbon expulsion .... *Pliocene–Pleistocene*
Present depth to top of source .... *≈11,369 ft (3465 m)*
Thickness .... *Unknown*
Potential yield .... *NA*

Appendix 2. Production Data

Field name .... *Maui*

Field size:

Proved acres .... *74,900 ac (303 km^2)*
Number of wells all years *20*
Current number of wells (as of year) *20*
Well spacing .... *NA*
Ultimate recoverable .... *5.3–8.4 tcf; 75–230 million bbl*
Cumulative production .... *0.626 tcf; 0.024 million bbl*
Annual production .... *Shut in*
Present decline rate .... *NA*
Initial decline rate .... *NA*
Overall decline rate .... *NA*
Annual water production .... *NA*
In place, total reserves .... *NA*
In place, per acre-foot .... *NA*
Primary recovery .... *NA*
Secondary recovery .... *NA*
Enhanced recovery .... *NA*
Cumulative water production .... *NA*

Drilling and casing practices:

Amount of surface casing set .... *506 ft (154 m)*
Casing program
30-in. to 506 ft (154 m); 20-in. to 1041 ft (317 m); 13½-in. to 3030 ft (924 m); 9⅝-in. to 8351 ft (2545 m); log, plug back to 10,200 (3109 m); 7-in. log 7865–10,200 ft (2397–3109 m)

Drilling mud .... *Salt water to 1100 ft (335 m); Freshwater Spersene XP-20 mud*
Bit program .... *Variable*
High pressure zones .... *None*

Completion practices:

Interval(s) perforated .... *NA*
Well treatment .... *NA*

Formation evaluation:

Logging suites .... *SP, Resistivity, Caliper, Acoustic Velocity, Density, FORXO, Dipmeter*
Testing practices .... *Typically production tested*
Mud logging techniques .... *NA*

Oil characteristics:

Type .... *Paraffinic*
(Tissot and Welte Classification in "Petroleum Formation and Occurrence," 1984, Springer-Verlag, p. 419)
API gravity .... *51.5°*

Base .... *NA*
Initial GOR .... *1450 cf/bbl*
Sulfur, wt% *0.01*
Viscosity, SUS .... *1.18 cp at 70°F (21°C)*
Pour point .... *Less than 15°C (60°F)*
Gas-oil distillate .... *NA*

Field characteristics:

Average elevation .... *-110 m MSL*
Initial pressure .... *1886 psi (13,105 kPa); overpressured zones reported*
Present pressure .... *NA*
Pressure gradient .... *NA*
Temperature .... *210°F (99°C)*
Geothermal gradient .... *5°C(?) 100 m/20°C*
Drive .... *Pressure depletion; water influx*
Oil column thickness .... *Oil, 32 ft (10 m); gas, 278 ft (85 m)*
Oil-water contact .... *2773.7 m; 3007.2 m*
Connate water .... *NA*
Water salinity, TDS .... *48 g/L*
Resistivity of water .... *0.12 ohm*
Bulk volume water (%) *20%*

Transportation method and market for oil and gas:

Pipeline.

Kapuni Field

W. O. ABBOTT
Occidental International Exploration and Production Company
Bakersfield, California, U.S.A.

FIELD CLASSIFICATION

BASIN: Taranaki
BASIN TYPE: Interarc Basin
RESERVOIR ROCK TYPE: Sandstone
RESERVOIR ENVIRONMENT OF DEPOSITION: Coastal and Paralic Coal Basin
RESERVOIR AGE: Upper Eocene
PETROLEUM TYPE: Gas and Condensate
TRAP TYPE: Anticline

LOCATION

There are two onshore commercial fields in New Zealand (Kapuni and McKee) and one noncommercial field (Motorua). Kapuni is the largest field and is located approximately 30 mi (48 km) south-southeast of New Plymouth, North Island (Figures 1 and 2). It is situated in the northern portion of the Taranaki graben, located on the Kupe block east of the Manaia fault (Figures 2 and 3). The coordinates for the discovery well (Kapuni #1) are New Zealand grid references: 1,753.34; 3,415.25. Estimated ultimate recovery is 9.91 bcf of gas and 41.3 million bbl of natural gas liquids (NGL).

The largest field in New Zealand is the offshore Maui field. In addition, the offshore Kupe South area is presently being evaluated as a potential gas condensate field.

HISTORY

Pre-Discovery

Several oil seeps and numerous gas seeps have been found in New Zealand, and they attracted attention to the country's oil and gas potential early in the nineteenth century. Exploration began in New Zealand as early as 1839 with the digging of pits near oil seeps. New Zealand's oil exploration effort began in the Taranaki basin in 1865, when the Alpha well was drilled adjacent to oil seeps then active at Motorua, near New Plymouth. This well produced 2 bbl/day. Between then and 1965, 46 wells were drilled in the vicinity of New Plymouth, of which nine were successful. Between 1934 and 1972, the Motorua field produced 216,000 bbl of oil. The field was abandoned in 1972.

Several unsuccessful wells were drilled elsewhere in Taranaki prior to 1955. Like those near New Plymouth, none of these wells reached Kapuni Group sediments; the deepest well drilled was Midhirst #1 (3330 m).

Discovery

A new era in exploration began in 1955 when the Shell-B.P.-Todd consortium was formed. Their efforts led to the Kapuni field discovery in 1959. The Kapuni #1 well spudded in on 27 January 1959 in the onshore South Taranaki graben. It was drilled to 3974.45 m total depth. Its location within the field is shown on Figure 4. The primary objective was the Mokau (Moki) Formation encountered at 2005 m. The Kapuni Group was the secondary objective, and the top of this group was drilled at 3245.6 m (Figures 5 and 6).

The "K3E" hydrocarbon reservoir was found 1400 ft (427 m) below the top of the Kapuni Formation (Figure 6). Two production tests of individual sandstones within the interval 12,366–12,395 ft (3769–3778 m) produced 1.9 million ft^3/d of gas and 41 m^3 (262 bbl) of condensate to prove the accumulation. The oil gravity was 43°–53° API and 15%–30% wax. The gas included 48% CO_2. The Kapuni #1 was suspended as a gas-condensate producer in October 1958.

Post-Discovery

Up to 1965, the Shell-B.P.-Todd consortium drilled several wildcat wells onshore, both in the Taranaki and neighboring North and South Wanganui basins, but without commercial success. During this period, three appraisal wells (Kapuni #2, #3, and #4) were drilled on the Kapuni structure. Following this development drilling, the Kapuni field commenced production in 1970.

After production began, with facilities capable of delivering up to 90 mcf/d of gas, a seismic survey was carried out early in 1971 to define more clearly the field and redetermine the original recoverable reserves. In 1974 and 1975, a further round of drilling was undertaken. Eight wells (#5 through #11), including a redrill of Kapuni #3 which had watered out, and an appraisal outstep, Kapuni #8, were staked on the structure. Two further wells, Kapuni #12, drilled to the east of Kapuni #8 (offstructure), and

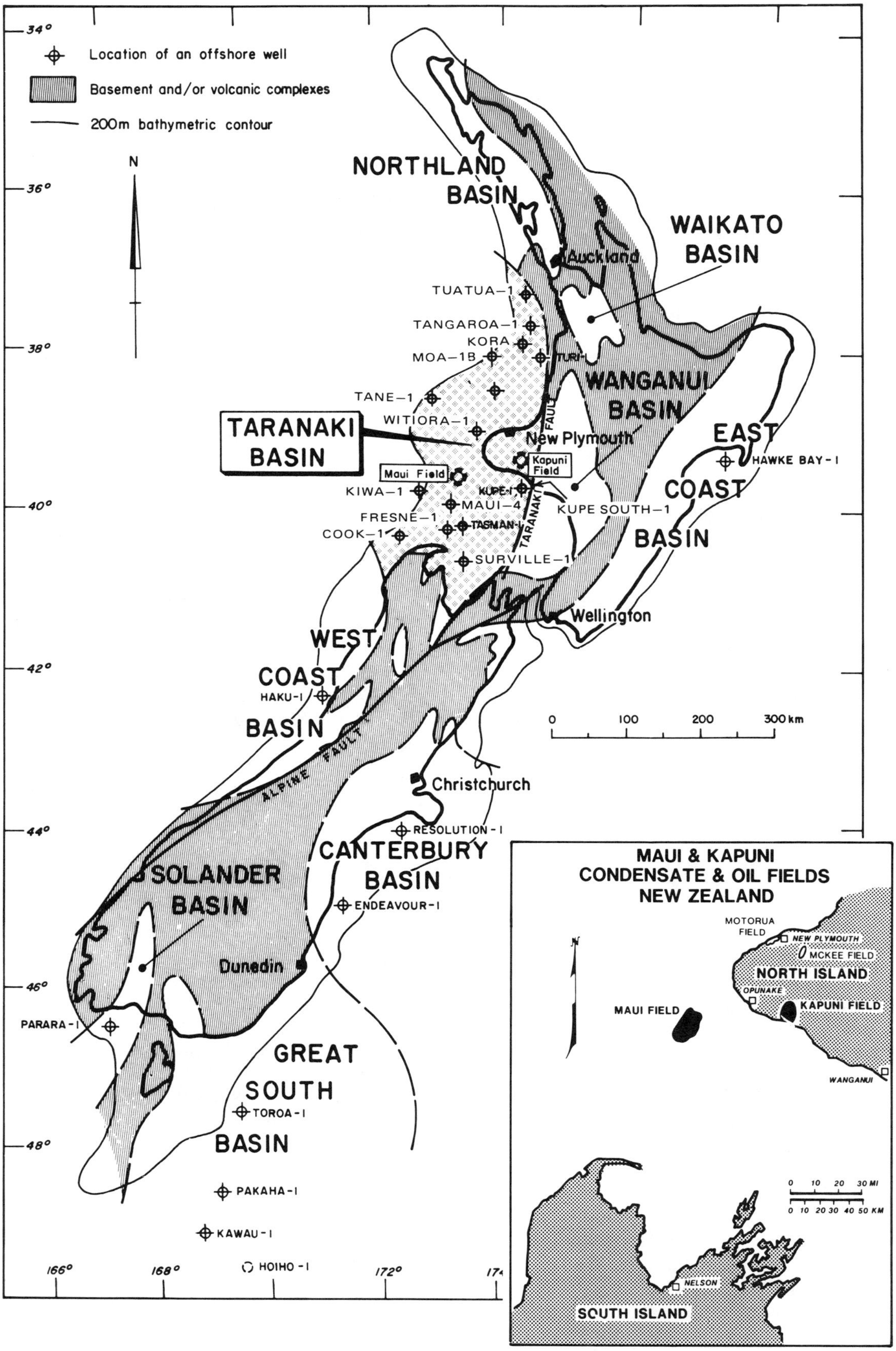
Location of an offshore well
Basement and/or volcanic complexes
200m bathymetric contour
N
NORTHLAND BASIN
WAIKATO BASIN
Auckland
TUATUA-1
TANGAROA-1
KORA
MOA-1B
TURI-1
TANE-1
WITIORA-1
TARANAKI BASIN
WANGANUI BASIN
New Plymouth
Kapuni Field
Maui Field
KIWA-1
KUPE-1
MAUI-4
FRESNE-1
TASMAN-1
COOK-1
SURVILLE-1
KUPE SOUTH-1
TARANAKI FAULT
EAST COAST BASIN
HAWKE BAY-1
Wellington
WEST COAST BASIN
HAKU-1
ALPINE FAULT
0 100 200 300 km
Christchurch
RESOLUTION-1
CANTERBURY BASIN
ENDEAVOUR-1
SOLANDER BASIN
Dunedin
PARARA-1
GREAT SOUTH BASIN
TOROA-1
PAKAHA-1
KAWAU-1
HOIHO-1
34°
36°
38°
40°
42°
44°
46°
48°
166°
168°
172°
MAUI & KAPUNI
CONDENSATE & OIL FIELDS
NEW ZEALAND
MOTORUA FIELD
NEW PLYMOUTH
MCKEE FIELD
NORTH ISLAND
OPUNAKE
MAUI FIELD
KAPUNI FIELD
WANGANUI
0 10 20 30 MI
0 10 20 30 40 50 KM
NELSON
SOUTH ISLAND

Figure 1. Index map and sedimentary basins in New Zealand (after Pilaar and Wakefield, 1984).

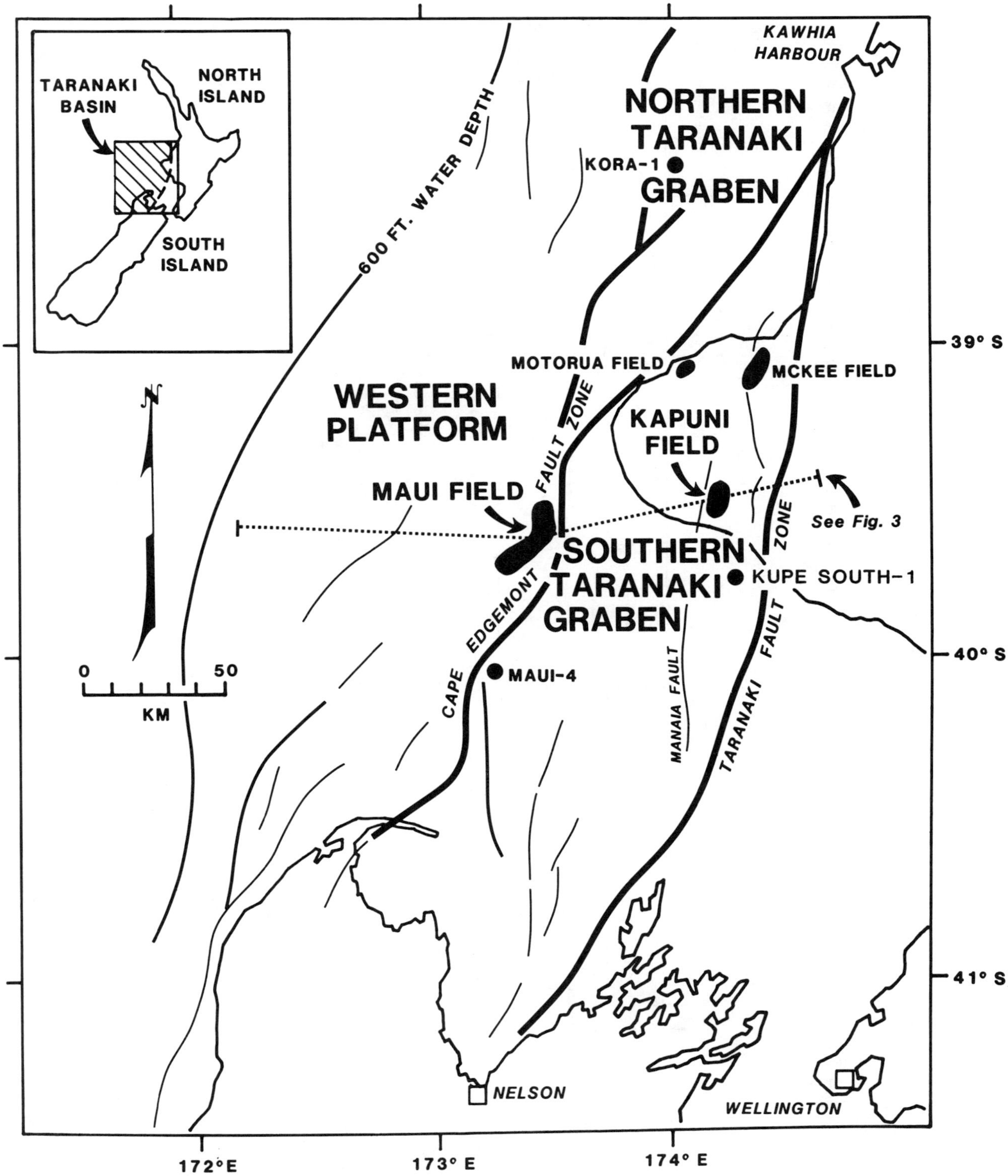
TARANAKI
BASIN
NORTH
ISLAND
SOUTH
ISLAND
KAWHIA
HARBOUR
NORTHERN
TARANAKI
GRABEN
KORA-1
600 FT. WATER DEPTH
MOTORUA FIELD
MCKEE FIELD
WESTERN
PLATFORM
KAPUNI
FIELD
MAUI FIELD
FAULT ZONE
See Fig. 3
SOUTHERN
TARANAKI
GRABEN
KUPE SOUTH-1
CAPE EDGEMONT
MAUI-4
MANAIA FAULT
TARANAKI FAULT ZONE
0
50
KM
NELSON
WELLINGTON
39° S
40° S
41° S
172°E
173° E
174° E

Figure 2. Taranaki basin, major structural elements (after King and Robinson, in press).

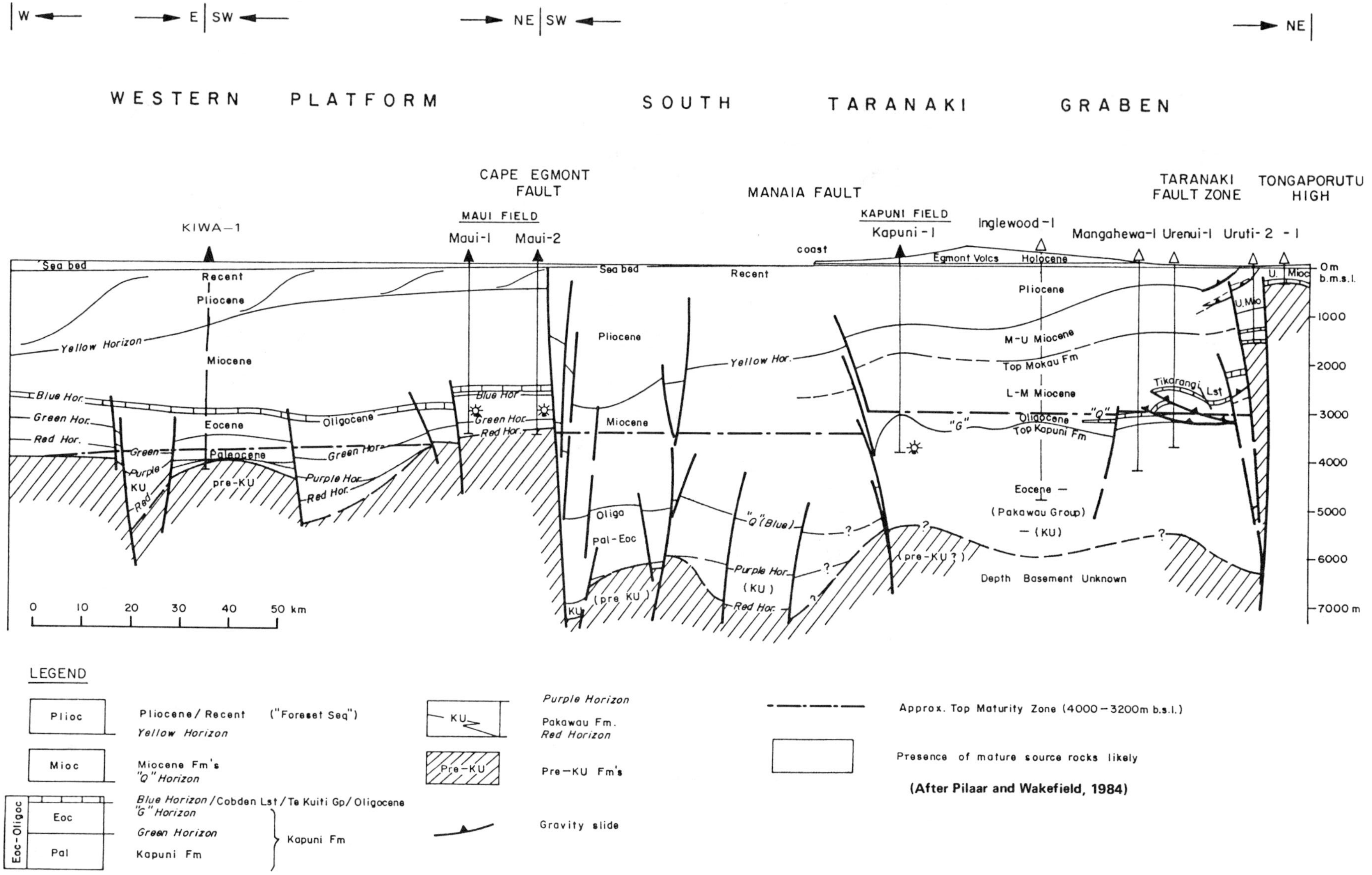
W
E
SW
NE
SW
NE
WESTERN PLATFORM
SOUTH TARANAKI GRABEN
CAPE EGMONT FAULT
MANAIA FAULT
TARANAKI FAULT ZONE
TONGAPORUTU HIGH
KIWA-1
MAUI FIELD
Maui-1
Maui-2
KAPUNI FIELD
Kapuni-1
Inglewood-1
Mangahewa-1
Urenui-1
Uruti-2 - 1
coast
Sea bed
Recent
Pliocene
Miocene
Eocene
Paleocene
Oligocene
Yellow Horizon
Blue Hor.
Green Hor.
Red Hor.
Purple Hor.
pre-KU
KU
Egmont Volcs
Holocene
M-U Miocene
Top Mokau Fm
L-M Miocene
Oligocene
Top Kapuni Fm
"G"
"Q"
Tikorangi Lst
Eocene
(Pakawau Group)
(KU)
Depth Basement Unknown
Oliga
Pal-Eoc
"Q"(Blue)
U. Mioc
0 m b.m.s.l.
1000
2000
3000
4000
5000
6000
7000 m
0
10
20
30
40
50 km
LEGEND
Plioc
Pliocene/Recent ("Foreset Seq")
Yellow Horizon
Mioc
Miocene Fm's
"Q" Horizon
Eoc-Oligoc
Eoc
Pal
Blue Horizon/Cobden Lst/Te Kuiti Gp/Oligocene
"G" Horizon
Green Horizon
Kapuni Fm
Kapuni Fm
Purple Horizon
Pakawau Fm.
Red Horizon
Pre-KU Fm's
Gravity slide
Approx. Top Maturity Zone (4000-3200m b.s.l.)
Presence of mature source rocks likely
(After Pilaar and Wakefield, 1984)

Figure 3. Taranaki basin, structural cross section. See Figure 2 for location.

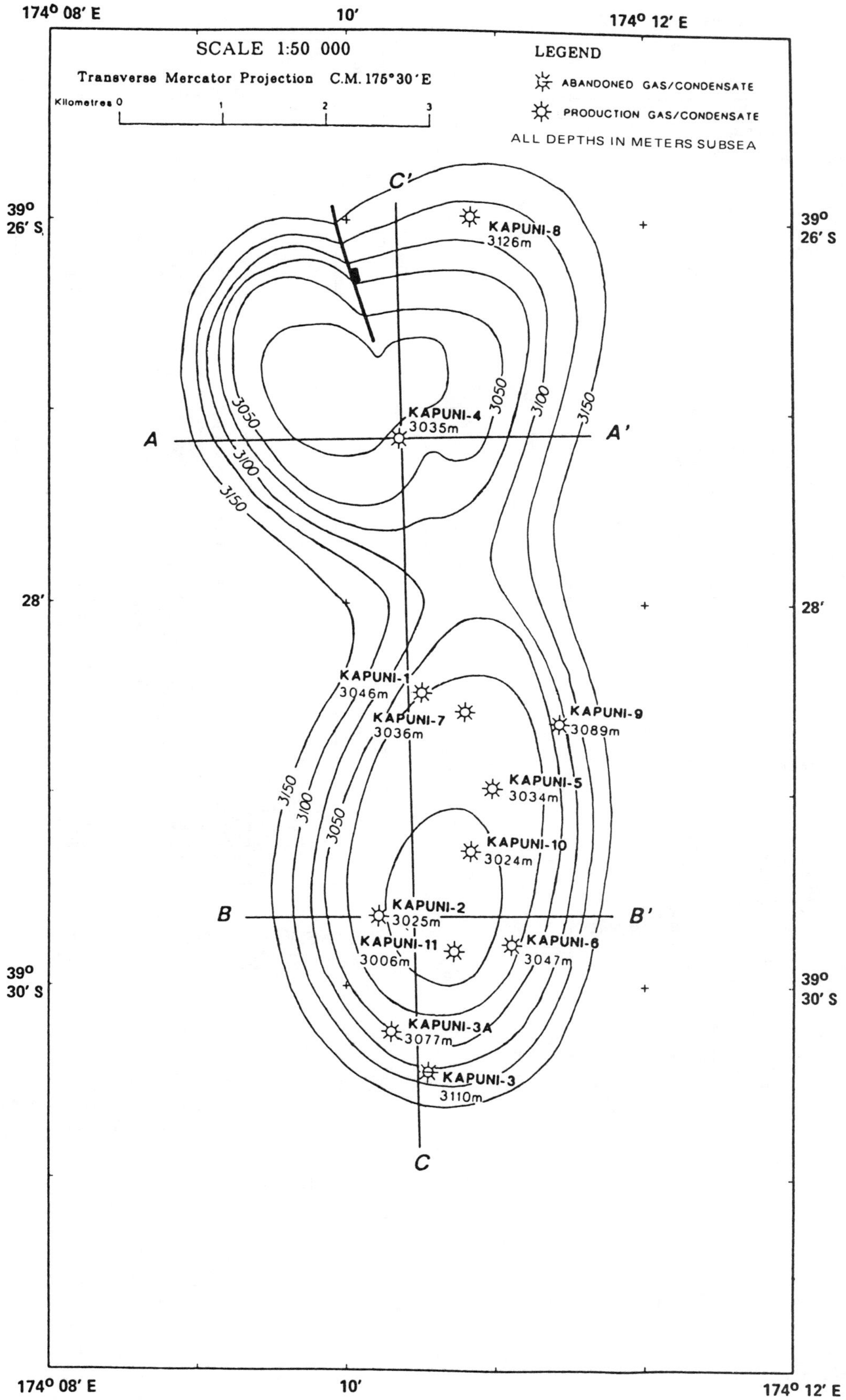
174° 08′ E
10′
174° 12′ E
SCALE 1:50 000
Transverse Mercator Projection C.M. 175°30′E
Kilometres 0 1 2 3
LEGEND
ABANDONED GAS/CONDENSATE
PRODUCTION GAS/CONDENSATE
ALL DEPTHS IN METERS SUBSEA
39° 26′ S
28′
39° 30′ S
C′
KAPUNI-8 3126m
KAPUNI-4 3035m
3050
3100
3150
A
A′
KAPUNI-1 3046m
KAPUNI-7 3036m
KAPUNI-9 3089m
KAPUNI-5 3034m
KAPUNI-10 3024m
KAPUNI-2 3025m
B
B′
KAPUNI-11 3006m
KAPUNI-6 3047m
KAPUNI-3A 3077m
KAPUNI-3 3110m
C

Figure 4. Structural contours on top Kapuni Group. Contour interval, 25 m. See Figure 9 for structural cross sections. (After Haskell, 1986.)

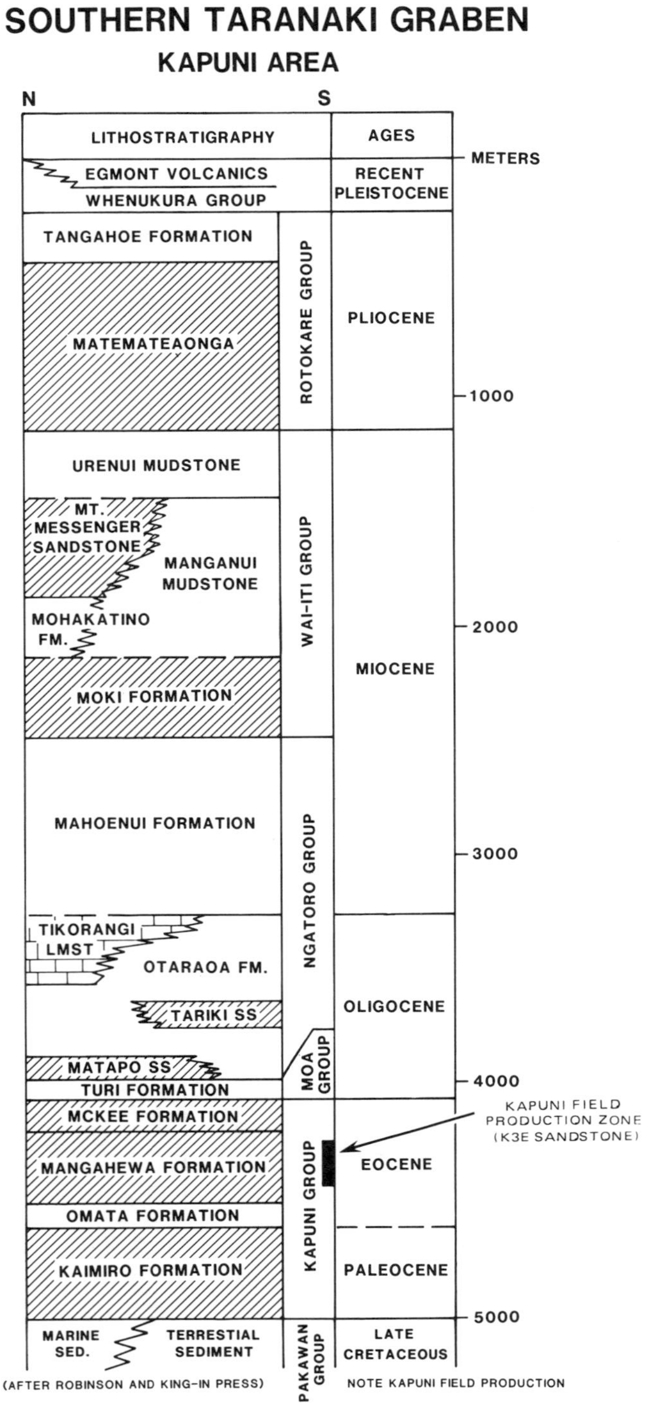
SOUTHERN TARANAKI GRABEN
KAPUNI AREA
N
S
LITHOSTRATIGRAPHY
AGES
METERS
EGMONT VOLCANICS
WHENUKURA GROUP
RECENT
PLEISTOCENE
TANGAHOE FORMATION
MATEMATEAONGA
ROTOKARE GROUP
PLIOCENE
1000
URENUI MUDSTONE
MT. MESSENGER SANDSTONE
MANGANUI MUDSTONE
MOHAKATINO FM.
WAI-ITI GROUP
2000
MIOCENE
MOKI FORMATION
MAHOENUI FORMATION
NGATORO GROUP
3000
TIKORANGI LMST
OTARAOA FM.
TARIKI SS
OLIGOCENE
MATAPO SS
MOA GROUP
4000
TURI FORMATION
MCKEE FORMATION
KAPUNI FIELD PRODUCTION ZONE (K3E SANDSTONE)
MANGAHEWA FORMATION
EOCENE
OMATA FORMATION
KAPUNI GROUP
KAIMIRO FORMATION
PALEOCENE
5000
MARINE SED.
TERRESTIAL SEDIMENT
PAKAWAN GROUP
LATE CRETACEOUS
(AFTER ROBINSON AND KING-IN PRESS)
NOTE KAPUNI FIELD PRODUCTION

Figure 5. Generalized stratigraphic section, South Taranaki graben.

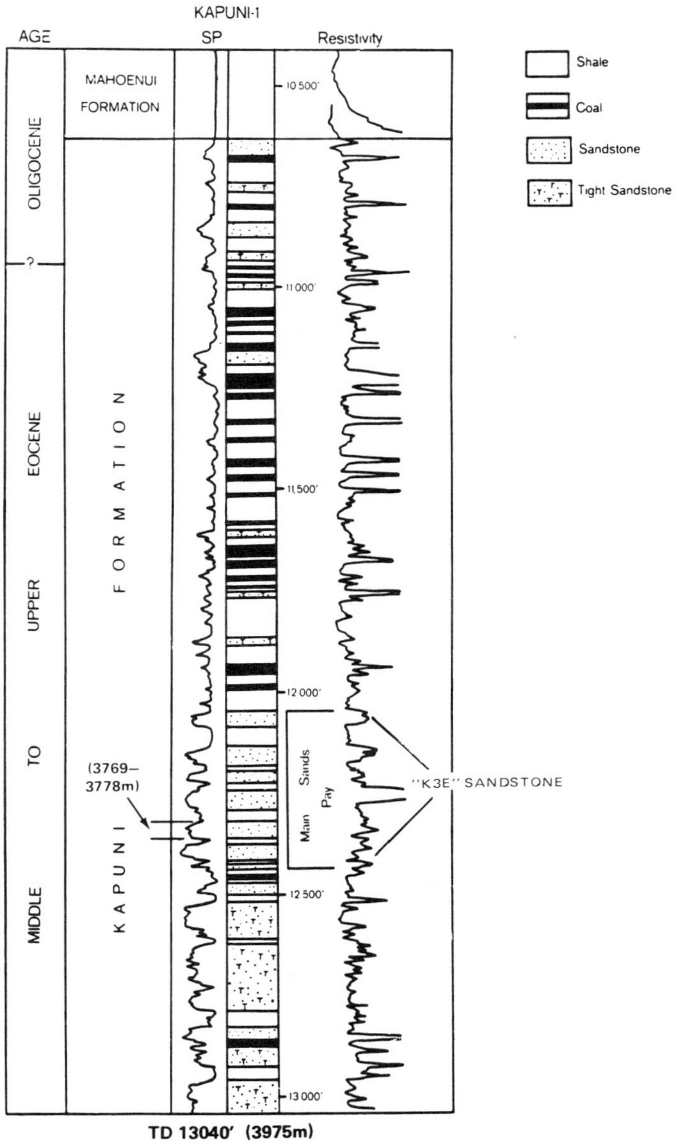
KAPUNI-1
AGE
SP
Resistivity
Shale
Coal
Sandstone
Tight Sandstone
MAHOENUI FORMATION
OLIGOCENE
10 500'
11 000'
11 500'
12 000'
12 500'
13 000'
MIDDLE TO UPPER EOCENE
KAPUNI FORMATION
(3769–3778m)
Main Sands Pay
"K3E" SANDSTONE
TD 13040' (3975m)

Figure 6. Well log of the Kapuni Formation from Kapuni-1 well. SP, spontaneous potential. (After McBeath, 1977.)

Kapuni Deep #1, located north-northeast of Kapuni #2, were drilled in 1983.

In 1969, the Shell consortium made a significant offshore discovery when they found the Maui gas-condensate field with estimated ultimate recovery of 5.3 tcf of gas, 230 million bbl NGL, and 300 million bbl of oil. This was followed a year later by the discovery of a subcommercial oil accumulation in Maui #4. In onshore Taranaki, only one well (Urenui #1; Aquitaine) was drilled between 1965 and 1977. This well drilled a structure that was then poorly understood but is now known to be part of the Tarata overthrust trend. Urenui #1 was tested at 1.3 MCFGD and 44 bbl/d condensate, then abandoned as noncommercial.

In 1978, the newly formed Petroleum Corporation of New Zealand (Petrocorp) took over the license to prospect the entire Taranaki Peninsula. Their second well (McKee #1) was located near the onshore northern end of the Tarata thrust zone. The well was noncommercial, but it was the discovery well of the McKee oil field, which through 12 stepout wells now produces 8000 BOPD.

Following the discovery of the McKee field, Petrocorp drilled 19 exploratory wells in onshore

Taranaki and discovered 5 smaller fields. To date, 5500 km of seismic reflection data have been collected in onshore Taranaki.

Two recent significant tests in the offshore Taranaki basin have sparked renewed interest in the area. The TCPL Resources, Ltd., 3B Kupe South appraisal well flowed 10,000 bbl/d during three drill-stem tests. It is located 60 km southeast of the Maui field in the South Taranaki graben. The Kora #1 well, drilled by Arco Petroleum New Zealand, Inc., 115 km north-northeast of the Maui field in the North Taranaki graben, has reported possible commercial hydrocarbons.

DISCOVERY METHOD

Geologic evaluation of the Kapuni field area was initiated with structural and stratigraphic surface mapping. In March 1958, a seismic survey defined the Kapuni anticline, and this structure was proposed for drilling. The Midhirst #1 well, 24 km north-northeast of the Kapuni structure, provided a measure of subsurface control to which the seismic results were tied, but it did not reach the most prominent reflector in the Kapuni area. This reflector was thought to represent the Oligocene Te Kuiti limestone or the Eocene-lower Oligocene coal measures known to be present on the South Island. However, it was thought that similar sediments would be present in the Kapuni area and that they must be considered oil source and reservoir rocks.

STRUCTURE

Tectonic History

During late Paleozoic through Jurassic time, the New Zealand geosyncline was rapidly subsiding and filling with turbidites. The geosyncline essentially was filled, then uplifted and partially stripped during the Rangitata orogeny in Late Jurassic and Early Cretaceous time. New Zealand began to separate from Australia in the Late Cretaceous. At this time the proto-Taranaki basin existed as the southeastern-most limit of the New Caledonia basin, which extended into the New Zealand continental platform (Figure 7). In response to the Tasman Sea rifting, a series of north-south-trending subbasins and half-grabens developed in the Taranaki region. Subsidence in these depocenters was initially rapid but gradually slowed as the rate of crustal cooling diminished. The Eocene and early Oligocene was marked by a period of tectonic stability interrupted in late Oligocene by the onset of rapid subsidence in the Taranaki region (Hayward, 1987). This event is attributed to oblique extension and westerly downthrow on the Taranaki fault, which in turn is probably related to the development of a transform plate boundary through New Zealand and general foundering of the New Zealand platform. Increasing oblique compression on the plate boundary caused several episodes of fault reversal and subbasin inversion.

Emplacement of low-angle overthrusts along the eastern margin of the Taranaki basin occurred from early to late Miocene time. Thereafter, the South Taranaki graben region continued to subside while the area south of the graben experienced pronounced uplift.

Conversely, to the north, extensional tectonics dominated, and movement along an en echelon series of antithetic normal faults produced the North Taranaki graben.

The Western platform remained a relatively stable block throughout most of the Tertiary and was affected only by Upper Cretaceous to Eocene normal block faulting.

Increased convergence along the plate boundary during Pleistocene time resulted in the Taranaki Peninsula region being uplifted and tilted. Volcanic extrusions were emplaced during this period.

Regional Structure

New Zealand is a fragmented block of continental crust isolated by the Tasman Sea from its parent continent, Australia, and is in an active seismic-volcanic belt.

The region northwest of the Alpine fault (Figure 1) and west of the Taranaki fault, which includes the Taranaki basin, is a separate tectonic province from southeastern New Zealand. This northwestern region has generally been more stable in the post-rift period than its southeastern counterpart, remaining essentially fixed relative to the Challenger plateau and Lord Howe rise. The subsidence and sedimentary history of the Taranaki basin reflects three predominant tectonic regimes: extensional, obliquely extensional, and obliquely compressional. These influences evolved in relation to the changing nature and position of the New Zealand continental block with respect to the Pacific and Indonesian-Australian plate boundary.

The Taranaki basin can be defined by the Taranaki fault and the continental shelf break (approximately 200 m isobath). To the north, the basin merges arbitrarily with the southwestward offshore portions of the Northland basin, while to the south, it overlaps basin and range provinces of the northwestern South Island. Two distinct elements of the Taranaki basin are recognized: the Taranaki graben to the east and the Western platform. The Cape Egmont fault zone separates these elements. The graben is divided into northern and southern segments, differentiated mainly by the greater depths of the northern graben. The two areas are separated by a complex zone of normal faults, downthrown to the northwest (Cook and King, 1987).

Two main fault trends are found within the basin: north-south and northeast-southwest. These tectonic

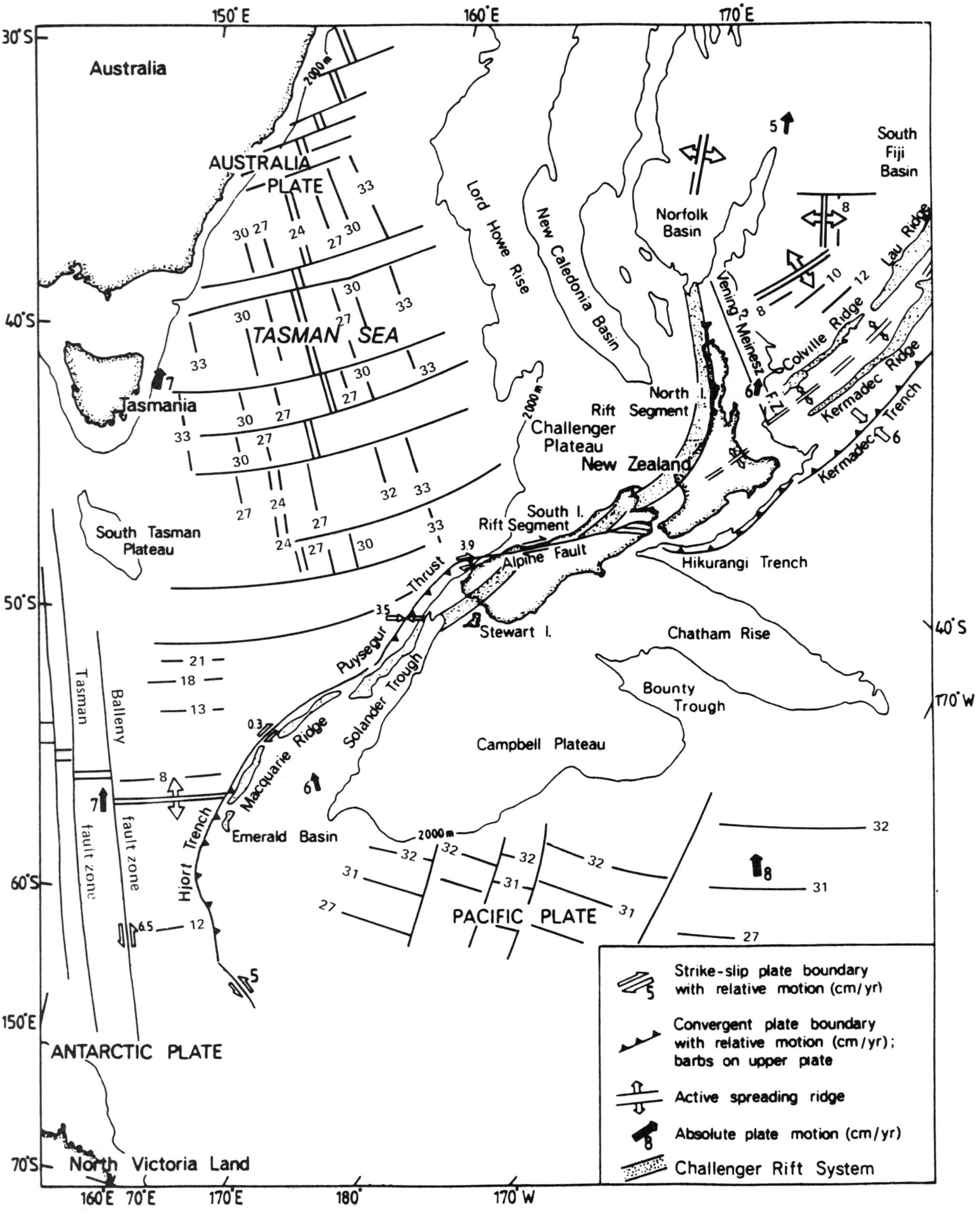
150°E
160°E
170°E
30°S
40°S
50°S
60°S
70°S
Australia
AUSTRALIA PLATE
TASMAN SEA
Tasmania
South Tasman Plateau
Lord Howe Rise
New Caledonia Basin
Norfolk Basin
South Fiji Basin
Lau Ridge
Colville Ridge
Kermadec Ridge
Kermadec Trench
Vening Meinesz F.Z.
North I.
Rift Segment
Challenger Plateau
New Zealand
South I.
Rift Segment
Alpine Fault
Puysegur Thrust
Hikurangi Trench
Stewart I.
Chatham Rise
Bounty Trough
Campbell Plateau
Solander Trough
Macquarie Ridge
Emerald Basin
Hjort Trench
Tasman fault zone
Balleny fault zone
PACIFIC PLATE
ANTARCTIC PLATE
North Victoria Land
40°S
170°W
150°E
160°E 70°E
170°E
180°
170°W
Strike-slip plate boundary with relative motion (cm/yr)
Convergent plate boundary with relative motion (cm/yr); barbs on upper plate
Active spreading ridge
Absolute plate motion (cm/yr)
Challenger Rift System

Figure 7. Plate reconstruction and tectonic character of the Australia-Pacific plate boundary showing Alpine fault dislocation of the Challenger rift system through western New Zealand. (Modified after Kamp, 1986; copyright the American Association of Petroleum Geologists.)

trends control the structural configuration of the Taranaki basin. At their intersections, the north-south trend is usually offset by the younger, northeast-southwest oblique trend. This system of faults is interpreted as a first-order wrench system (Pilaar and Wakefield, 1984).

Taranaki Fault

The Taranaki fault was originally a normal fault, downthrown to the west. Compression during the Miocene caused it to become a transform fault, with a right-lateral sense of movement presumed to have been predominant. Oversteepening of the fault plane also produced a reverse sense of throw, still downthrown to the west. The angle of the fault plane increases from north to south (Cook and King, 1987).

Taranaki Graben

From Late Cretaceous to early Miocene, the Taranaki graben in the vicinity of the Kapuni field was an asymmetric feature, with maximum downthrow to the east, alongside the downthrown Taranaki fault. This was especially pronounced in late Oligocene and early Miocene. Subsequent events contributed to reversing the direction of asymmetry in the graben. These include overthrusting and doming in the early to late Miocene, Pliocene development of the North Taranaki graben, and Pleistocene tilting up to the northeast. As a result, the Pliocene-Pleistocene sequence is thickest in the west and southwest.

Local Structure

The Kapuni #1 well was drilled on the culmination of the seismically defined Kapuni (Manaia) structure. A detailed survey of the structure by the Shell consortium revealed a well-developed, rather steeply flanked north-south-trending asymmetrical structure with the Manaia fault on the west flank. Closure is provided by a saddle to the north. Closure at the middle Miocene layer was 180 ft (55 m). The top of the Eocene Kapuni coal measures shows a closure of approximately 2600 ft (792 m).

The Kapuni anticline is indicated on seismic to be a typical example of a growing structure. Folding movements started in the middle Oligocene and continued into the late Miocene. In the last period of folding, the structure rose above sea level.

The Kapuni structure at the top of the Kapuni K3E sandstone is shown on Figure 8. This structure map, based on well control (Figure 9) and seismic (Figure 10), defines a north-south anticlinal fold with a pronounced saddle between the Kapuni #4 and Kapuni #1 wells.

STRATIGRAPHY

The stratigraphy of the Taranaki graben cannot be closely correlated lithologically with that of the Western platform, and different formation names are used (Figures 5 and 11).

Upper Cretaceous (Pakawau Group)

The oldest sediments so far drilled in onshore Taranaki are Paleocene in age. However, Late Cretaceous sediments are inferred to form the base of the sedimentary sequence in this area, where they are most likely buried to approximately 6–9 km. Rocks of Late Cretaceous age have been widely encountered elsewhere in the basin, especially in the south and on the Western platform.

Initial Late Cretaceous Pakawau Group sedimentation in Taranaki was confined to localized depressions and faulted half-grabens on a well-weathered basement surface of otherwise subdued relief. Sediments were characteristically terrestrial, with sands, silts, carbonaceous mudstones, and coals commonly deposited, and conglomerates locally prevalent.

By the end of the Cretaceous, a shallow sea had transgressed from the northwest as far as the South Taranaki graben.

Paleocene–Eocene (Kapuni Group)

Throughout the Paleocene and Eocene, minor fluctuations in sea level with respect to the low-lying landscape in this area resulted in the alternating deposition of coastal plain, lagoonal, and inner shelf sediments along a paleoshoreline aligned northwest-southeast. These sediments are assigned to the Kapuni Group. Coastal sands of the McKee Formation were deposited along this trend during the late Eocene.

Oligocene–Lower Miocene (Moa/Ngatoro Groups)

By the start of the Oligocene, a greater marine influence on sedimentation developed both in the Taranaki Peninsula region and elsewhere in the basin where, up until this time, a gentle transgression from the northwest had been progressing.

Basin deepening, which initially resulted in the deposition of shallow-marine silts and muds (Turi Formation) over the coastal sand sequence in the South Taranaki graben, accelerated in the late Oligocene. Accompanying subsidence and inundation of land to the south and southeast reduced sediment supply, such that carbonates of the Tikorangi Formation were variously deposited over much of the Taranaki basin. However, terrigenous sedimentation dominated in the graben area, with sediment presumably being sourced from immediately east of the Taranaki fault. Although a relatively thick sequence of muds and silts resulted (Otaraoa and Mahoenui formations), subsidence rates greatly exceeded sediment supply, such that bathyal water

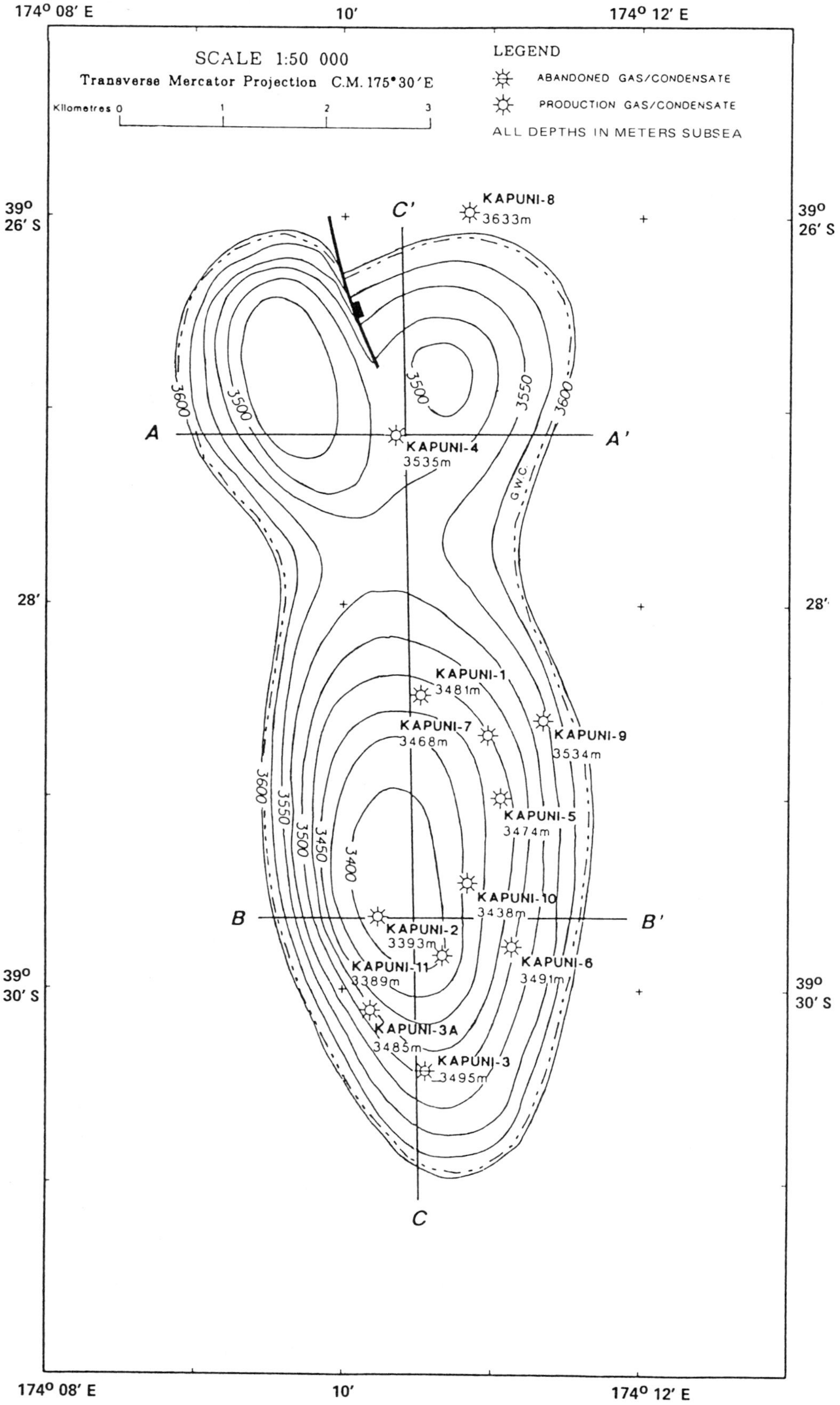
174° 08′ E
10′
174° 12′ E
SCALE 1:50 000
Transverse Mercator Projection C.M. 175°30′E
Kilometres 0 1 2 3
LEGEND
ABANDONED GAS/CONDENSATE
PRODUCTION GAS/CONDENSATE
ALL DEPTHS IN METERS SUBSEA
39° 26′ S
28′
39° 30′ S
C′
KAPUNI-8
3633m
A
A′
KAPUNI-4
3535m
3600
3500
3550
G.W.C.
KAPUNI-1
3481m
KAPUNI-7
3468m
KAPUNI-9
3534m
KAPUNI-5
3474m
3450
3400
KAPUNI-10
3438m
B
B′
KAPUNI-2
3393m
KAPUNI-11
3389m
KAPUNI-6
3491m
KAPUNI-3A
3485m
KAPUNI-3
3495m
C

Figure 8. Structural contours on Kapuni K3E sandstone (main sands). Contour interval, 25 m. See Figure 9 for structural cross section. (After Haskell, 1986.)

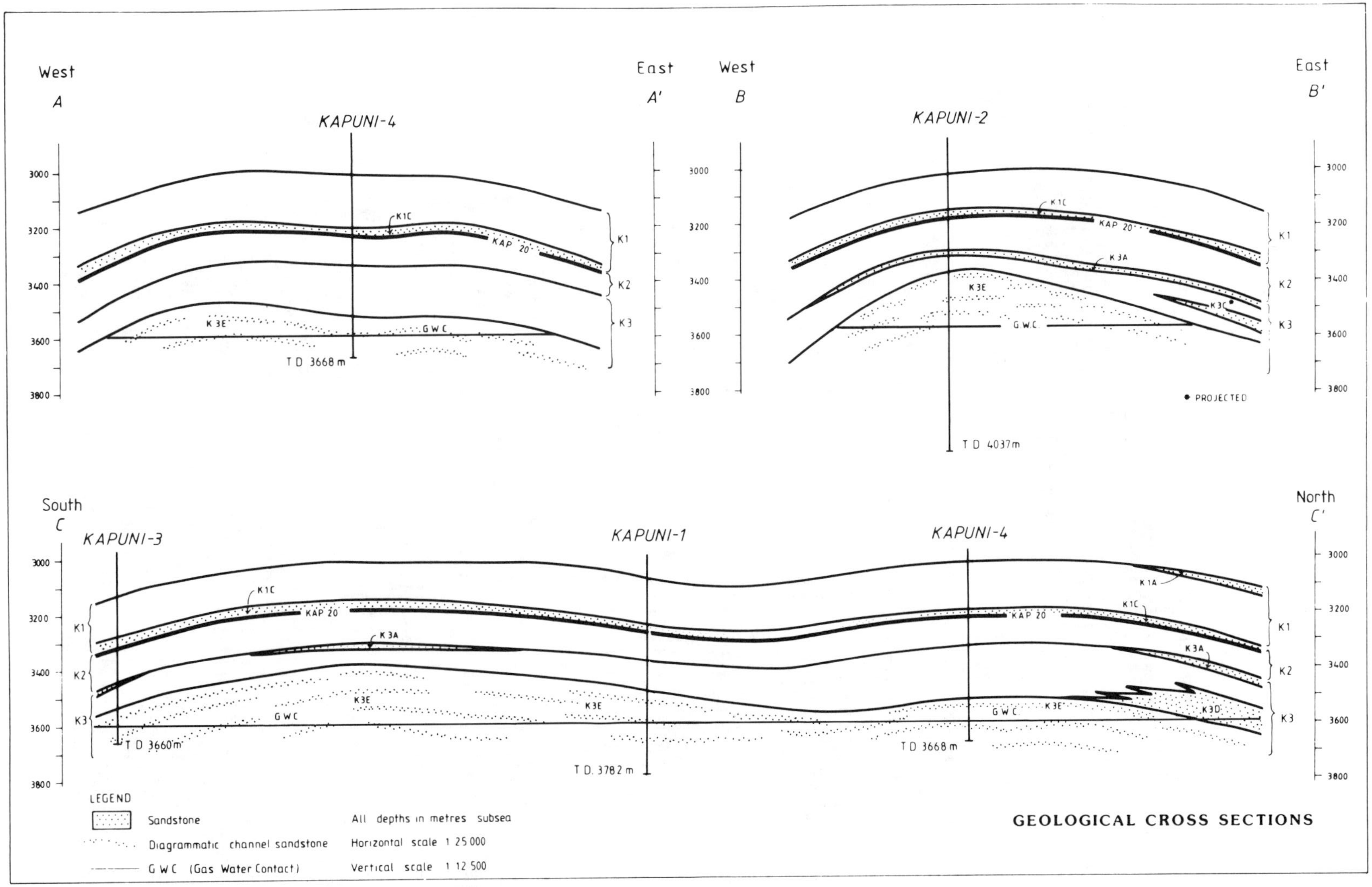
West
A
East
A'
KAPUNI-4
K1C
KAP 20
K1
K2
K3
K 3E
G W C
T D 3668 m
West
B
East
B'
KAPUNI-2
K 3A
K 3E
K 3C
T D 4037m
PROJECTED
South
C
North
C'
KAPUNI-3
KAPUNI-1
T D 3660m
T D. 3782 m
K 1A
K3D
3000
3200
3400
3600
3800
LEGEND
Sandstone
Diagrammatic channel sandstone
G W C (Gas Water Contact)
All depths in metres subsea
Horizontal scale 1 25 000
Vertical scale 1 12 500
GEOLOGICAL CROSS SECTIONS

Figure 9. Kapuni field structural and stratigraphic detail. See Figures 4 and 9 for cross section locations. (After Haskell, 1986.)

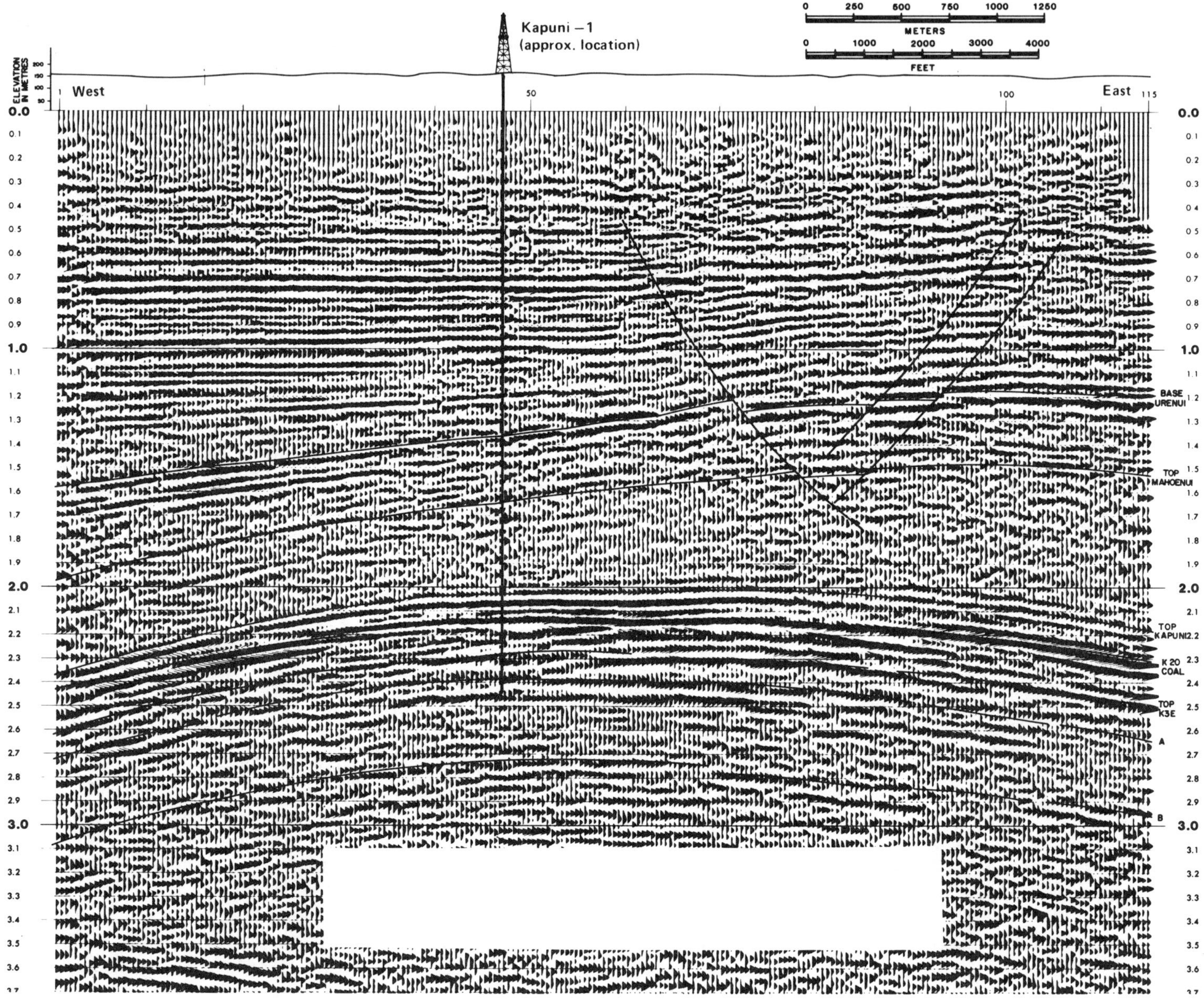
Kapuni –1
(approx. location)
0 250 500 750 1000 1250
METERS
0 1000 2000 3000 4000
FEET
ELEVATION IN METRES
West
East
50
100
115
BASE URENUI
TOP MAHOENUI
TOP KAPUNI
K 20 COAL
TOP K3E
A
B
0.0
1.0
2.0
3.0

Figure 10. East-west seismic profile across Kapuni structure in the vicinity of Kapuni-1. Note (1) domal structure at top K3E sand (2.5 sec); (2) marine onlap and unconformity at base Urenui; (3) flat-lying Pliocene sediments at 1.0 sec. Seismic line from Shell-B.P.-Todd.

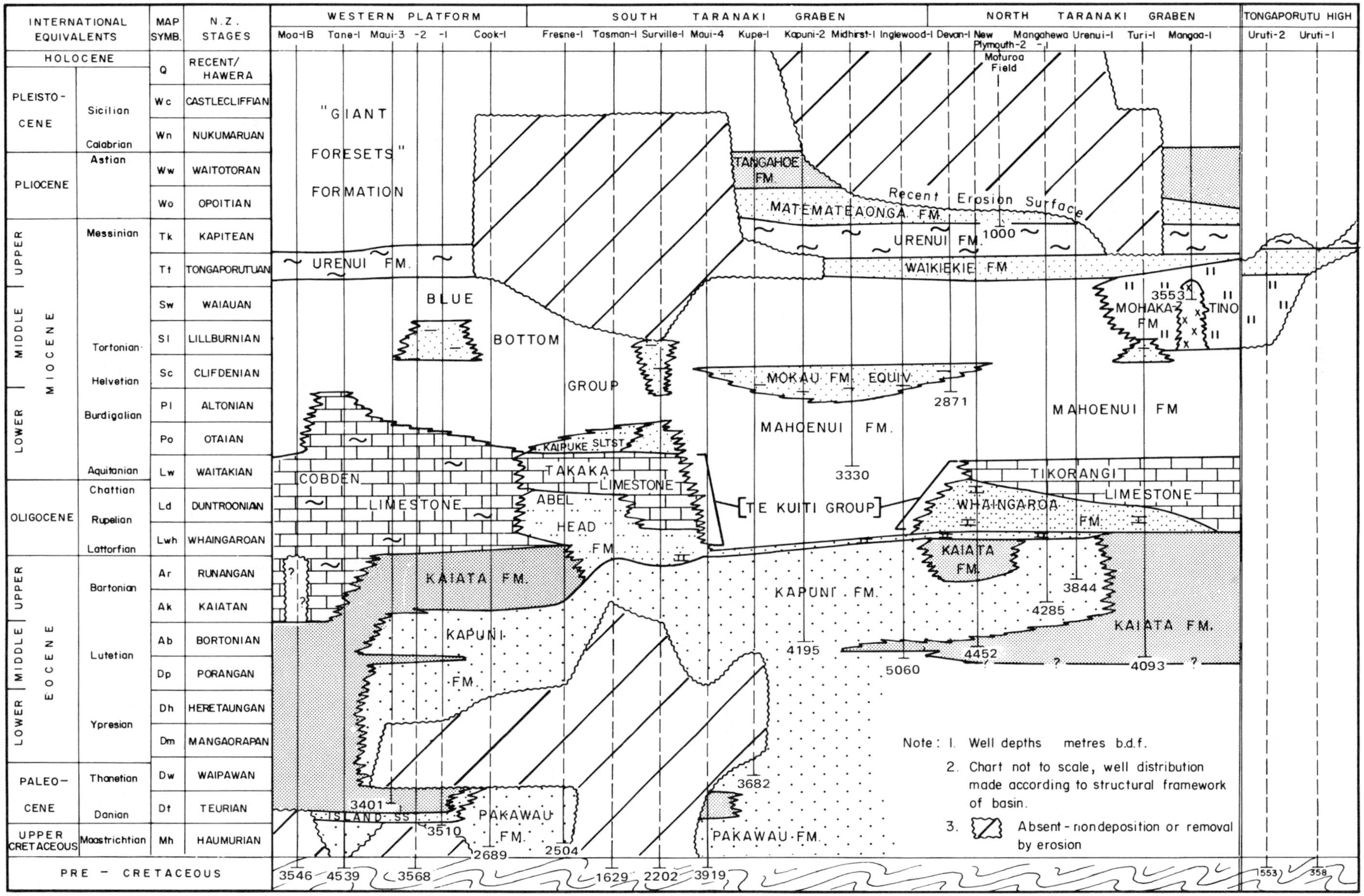
INTERNATIONAL EQUIVALENTS
MAP SYMB.
N.Z. STAGES
WESTERN PLATFORM
SOUTH TARANAKI GRABEN
NORTH TARANAKI GRABEN
TONGAPORUTU HIGH
"GIANT FORESETS" FORMATION
BLUE BOTTOM GROUP
URENUI FM.
TANGAHOE FM.
MATEMATEAONGA FM.
Recent Erosion Surface
WAIKIEKIE FM
MOHAKA FM
TINO
MOKAU FM. EQUIV.
MAHOENUI FM.
KAIPUKE SLTST
TAKAKA LIMESTONE
COBDEN LIMESTONE
ABEL HEAD FM
TE KUITI GROUP
TIKORANGI LIMESTONE
WHAINGAROA FM.
KAIATA FM.
KAPUNI FM.
ISLAND SS
PAKAWAU FM.
PRE - CRETACEOUS
Note: 1. Well depths metres b.d.f.
2. Chart not to scale, well distribution made according to structural framework of basin.
3. Absent - nondeposition or removal by erosion

Figure 11. Taranaki basin stratigraphic correlation chart (after Pilaar and Wakefield, 1984).

depths prevailed in this area into the early Miocene. Periodically, sands were channeled off the narrow shelf and locally deposited as submarine fans. These are referred to as the Tariki sandstone member of the Otaraoa Formation.

Middle-Upper Miocene (Wai-iti Group)

In the middle Miocene, sand distribution (Moki Formation) became more widespread, a result of increasing uplift in the hinterland. Sediment supply began to exceed subsidence rates, and large fans developed on the continental slope and slope base. This marked the onset of shelf progradation toward the northwest, a generally regressive sedimentary phase that has continued until the present day.

Subsidence and sedimentation patterns were interrupted along the eastern margin of the basin by compression and overthrusting that occurred episodically through the early middle Miocene and culminated in the late Miocene. Elsewhere in the South Taranaki graben region, subsidence continued largely unabated, and muds and silts of the Manganui Formation were deposited. Some sandy intervals of local distribution and presumed fan origin are found within this predominantly fine-grained middle to late Miocene sequence.

Shelf progradation did not become pronounced in the area until after the late Miocene, by which time the shelf edge ran roughly north-south through the center of the Taranaki graben.

About this time, andesitic volcanism in the north of the Taranaki basin produced considerable quantities of volcaniclastic sediment, mostly redeposited in deep water. Tuffaceous sandy and muddy sediments were deposited in the northern part of the graben but are best developed in the southern portion of the Northland basin. Following this, fan deposits of the Mt. Messenger Formation were laid down in the northern portion of the Taranaki graben. Thereafter, the latest Miocene was a period of relative quiescence, and fine-grained sediments of the Urenui Formation were deposited over most of the South Taranaki graben area.

Pliocene (Rotokare Group)

A major episode of uplift in the southeastern hinterland began supplying vast quantities of sediment to the Taranaki basin in the early Pliocene. Differential subsidence along the Taranaki fault allowed a thick sequence of shelf sediments to accumulate. The cyclic sedimentation exhibited by the sheet sands, coquina shell hash, and muds of this formation probably reflects the combined effects of tectonism, sea level changes, and storm activity.

Pleistocene

The onshore Taranaki region was uplifted and tilted in the Pleistocene. However, the peninsula's configuration also largely results from volcanic activity and associated sedimentation during this time (Cook and King, 1987).

TRAP

The trap in the Kapuni field is a compound anticlinal trap. It may owe its origin to the high-angle Manaia fault associated with local compression in an extensional setting during middle Miocene. The mudstones and shales of the Kapuni Formation seal the reservoir sandstones.

In all wells, a similar hydrocarbon-bearing zone was penetrated as in Kapuni #1, but in the Kapuni #2 and #3 wells, a second productive interval, the K1C sandstone, was encountered about 300 ft (91 m) above the K3E sandstone (Figure 9). Seventy-five feet (23 m) of microlog porosity was indicated to be hydrocarbon-bearing in the K1C sandstone in each of these two wells, but to date they remain undeveloped.

The gas-water contact was established at 11,825 ft (3604 m) subsea for the K3E sandstone, although it appeared to be somewhat higher at about 11,775 ft (3589 m) subsea in Kapuni #4 and 11,240 ft (3426 m) subsea for the K1C sandstone. For the K3E sandstone, the gross gas-bearing interval of 691 ft (211 m) was thickest in Kapuni #2 and thinnest in Kapuni #4 with 272 ft (83 m) (Figure 9). There are stratigraphic traps on the flanks of the structure where point bar sandstones terminate (K3D) against flood basin siltstones. Additional isolated point bar sandstones occur on the crest and flanks of the Kapuni structure (K1A, K3A, and K3C sandstone).

RESERVOIR

Reservoir pressures in the onshore Kapuni field are somewhat above hydrostatic, the pressure gradient being 0.526 psi/ft (6500 psi at 12,360 ft). The porosity and permeability of the sandstone comprising the K3E reservoir is highly variable, as indicated in a core cut from the Kapuni #3 well. The permeabilities are relatively low, averaging 15 md, although a few beds reach 500 md and account for much of the production. Intergranular porosity ranges up to around 21%, with an average of 15%.

The pressure gradients over the various sandstones indicate that there are two reservoir systems in the Kapuni field. The major reservoir is the K3E sandstone system. However, pressure data indicate that the K1A, K1C, and K3A sandstones are all within the same pressure regime, even though separated by mudstones. Fracture permeability provides the conduit for the latter system.

Production from the K3E sandstone wells varies from approximately 1.6 million m^3 (60 million standard ft^3) gas (CO_2 and hydrocarbons) and 500 m^3 (3120 bbl) condensate per day in Kapuni #6 to 500,000 m^3 (18 million standard ft^3) gas and 150 m^3 (936 bbl) condensate per day from Kapuni #4. The K1A sandstone in Kapuni #8 produced 320,000 m^3 (12 million standard ft^3) gas and 100 m^3 (626 bbl) condensate. The CO_2 and gas/oil ratios are constant over the field and its two reservoirs.

Ultimate recovery is estimated to be 28×10^9 m^3 of total gas and 5.8×10^6 m^3 condensate (1000×10^9 ft^3 gas and 35×10^6 bbl).

SOURCE

The Late Cretaceous and Eocene coal measure sequences are the only proven source rocks in the Taranaki basin, and their presence can be established over much of the basin from seismic and well data. Despite their common abundance of leaf cuticles, pollen, spores, and resins relative to woody material, the coals are classified as hydrogen-poor or gas-generative (Pilaar and Wakefield, 1984). However, high volatile perhydrous coals, which could be a source for oil, do occur in the West Coast basin of the South Island and might be found to occur in the Taranaki basin.

Similarities in the produced liquids from Maui, Kapuni, and Motorua fields suggest a common source for the oil and condensate. The oils of the Taranaki basin are high wax and paraffinic. The chemical and biomarker parameters consistently indicate that the oils are mainly derived from nonmarine source rocks deposited under anoxic and low bacterial freshwater swamp conditions. An exception is oil from the Tangaroa #1 well, which has a strong marine influence and represents a different source in the North Taranaki graben.

Pilaar and Wakefield (1984) have plotted the vitrinite reflection values measured in the various wells against depth. Maturation curves were drawn; one for wells drilled on the Western platform and another for those drilled in the Taranaki graben (Figure 12). From these data it can be assumed that top maturity level and, consequently, the zones of generation and expulsion of hydrocarbons lie at a greater depth on the Western platform than in the graben complex.

Pilaar and Wakefield (1984) indicate that this difference in depth cannot be readily explained by the small difference in geothermal gradients (16°–41°C/km) calculated for the two provinces. It seems more likely to be the result of the age difference of the "effective" overburden in the graben being Miocene while it is Pliocene to Pleistocene on the platform.

Pilaar and Wakefield have also constructed burial graphs giving the relationship between the depth of burial and geologic time for each vitrinite reflectance value measured in selected graben complex wells. The results are given on Figure 13. These curves indicate that present-day depths are close to maximum depths.

The effect of geologic time on the maturation process in the Taranaki basin is illustrated on Figure 14, in which the expulsion areas are outlined as they are thought to be today (Pilaar and Wakefield, 1984).

It is theorized that hydrocarbon charge for the Kapuni field is from the deeper portion of the South Taranaki graben. In particular, the Kapuni field appears well placed relative to the oil and gas expulsion area (Figures 12, 13, and 14).

The high CO_2 content (48%) in Kapuni field is of interest. Such concentrations usually are attributed to an association of carbonate rocks and high temperatures in proximity to volcanic centers. The volcanoes are present but the carbonates are lacking near Kapuni field. This suggests a direct contribution from the volcanoes.

EXPLORATION CONCEPTS

A study of Kapuni field and its neighbors, Maui and Motorua, did not reveal anything surprising. Seeps provide evidence of effective source beds; source bed quality deposits were confirmed by drilling through thick shales, mudstones, and coals that are rich in organic matter. The faults, especially the major Cape Egmont system, are logical migration avenues to feed the gas and oil to the traps, which are of a size to hold major reserves, given adequate reservoir capacity. Neither Maui nor Kapuni are filled to spill point. This suggests inadequate source, partial isolation from the source, inadequate seals, and escape migration, or a combination of these factors.

Using upgraded seismic and well data, the original recoverable reserves were redetermined at 460 million million BTU of dry gas, an increase of 84%. Estimated condensate recovery was 34 billion bbl.

If this field were to be evaluated today using present techniques, 3D seismic might be employed to define the discontinuous fluvial sand bodies in the upper Kapuni interval and more precisely map the fault trends and fault trend geometries.

ACKNOWLEDGMENTS

Occidental International Exploration and Production Company offered assistance in preparation of this manuscript. Special assistance was given by L. M. Seibert, typist; H. L. Scott and B. D. De La Cruz, graphics; and F. R. Abbott, research and editing. Conversations with J. M. Winterman, Occidental International Exploration and Production Company, and published data by Shell, B.P., and Todd Oil

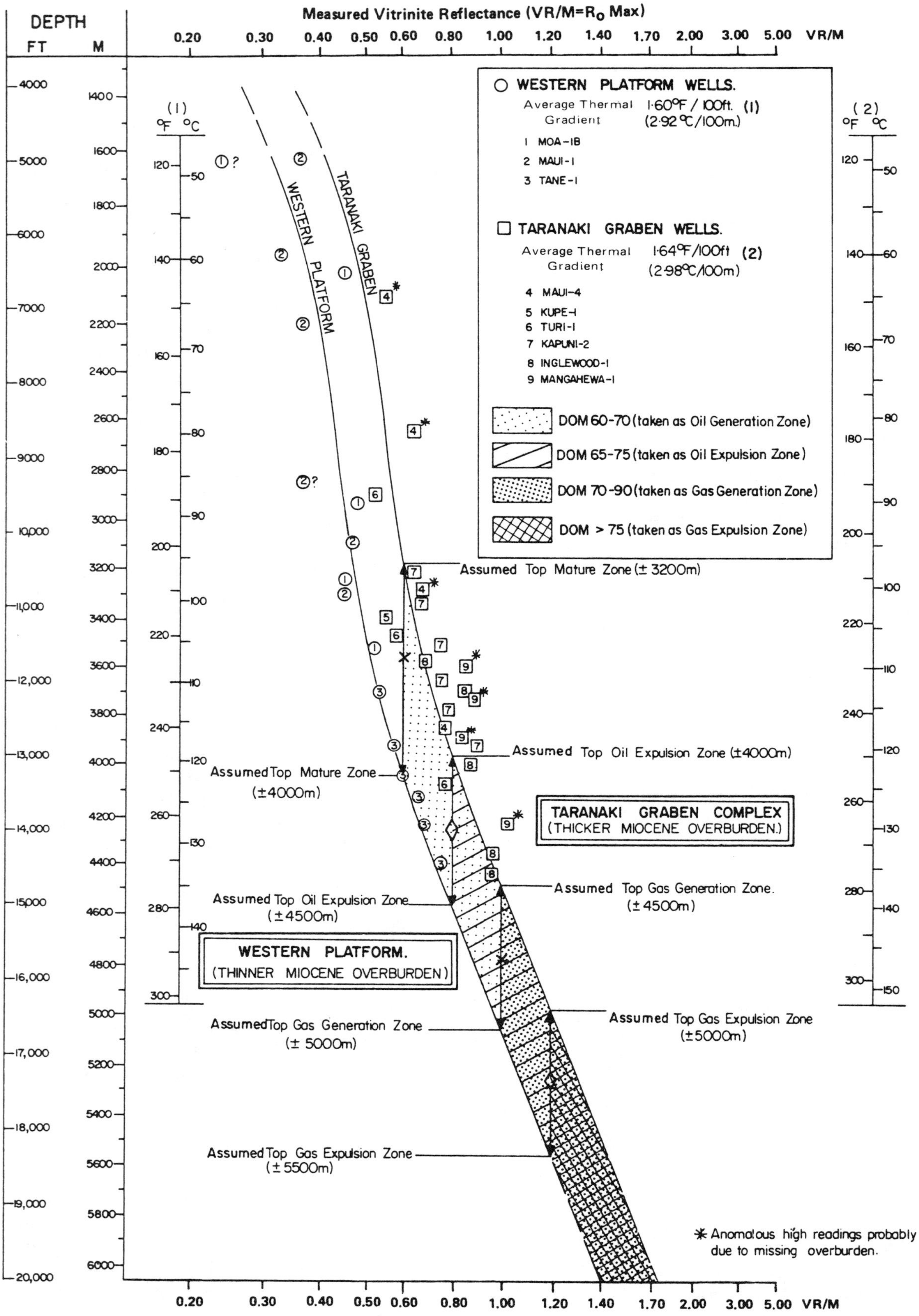

DEPTH
FT
M
Measured Vitrinite Reflectance (VR/M=Ro Max)
0.20 0.30 0.40 0.50 0.60 0.80 1.00 1.20 1.40 1.70 2.00 3.00 5.00 VR/M
WESTERN PLATFORM WELLS.
Average Thermal Gradient 1·60°F / 100ft. (1) (2·92°C/100m.)
1 MOA-1B
2 MAUI-1
3 TANE-1
TARANAKI GRABEN WELLS.
Average Thermal Gradient 1·64°F/100ft (2) (2·98°C/100m)
4 MAUI-4
5 KUPE-1
6 TURI-1
7 KAPUNI-2
8 INGLEWOOD-1
9 MANGAHEWA-1
DOM 60-70 (taken as Oil Generation Zone)
DOM 65-75 (taken as Oil Expulsion Zone)
DOM 70-90 (taken as Gas Generation Zone)
DOM > 75 (taken as Gas Expulsion Zone)
WESTERN PLATFORM
TARANAKI GRABEN
Assumed Top Mature Zone (± 3200m)
Assumed Top Oil Expulsion Zone (±4000m)
Assumed Top Mature Zone (±4000m)
TARANAKI GRABEN COMPLEX (THICKER MIOCENE OVERBURDEN.)
Assumed Top Gas Generation Zone. (± 4500m)
Assumed Top Oil Expulsion Zone (± 4500m)
WESTERN PLATFORM. (THINNER MIOCENE OVERBURDEN)
Assumed Top Gas Expulsion Zone (±5000m)
Assumed Top Gas Generation Zone (± 5000m)
Assumed Top Gas Expulsion Zone (± 5500m)
✱ Anomalous high readings probably due to missing overburden.

Figure 12. Taranaki basin vitrinite reflectance versus depth (after Pilaar and Wakefield, 1984).

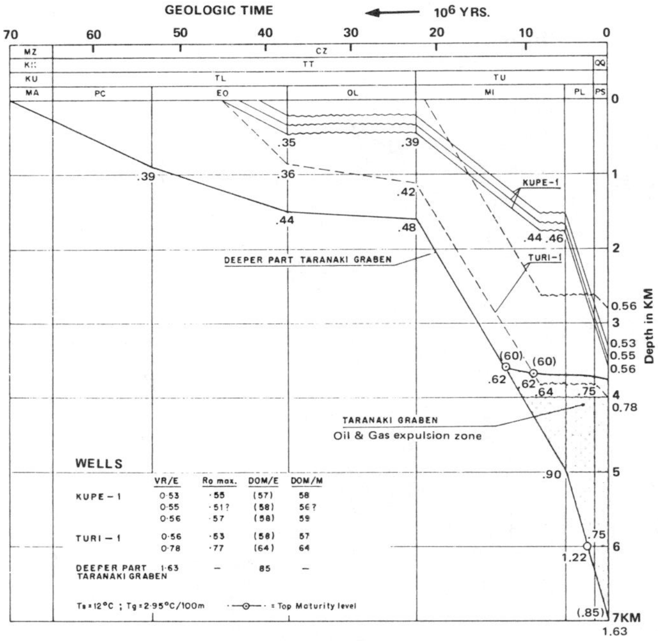
GEOLOGIC TIME
10^6 YRS.
70 60 50 40 30 20 10 0
MZ
KU
MA
PC
TL
EO
OL
CZ
TT
TU
MI
PL
PS
KUPE-1
DEEPER PART TARANAKI GRABEN
TURI-1
TARANAKI GRABEN
Oil & Gas expulsion zone
Depth in KM
7KM
WELLS
VR/E Ro max. DOM/E DOM/M
KUPE-1 0·53 ·55 (57) 58
0·55 ·51? (58) 56?
0·56 ·57 (58) 59
TURI-1 0·56 ·53 (58) 57
0·78 ·77 (64) 64
DEEPER PART TARANAKI GRABEN 1·63 – 85 –
Ts=12°C ; Tg=2·95°C/100m
= Top Maturity level

Figure 13. Burial graph, Taranaki graben complex (after Pilaar and Wakefield, 1984). DOM, degree of organic metamorphism; DOM/E, estimated; DOM/M, measured; VR/E, vitrinite reflectance equivalent; R.O., vitrinite reflectance of oil.

Services, Ltd., contributed greatly to this publication.
This paper draws heavily on the published work and illustrations of W.F.H. Pilaar and L. L. Wakefield as well as Haskell Exploration Services.

REFERENCES

Alexander, R., R. I. Kagi, G. W. Woodhouse, and J. K. Volkman, 1983, The geochemistry of some biodegraded Australian oils: APEA Journal, v. 23, n. 1, p. 53-63.

Analabs (Oil and Gas Division), 1984, Petroleum geochemistry of the Taranaki Basin: New Zealand Geological Survey unpublished open file petroleum report 1013.

Austin, P. M., et al., 1973, Structure and petroleum potential of eastern Chatham Rise, New Zealand: American Association of Petroleum Geologists Bulletin, v. 57, n. 3, p. 477-497.

Barnard, E. J., 1969, Design criteria for Kapuni Field facilities: N.Z. Engineering, v. 24, p. 15-22.

Beddoes, L. R., Jr. (ed.), 1973, Oil and gas fields of Australia, Papua New Guinea and New Zealand: Tracer Petroleum & Mining Publications Pty Ltd., 382 p.

Blake, M. C., et al., 1974, Active continental margins; comparisons between California and New Zealand, *in* C. A. Burke and C. L. Drake, eds., The geology of continental margins: Springer-Verlag New York, Inc., p. 853-872.

Carmalt, S. W., and B. St. John, 1986, Giant oil and gas fields, *in* M. T. Halbouty, ed., Future petroleum provinces of the world: American Association of Petroleum Geologists Memoir 40, p. 11-54.

Carter, R. M., and R. J. Morris, 1976, Cainozoic history of southern New Zealand; an accord between geological observations and plate tectonic predictions: Earth Science and Planetary Letters, v. 3, n. 1, p. 85-94.

Connan, J., 1974, Time-temperature relation in oil genesis: American Association of Petroleum Geologists Bulletin, v. 58, n. 12, p. 2516-2521.

Cook, R. A., 1987, The geology and geochemistry of the crude oils and source rocks of western New Zealand: N.Z. Geological Survey Report PR 1250, 380 p.

Cook, R. A., and P. R. King, 1987, Summary of petroleum geology of onshore Taranaki: Petroleum Exploration News, November, p. 6-11.

Cope, R. N., and I. J. Reed, 1967, The Cretaceous paleogeology of the Taranaki-Cook Strait area: Proceedings Australian Institute of Mining and Metallurgy.

Editorial Staff, 1975, Work starts to develop big Maui Field: Oil and Gas Journal, September 1, p. 52-53.

Editorial Staff, 1975, Maui Field soon to hit full stride: Oil and Gas Journal, October 27, p. 140-155.

Elphick, J. O., and R. P. Suggate, 1964, Depth/rank relations of high volatile bituminous coals: N.Z. Journal of Geology and Geophysics, v. 7, n. 3, p. 594-601.

Fairburn, S. G., 1980, Diagenesis of the Kapuni Formation—offshore South Taranaki: BSc thesis, Victoria University, Wellington, New Zealand, 33 p.

Furzey, D. G., 1970, Kapuni plant treats high CO_2 content natural gas: World Petroleum, September 1970, p. 48-54.

Gibbons, M. J., and S. Fry, 1983, A compilation of the geochemical characteristics of the oils and condensates of New Zealand: New Zealand Geological Survey unpublished open file report 989.

Griffiths, J. R., 1971, Reconstruction of the southwest Pacific margin of Gondwanaland: Nature, v. 234, November 26, p. 203-207.

Harding, T. P., 1974, Petroleum traps associated with wrench faulting: American Association of Petroleum Geologists Bulletin, v. 57, p. 74-96.

Haskell, T. R., 1986, A study of structural development, sandstone depositional systems, and hydrocarbon accumulation: Haskell Exploration Services Ltd., Monograph-1, p. 1-96.

Hayward, B. W., 1987, Paleobathymetry and structural and tectonic history of Cenozoic drillhole sequences in Taranaki Basin: N.Z. Geological Survey Report PAL 122, 63 p.

Hill, P. J., and J. D. Collen, 1978, The Kapuni sandstones from Inglewood-1 well, Taranaki—petrology and the effects of diagenesis on reservoir characteristics: N.Z. Journal of Geology and Geophysics, v. 21, n. 2, p. 215-228.

Hogan, J. A., 1979, Stratigraphy and sedimentology of the Kapuni Formation, Taranaki, New Zealand: MSc thesis, Victoria University, Wellington, New Zealand, 189 p.

Hood, A., C. C. M. Gutjahr, and R. I. Heacock, 1975, Organic metamorphism and the generation of petroleum: American Association of Petroleum Geologists Bulletin, v. 59, p. 986-996.

Kamp, P. J. J., 1986, The mid-Cenozoic Challenger Rift system of western New Zealand and its implications for the age of alpine fault inception: Geological Society of America Bulletin, v. 97, p. 255-281.

Katz, H. R., 1968, Potential oil formations in New Zealand and their stratigraphic position as related to basin evolution: New Zealand Journal of Geology and Geophysics, v. 11, n. 5, p. 1077-1133.

Katz, H. R., 1974, Margins of the southwest Pacific, *in* C. A. Burke and C. L. Drake, eds., The geology of continental margins: Springer-Verlag New York Inc., p. 549-565.

Katz, H. R., 1974, Offshore petroleum potential in New Zealand: APEA Journal, v. 14, n. 1, p. 3-13.

Katz, H. R., 1976, Sedimentary basins and petroleum prospects, onshore and offshore New Zealand: American Association of Petroleum Geologists Memoir 25, Circum-Pacific Energy and Mineral Resources, p. 217-228.

Kear, D. 1965, The Kapuni gas-condensate field, New Zealand—a case study: Third Symposium on the Development of Petroleum Resources of Asia and the Far East: Mineral Resources Development Series No. 29, p. 86-91.

King, P. R., and P. H. Robinson, in press, An overview of the Taranaki region geology, New Zealand: New Zealand Geological Survey DSIR, Lower Hutt, New Zealand.

Kingma, J. T., 1974, The geological structure of New Zealand: New York, John Wiley & Sons, Inc., 407 p.

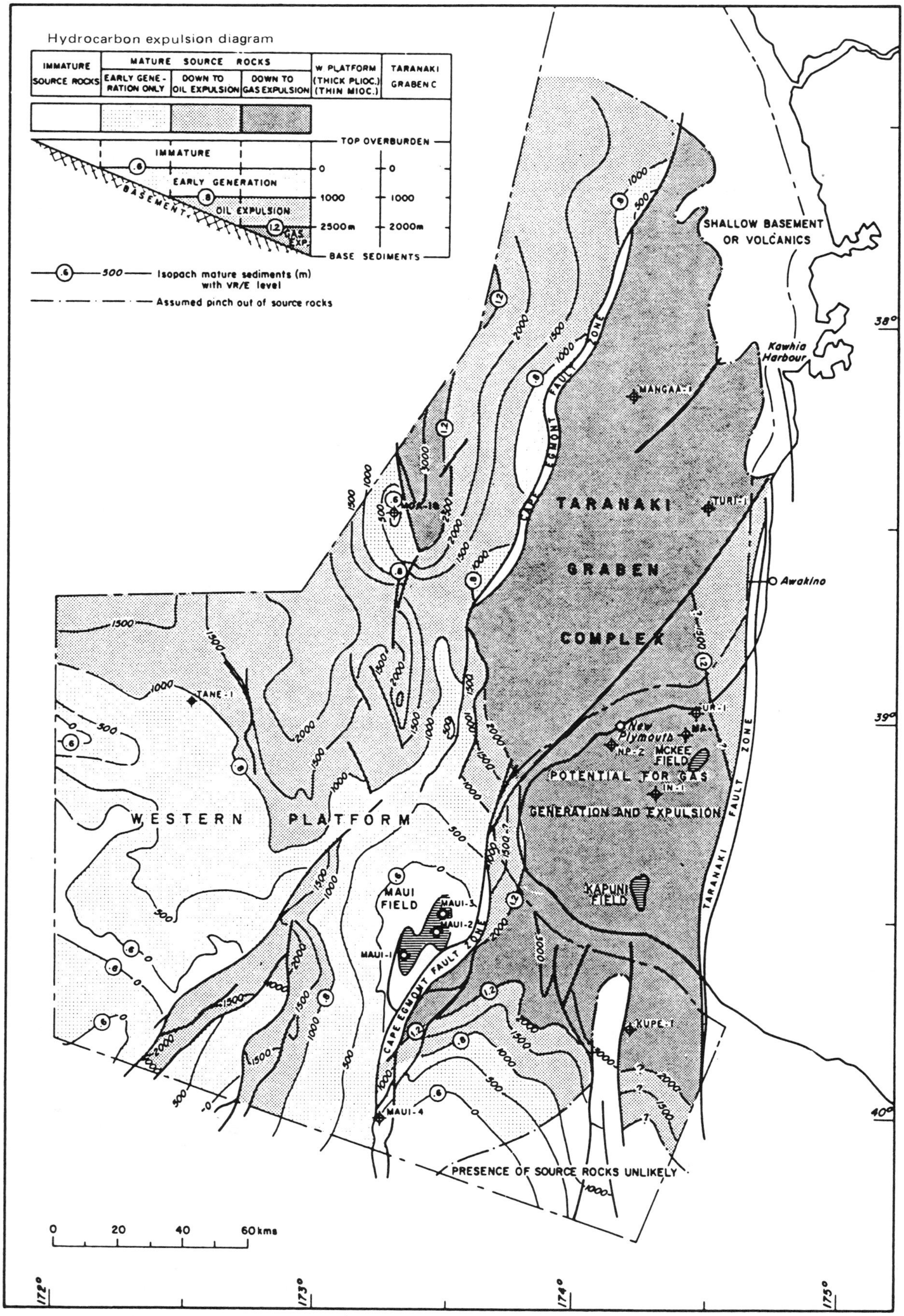
Hydrocarbon expulsion diagram
IMMATURE SOURCE ROCKS
MATURE SOURCE ROCKS
EARLY GENERATION ONLY
DOWN TO OIL EXPULSION
DOWN TO GAS EXPULSION
W PLATFORM (THICK PLIOC.) (THIN MIOC.)
TARANAKI GRABEN C
TOP OVERBURDEN
IMMATURE
EARLY GENERATION
OIL EXPULSION
GAS EXP.
BASEMENT
BASE SEDIMENTS
0
1000
2500m
2000m
Isopach mature sediments (m) with VR/E level
Assumed pinch out of source rocks
SHALLOW BASEMENT OR VOLCANICS
Kawhia Harbour
Awakino
New Plymouth
TARANAKI GRABEN COMPLEX
CAPE EGMONT FAULT ZONE
TARANAKI FAULT ZONE
MANGAA-1
TURI-1
TANE-1
WESTERN PLATFORM
POTENTIAL FOR GAS GENERATION AND EXPULSION
MCKEE FIELD
KAPUNI FIELD
MAUI FIELD
MAUI-1
MAUI-2
MAUI-3
MAUI-4
KUPE-1
PRESENCE OF SOURCE ROCKS UNLIKELY
0 20 40 60kms
172°
173°
174°
175°
38°
39°
40°

Figure 14. Taranaki basin, tentative hydrocarbon generation and expulsion provinces at present (after Pilaar and Wakefield, 1984).

Knox, G. J., 1982, Taranaki Basin, structural style and tectonic setting: New Zealand Journal of Geology and Geophysics, v. 25, n. 2, p. 125-140.

Knox, G. J, 1982, Taranaki Basin, structural style and tectonic setting: New Zealand Journal of Geology and Geophysics, v. 28, n. 2, p. 197-216.

Krebs, W., 1975, Formation of southwest Pacific island arc-trench and mountain systems; plate or global-vertical tectonics: American Association of Petroleum Geologists Bulletin, v. 59, n. 9, p. 1639-1666.

Lopatin, N. V., 1971, Temperature and geological time as factors in coalification: Akademiya Nauk SSSR, Izviesta, Seriya Geologicheskaya, n. 3 (English translation by N. H. Bosteck, 1972, Illinois State Geological Survey, p. 96-106).

Lowery, J. H., 1986, Vitrinite reflectance and maturation in the 5 km deep Inglewood-1 prospecting well, Taranaki Basin: New Zealand Geological Survey Report M149.

McBeath, D. M., 1976, Kapuni and Maui gas-condensate fields of New Zealand—summary: American Association of Petroleum Geologists Memoir 25, Circum-Pacific Energy and Mineral Resources, p. 211-216.

McBeath, D. M., 1977, Gas-condensate fields of the Taranaki Basin, New Zealand: New Zealand Journal of Geology and Geophysics, v. 20, n. 1, p. 99-127.

Nathan, S., H. J. Anderson, R. A. Cook, R. H. Herzer, R. H. Hoskins, J. I. Raine, and D. Smale, 1986, Cretaceous and Cenozoic sedimentary basins of the west coast region, South Island, New Zealand: Geological Survey of New Zealand and Basin Studies 1.

Nelson, C. S., 1978, Temperate shelf carbonate sediments in the Cenozoic of New Zealand: Sedimentology, v. 15, p. 737-771.

New Zealand Government, 1964, Report on utilisation of Kapuni natural gas: Government Printer, Wellington, 29 p.

New Zealand Government, 1965, Second report on utilisation of Kapuni natural gas: Government Printer, Wellington, 14 p.

New Zealand Government, 1967, Gas purchase contract between Shell (petroleum mining) Company Limited, BP (oil exploration) Company of New Zealand Limited, Todd Petroleum Mining Company Limited, sellers, and the Minister of Mines, buyer: Government Printer, Wellington, 36 p.

New Zealand Government, 1973, Development of the Maui gas field: Government Printer, Wellington, 316 p.

Palmer, J., 1985, Pre-Miocene lithostratigraphy of Taranaki Basin, New Zealand: New Zealand Journal of Geology and Geophysics, v. 28, n. 2, p. 197-216.

Pilaar, W. F. H., and L. L. Wakefield, 1978, Structural and stratigraphic evolution of the Taranaki Basin, offshore North Island, New Zealand: Australian Petroleum Exploration Association Journal, p. 93-101.

Pilaar, W. F. H., and L. L. Wakefield, 1984, Hydrocarbon generation in the Taranaki Basin, New Zealand, *in* G. Demaison and R. J. Murris, eds., Petroleum geochemistry and basin evaluation: American Association of Petroleum Geologists Memoir 35, p. 405-423.

Robinson, P. H., P. R. King, and G. P. Thrasher, in preparation, Lithostratigraphic nomenclature for Taranaki region, New Zealand.

Roux, B., et al., 1980, An improved approach to estimating true reservoir temperature from transient temperature data: Society of Petroleum Engineers of AIME, n. 8888.

Shell, B.P., Todd Oil Services Ltd., 1975, Exploration well proposal structure D (Kupe) PPL 682: New Zealand Geological Survey unpublished open-file petroleum report 643.

Short, K. C., 1962, The stratigraphy of the Taranaki Peninsula structure: Shell, B.P., and Todd Oil Services Limited unpublished report.

Smale, D., and A. C. Morton, in press, Heavy mineral suites of core samples from the McKee Formation (Eocene-Lower Oligocene), Taranaki; implications for provenance and diagenesis: New Zealand Journal of Geology and Geophysics, v. 30, n. 3.

Sprigg, R. C., et al., 1969, New Zealand Basin offers promise: Oil and Gas Journal, June 16, p. 124-129.

Stach, E., et al., 1975, Stach's textbook on coal petrology, 2nd revised edition: Berlin-Stuttgart, Gebr. Borntraeger.

Suggate, R. P., 1950, Quartzose coal measures of west Nelson and north Westland: New Zealand Journal of Science and Technology, Bulletin 31, n. 4, p. 1-14.

Suggate, R. P., 1956, Puponga coal field: New Zealand Journal of Science and Technology, Bulletin 37, n. 5, p. 539-559.

Suggate, R. P., 1973, Coal ranks in relation to depth and temperature in Australia and New Zealand oil and gas wells: New Zealand Journal of Geology and Geophysics, v. 17, n. 1, p. 149-167.

Thompson, J. G., 1982, Hydrocarbon source rock analyses of Pakawau Group and Kapuni Formation sediments, northwest Nelson and offshore South Taranaki, New Zealand: New Zealand Journal of Geology and Geophysics, v. 25, n. 2, p. 141-148.

Van der Lingen, G. J., D. Smale, G. A. Challis, W. A. Watters, and P. H. Robinson, in press, Siliciclastic diagenesis in Paleocene-Eocene reservoir sandstones of the Taranaki Basin: New Zealand Geological Survey DSIR, Lower Hutt, New Zealand.

Van de Watering, W. P. M., 1976, Planning pays in setting Maui structure: Oil and Gas Journal, March 1, p. 107-113.

Walcott, R. I., 1987, Geodetic strain and the deformational history of the North Island of New Zealand during the Late Cenozoic: Philosophical Transactions of the Royal Society of London, A321, p. 163-181.

Waples, D. W., 1980, Time and temperature in petroleum formation; application of Lopatin's method to petroleum exploration, American Association of Petroleum Geologists Bulletin, v. 64, p. 916-926.

Watt, D. S., 1965, Natural gas and oil in western New Zealand: Publications 8th Commonwealth Mining and Metallurgical Congress 5, p. 77-88.

Wilcox, R. E., T. P. Harding, and D. R. Seely, 1973, Basic wrench faulting: American Association of Petroleum Geologists Bulletin, v. 57, p. 74-96.

Williams, H. R., 1968, Production of the Kapuni Field and separation of gas and condensate: New Zealand Engineering 23, p. 458-462.

Wodzicki, A., 1974, Geology of the pre-Cenozoic basement of the Taranaki-Cook Strait-Westland area, New Zealand, based on recent drillhole data: New Zealand Journal of Geology and Geophysics, v. 17, n. 4, p. 747-757.

SUGGESTED READING

Shell Oil Company, 1987, Atlas of seismic stratigraphy: American Association of Petroleum Geologists Studies in Geology 27, v. 1, p. 15-71.

Appendix 1. Field Description

Field name .... *Kapuni*

Ultimate recoverable reserves .... *Gas, 991 bcf (28 m^3)(16.3 $\times$ 10^9 m^3 gas; 4.6 $\times$ 10^6 m^3 cond.)*

Field location:

- **Country** .... *New Zealand*
- **State** .... *North Island*
- **Basin/Province** .... *Taranaki basin*

Field discovery:

- **Year field discovered** *1959*
- **Year second pay discovered** .... *NA*
- **Third pay** .... *NA*

Discovery well name and general location
(i.e., Jones No. 1, Sec. 2T12NR5E; or Smith No. 1, 5 mi west of Sheridan, Wyoming):

- **First pay** .... *Kapuni #1, 30 mi (48 km) south-southeast of New Plymouth*
- **Second pay** .... *NA*
- **Third pay** .... *NA*

Discovery well operator

Shell Co. Ltd., B.P. of New Zealand Ltd., and Todd Petroleum Mining Co. Ltd.

(if more than one pay in field, list operators of discovery well in other pays)

- **Second pay** .... *NA*
- **Third pay** .... *NA*

IP in barrels per day and/or cubic feet or cubic meters per day:

- **First pay** .... *NA*
- **Second pay** .... *NA*
- **Third pay** .... *NA*

All other zones with shows of oil and gas in the field:

Age	Formation	Type of Show
None		

Geologic concept leading to discovery and method or methods used to delineate prospect, e.g., surface geology, subsurface geology, seeps, magnetic data, gravity data, seismic data, seismic refraction, nontechnical:

The field was defined by surface geology and seismic reflection data and confirmed by drilling.

Structure:

Province/basin type (see St. John, Bally, and Klemme, 1984) .... *Interarc basin*

Tectonic history

After rifting away from Australia in late Paleozoic(?) time, Taranaki basin was part of extensive New Zealand geosyncline until Late Jurassic-Early Cretaceous Rangitata orogeny, during which the geosyncline was uplifted and eroded. Simultaneous movement on nearby transcurrent Alpine fault may have initiated Kapuni structural growth. A major rifting phase began after the Rangitata orogeny and transcurrent fault movement continued. The Taranaki basin began to form in Late Cretaceous time.

Regional structure

The Taranaki basin is composed of two main structural units: a stable western platform area to the west, the Western platform, and a graben system to the east, the Taranaki graben complex. The Cape Egmont fault separates these elements. The graben is divided into northern and southern segments. Subsidence reflects three tectonic regimes: extensional, oblique extensional, and oblique compressional.

Local structure

The Kapuni structure is a compound north-south-trending, slightly asymmetrical anticline with a faulted west flank located in the Taranaki graben, on the Kupe block east of the Manaia fault.

Trap

Trap type(s)

The Kapuni structure is a north-south elongate anticlinal trap with two main closures. There are stratigraphic traps on the flanks of the structure where point bar sandstones terminate against flood basin siltstones.

Basin stratigraphy (major stratigraphic intervals from surface to deepest penetration in field):

Age	Formation	Depth to Top in ft (m)
Recent	*Egmont Volcanics*	*Surface*
Pliocene	*Tangahoe*	*465 (142)*
Pliocene-Miocene	*Matemateaonga*	*1200 (366)*
Miocene	*Urenui*	*Absent*
Miocene	*Waikiekie-Mt. Messenger s.s.*	*Absent*
Miocene	*Mahakatino*	*6230 (1899)*
Miocene	*Mokau-Moki*	*6580 (2006)*
Oligocene	*Mahoenui*	*7870 (2399)*
Oligocene	*Te Kuiti-Tikorangi*	*Absent*
Oligocene/Eocene	*Mangaotaki-Kapuni*	*10,645 (3245)*

Location of well in field .... *NA*

Reservoir characteristics:

Number of reservoirs .... *1 primary and 4 secondary*

Formations .... *Kapuni Fm., K3E sandstone; K3D, K1A, K1C, and K3A sandstones*

Ages .... *Upper Eocene*

Depths to tops of reservoirs .... *12,268 ft (3739 m)*

Gross thickness (top to bottom of producing interval) .... *3000 ft (914 m)*

Net thickness—total thickness of producing zones

Average .... *± 500 ft (± 154 m)*

Maximum ....

Average ....

Maximum ....

Lithology

Sandstone: quartzose with varying amounts of argillaceous matrix, medium to fine grained, angular to subangular and friable. Kaolinite, illite, and chlorite present with some quartz overgrowths.

Porosity type .... *Intergranular*

Average porosity *15-21%*

Average permeability .... *15-500 md*

Seals:

Upper

Formation, fault, or other feature .... *Shale member, Kapuni Formation*

Lithology .... *Impervious shale*

Lateral

Formation, fault, or other feature .... *None*

Lithology ....

Source:

Formation and age .... *Lower Cretaceous Pakawau and Paleocene-Eocene Kapuni*

Lithology .... *Coal*

Average total organic carbon (TOC) .... *NA*

Maximum TOC .... *NA*

Kerogen type (I, II, or III) .... *II and III*

Vitrinite reflectance (maturation) .... $R_o = 0.95-1.0$

Time of hydrocarbon expulsion .... *Pliocene-Pleistocene*

Present depth to top of source .... *~12,500 ft (3810 m)*

Thickness .... *Unknown*
Potential yield .... *NA*

Appendix 2. Production Data

Field name .... *Kapuni*

Field size:

Proved acres .... *6400 ac (2590 ha.)*
Number of wells all years *12*
Current number of wells (as of year) *11*
Well spacing .... *1-1.5 mi (1.6-2.4 km)*
Ultimate recoverable .... *991 bcf gas*
Cumulative production .... *3,699,346 bbl condensate; 49,264 mmcf gas*
Annual production .... *3500 bbl condensate; 46.3 mmcf gas*
Present decline rate .... *Not on production*
Initial decline rate .... *NA*
Overall decline rate .... *NA*
Annual water production .... *NA*
In place, total reserves .... *NA*
In place, per acre-foot .... *NA*
Primary recovery .... *NA*
Secondary recovery .... *NA*
Enhanced recovery .... *NA*
Cumulative water production .... *NA*

Drilling and casing practices:

Amount of surface casing set .... *1019 ft (311 m)*
Casing program
16-in. to 1019 ft (311 m); 10¾-in. to 4344 ft (1324 m); 7⅝-in. cemented at 9998 ft (3047 m); 4½-in. to 12,827 ft (3910 m)

Drilling mud .... *Clay base to 2560 ft (777 m); gas-oil emulsion to 13,040 ft (3955 m)*
Bit program .... *Variable; mostly Hughes OSC-3*
High pressure zones .... *None*

Completion practices:

Interval(s) perforated .... *12,268-12,280 ft (3739-3743 m)*
Well treatment .... *NA*

Formation evaluation:

Logging suites .... *IES, Micro Caliper, Dipmeter, Temperature, Gamma Ray/Neutron*
Testing practices .... *Typically production tested*
Mud logging techniques .... *NA*

Oil characteristics:

Type .... *Paraffinic*
(Tissot and Welte Classification in "Petroleum Formation and Occurrence," 1984, Springer-Verlag, p. 419)
API gravity .... *43-53°*
Base .... *NA*
Initial GOR .... *1320 cf/bbl*
Sulfur, wt% *0.01*
Viscosity, SUS .... *1.49 cp at 70°F (21°C)*

Pour point .... *15°C (60°F)*
Gas-oil distillate .... *NA*

Field characteristics:

Average elevation .... *629 ft (192 m)*
Initial pressure .... *6500 psi (44,917 kPa) at 12,360 ft (3767 m)*
Present pressure .... *NA*
Pressure gradient .... *0.562 psi/ft*
Temperature .... *224°F (107°C)*
Geothermal gradient .... *69°F/100 ft (3.6°/100 m)*
Drive .... *Pressure depletion; water influx*
Oil column thickness .... *±1500 ft (457 m)*
Oil-water contact .... *12,000 ft (3658 m)*
Connate water .... *NA*
Water salinity, TDS .... *24 g/L*
Resistivity of water .... *5.26 ohm*
Bulk volume water (%) *14%*

Transportation method and market for oil and gas:

Gas sold to Natural Gas Corp. and condensate piped to storage in New Plymouth, then via tanker to refinery at Whangarei.

Leman Field

ALEC P. HILLIER
Shell U.K. Exploration and Production,
Lowestoft, United Kingdom

FIELD CLASSIFICATION

BASIN: North Sea
BASIN TYPE: Rift
RESERVOIR ROCK TYPE: Sandstone
RESERVOIR ENVIRONMENT OF DEPOSITION: Eolian and Fluvial
RESERVOIR AGE: Permian
PETROLEUM TYPE: Gas
TRAP TYPE: Anticline

LOCATION

The Leman field is situated in the southern part of the British sector of the North Sea some 50 km (30 mi) offshore from the small town of Bacton in the county of Norfolk (Figure 1).

Other important gas fields in the area are West Sole, Indefatigable, the "V" fields (Viking, Victor, Vanguard, Vulcan, and Valiant), Hewett, Dotty, Rough, Bure, Thames, Yare, and Audrey.

The Leman field is situated in the part of the North Sea that separates East Anglia from Holland. Geologically, the area is known as the Southern Permian basin of northwest Europe.

The field is a giant, ranked 130 by Carmalt and St. John (1986). It is the twenty-seventh largest gas field in the ranking. The estimated ultimate recovery from the field is 11,500 bcf of gas and an associated 9.7 million bbl of condensate.

HISTORY

Pre-Discovery

Interest in exploration in the southern North Sea was first generated by the discovery in 1959 of the giant Groningen gas field in North Holland. This field was the first major discovery in the area. It is now known to contain recoverable reserves of 86,000 bcf of gas in the Rotliegend sandstone. Extrapolation of geological data from onshore wells in Holland and Germany to the exposures of similar sandstone in Durham, England, led to the belief that Rotliegend sandstone may be present beneath the southern North Sea.

The Rotliegend, so called from the German miner's term for the redbeds beneath the Zechstein, is a series of continental red clastic sediments of Permian age. In the North Sea, the Rotliegend is given the status of a group composed of two formations, the Leman Sandstone Formation, which is the reservoir, and the Silverpit Formation (Rhys, 1975). By common usage, the name Rotliegend has been adopted for the Leman Sandstone Formation.

Exploration could not take place until the question of sovereignty over the continental shelf was settled. This occurred with the completing of bilateral agreements extending sovereignty over mineral rights on the continental shelf to a median line between the two countries concerned. This agreement gave the government control of the continental shelf of the United Kingdom. In 1964, a number of licenses were granted giving companies the right to explore for and, if successful, develop hydrocarbon accumulations on the continental shelf. Since this first round in 1964, a further ten rounds of licensing have taken place.

In 1964, seismic data were acquired over the area on a fourfold coverage. Based on these data, structures were defined and drilling commenced in the U.K. sector of the North Sea with well 38/29-1 in December 1964. A further nine wells were drilled in the U.K. sector prior to the Leman field discovery. Wells 38/29-1, 44/2-1, 44/21-1, and 41/20-1 found the Rotliegend Group to be developed as the Silverpit Formation with the Leman Sandstone Formation absent. The other six all found the Rotliegend developed as the Leman Sandstone Formation and included 48/6-1, the discovery well for West Sole field; well 49/17-1, later confirmed as the discovery well for Viking field, and noncommercial gas shows in the Rotliegend in wells 49/13-1 and 49/6-1. Wells 53/10-1 and 49/19-1 were dry holes. The eleventh well drilled in the southern North Sea, 49/26-1, discovered the Leman field.

Very little information from these wells was available to influence the choice of locations for further drilling as they were drilled by seven different operators each testing seismically defined structures within their own license areas. However, as time went on well data were traded between operators and a geological picture of the southern North Sea was built up.

A brief history of gas discoveries in the southern North Sea is given in Figure 2.

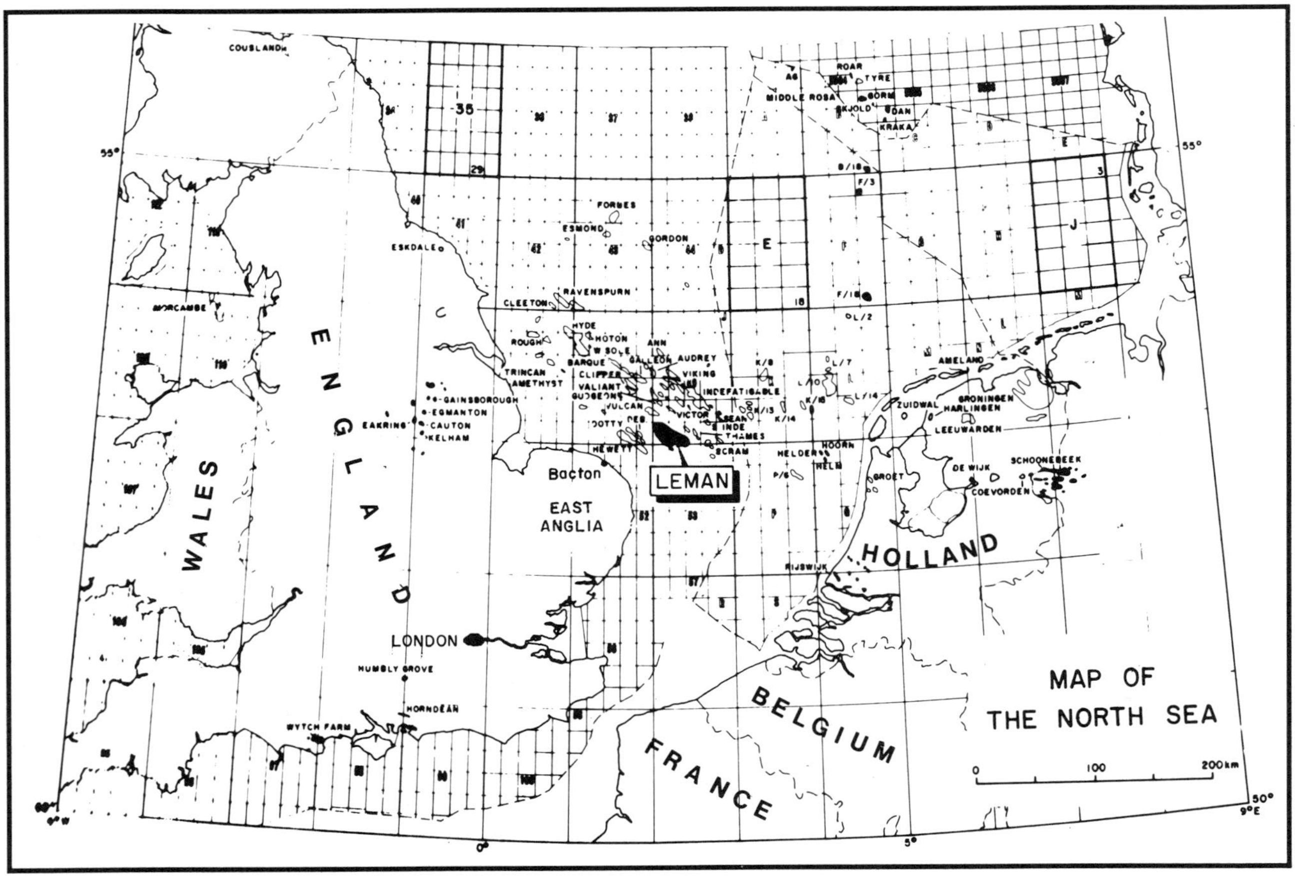
WALES
ENGLAND
LONDON
EAST ANGLIA
Bacton
LEMAN
HOLLAND
BELGIUM
FRANCE
MAP OF THE NORTH SEA
0
100
200km
55°
50°
9°E
5°
0°
WYTCH FARM
HUMBLY GROVE
HORNDEAN
EAKRING
GAINSBOROUGH
EGMANTON
CAUTON
KELHAM
ESKDALE
MORECAMBE
FORBES
ESMOND
GORDON
RAVENSPURN
CLEETON
HYDE
ROUGH
HOTON
W SOLE
ANN
AUDREY
BARQUE
CLIPPER
VIKING
TRINCAN
AMETHYST
VALIANT
VULCAN
VICTOR
INDEFATIGABLE
SEAN
INDE
THAMES
SCRAM
HEWETT
DOTTY
HELDER
HOORN
HELM
GROET
RIJSWIJK
AMELAND
GRONINGEN
HARLINGEN
LEEUWARDEN
ZUIDWAL
DE WIJK
COEVORDEN
SCHOONEBEEK
ROAR
TYRE
GORM
SKJOLD
DAN
KRAKA
MIDDLE ROSA
E
J
F/18
K/18
K/14
L/10
L/7
L/14
F/3

Figure 1. North Sea location map showing major fields.

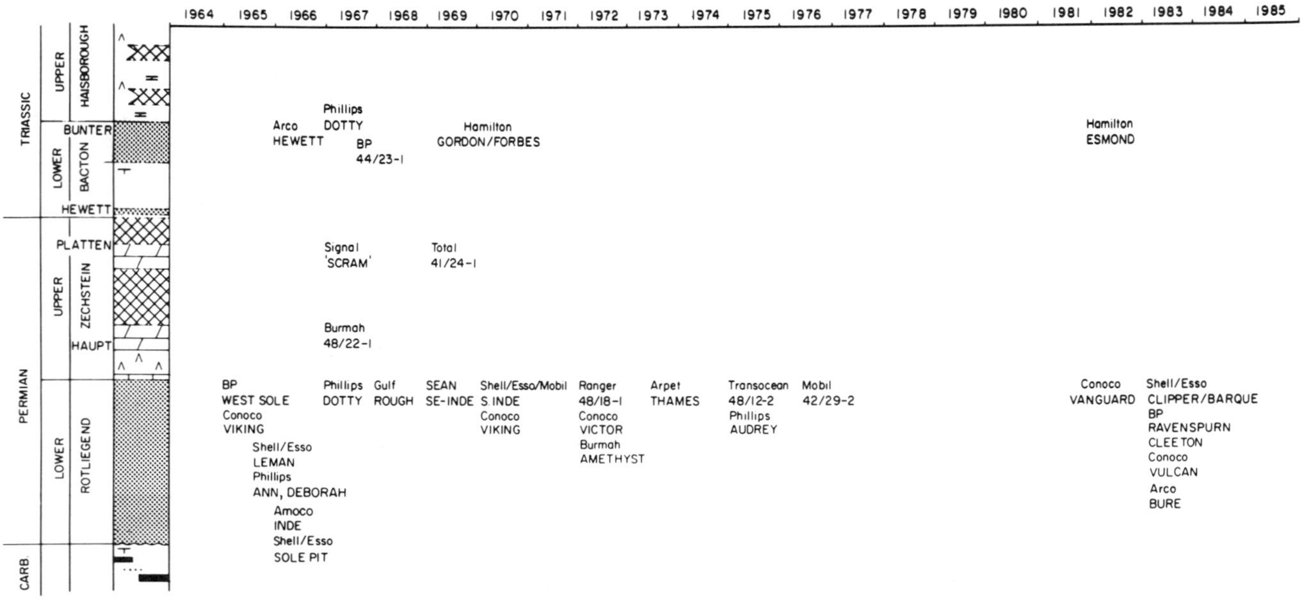
1964 1965 1966 1967 1968 1969 1970 1971 1972 1973 1974 1975 1976 1977 1978 1979 1980 1981 1982 1983 1984 1985
TRIASSIC
UPPER
HAISBOROUGH
LOWER
BUNTER
BACTON
HEWETT
PERMIAN
UPPER
PLATTEN
ZECHSTEIN
HAUPT
LOWER
ROTLIEGEND
CARB.
Phillips DOTTY
Arco HEWETT
BP 44/23-1
Hamilton GORDON/FORBES
Hamilton ESMOND
Signal 'SCRAM'
Total 41/24-1
Burmah 48/22-1
BP WEST SOLE
Conoco VIKING
Shell/Esso LEMAN
Phillips ANN, DEBORAH
Amoco INDE
Shell/Esso SOLE PIT
Phillips DOTTY
Gulf ROUGH
SEAN SE-INDE
Shell/Esso/Mobil S INDE
Conoco VIKING
Ranger 48/18-1
Conoco VICTOR
Burmah AMETHYST
Arpet THAMES
Transocean 48/12-2
Phillips AUDREY
Mobil 42/29-2
Conoco VANGUARD
Shell/Esso CLIPPER/BARQUE
BP RAVENSPURN
CLEETON
Conoco VULCAN
Arco BURE

Figure 2. Chronology of gas discoveries U.K. southern North Sea. Note lack of activity 1976-1982 as all eyes, and rigs, turn to the northern North Sea oil province.

Discovery

Shell U.K. Exploration & Production (Shell Expro), acting as operator for a 50/50 joint venture with Esso Exploration and Production U.K. Limited in the U.K. North Sea, took up several of the first-round licenses and drilled two wells, 44/2-1 and 49/19-1, during 1965. These two wells had secondary objectives of helping to assess Netherlands offshore acreage. It is impossible to single out individual names involved in the original idea that led to the discovery because it was part of an industry-wide move into a new area.

The discovery well, 49/26-1, located at 53°5′17″N, 2°7′46″E, was spudded on 17 December 1965 and abandoned on 18 April 1966 after testing gas from the Rotliegend sandstone.

The play concept tested was simple. A seismic survey on a 5 km grid with fourfold coverage had been acquired over block 49/26 in 1964. The deepest seismic reflection recorded corresponded to the base of the upper salt (BUS) in the Zechstein at which level a very large domal structure some 30 by 15 km was mapped. By extrapolation from the Dutch onshore geology, the Rotliegend was expected to lie some 300 m (1000 ft) below the BUS reflector and to be conformable with it; therefore, there should be a similar structure in the Rotliegend.

The Permian Rotliegend Leman sandstone reservoir was found at a depth of 1803 m (5914 ft) subsea. The reservoir was 281 m (921 ft) thick with 245 m (804 ft) of net gas-bearing sandstone with a porosity of 14% above a gas-water contact at 2048 m ss (6718 ft). A production test over the interval 1931–1967 m ss (6334–6454 ft) produced good-quality gas at a rate of 17 million standard ft^3/d through a $^9/_{16}$-in. choke.

Post-Discovery

Early estimates of the field's reserves were equivalent to supplying the entire gas consumption of the United Kingdom for a considerable number of years, and hence a rapid assessment of the true size of the field was essential. The first estimates were based on the initial structure map (Figure 3), but the actual figures have been lost. From the initial seismic data, it seemed certain that the field, named Leman after a nearby sand bank, extended into Amoco's block 49/27 to the east, the Signal block 49/21 to the north, Phillip's block 48/30 to the west, and Mobil's block 53/2 to the south. The initial structure map is shown in Figure 3.

To encourage the neighboring license holders to evaluate the acreage quickly, Shell/Esso pretraded the results of the discovery well with these operators in exchange for well data from their first wells in the adjacent blocks.

Shell/Esso drilled four stepout wells in the block, 49/26-2, 3, 4, and 5, which confirmed the extent of the field within the block. To the west Phillips' well 48/30-2 encountered a thin, 44 m (146 ft), tight Rotliegend that was thought to be on a separate structure. A later well, 48/30-4, drilled close to the block boundary, found a full Rotliegend sequence below the gas-water contact, thus excluding Phillips from the Leman field. To the north, the Signal well 49/21-1 found the Rotliegend at 2201 m ss (7220 ft) and penetrated 210 m (690 ft) of water-bearing sandstone that excluded Signal. Amoco's well 49/27-1 proved the crestal extension to the east finding the top Rotliegend at 1840 m ss (6037 ft) and testing gas at 25.5 million standard ft^3/d on a 1-in. choke. A further easterly extension was proved by Arpet well 49/28-1, and the southern extension was confirmed by Mobil's 53/2-1 well. The southeast limit was proven by the Mobil well 53/2-2, which found the Rotliegend at 2058 m ss (6751 ft). The appraisal campaign confirmed the extent of the field over the two blocks, 49/26 and 49/27, with minor extensions to the east and southeast. Infill seismic was shot, reducing the spacing to 2 km over the field.

Because the field was so large and strategically important, it was decided that it should have two operators, Amoco operating the eastern half on behalf of the East Leman Unit and Shell Expro as operator for the Shell/Esso joint venture in the western half, rather than unitizing the field and having a single group controlling potentially the entire country's gas supply. Platform development drilling started in 1967, and production commenced in 1968. Development drilling continued into 1974 when activity was much reduced as attention turned to the northern North Sea oil province. By this time, the structural picture of the field had changed to that shown in Figure 4 but was still based on the seismic picture of BUS and an isopach derived from well data to reach the top reservoir level. With the development of marine seismic data acquisition and processing by 1980 it was possible to image directly the reservoir level in the southern North Sea, and so a new grid of seismic data was shot on a 250 m by 500 m rectangular grid. Mapping from these data showed the reservoir to be far more faulted than had been believed and hence development wells could be targeted to potentially undrained fault blocks. Development drilling was recommenced after this new seismic was interpreted and is still ongoing in the field today.

DISCOVERY METHOD

Then

The initial coarse grid of fourfold seismic data shot over the southern North Sea delineated structural highs at the deepest visible reflector within the Zechstein. From onshore data, the Rotliegend was expected to lie more or less conformably below this reflector. Hence the discovery well was located on the crest of the seismic time high.

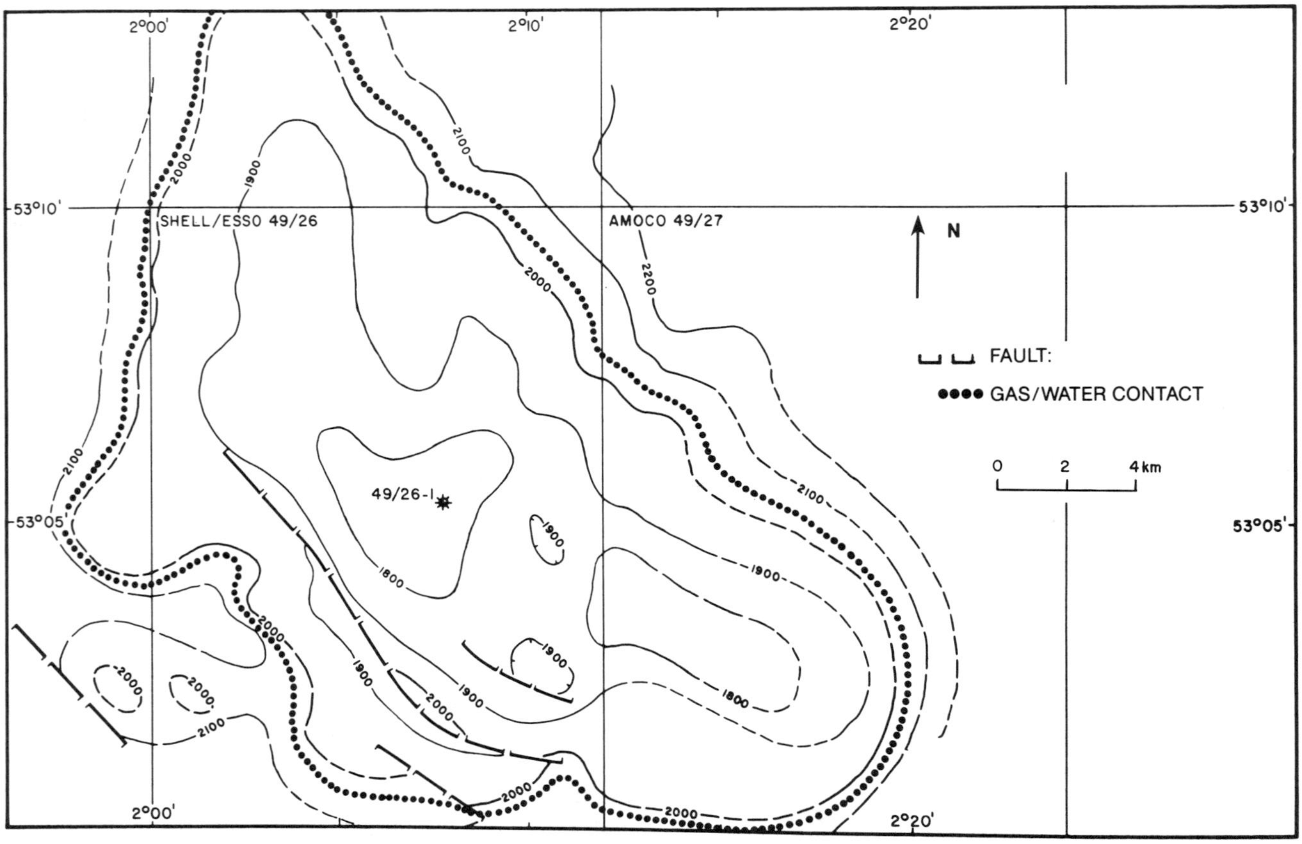
2°00'
2°10'
2°20'
53°10'
53°05'
SHELL/ESSO 49/26
AMOCO 49/27
N
FAULT:
GAS/WATER CONTACT
0 2 4km
49/26-1
1800
1900
2000
2100
2200

Figure 3. Top reservoir structure map of Leman field at discovery, 1965, based on map from 5 km seismic grid interpreted at BUS with isochore added to reach top reservoir.

Now

The higher multiplicity and better sampling and processing of seismic data available today would enable the Rotliegend to be directly mapped from seismic. However, the basic concept used in the original exploration well of extrapolation of data from a known area to an unknown area remains the same.

STRUCTURE

Tectonic History

The southern North Sea Permian sedimentary basin developed in the northern foreland of the Hercynian orogenic belt. This mountain belt was created by the northward movement of Gondwana and its subsequent collision with Laurasia. As a result of the compressional forces, the foreland was shortened, uplifted, and eroded in the late Carboniferous.

During the Permian, the orogen ceased to develop. The mountain belt and its foreland were modified as a result of east-west relative movement between Gondwana and Laurasia.

Sedimentary basins developed on and adjacent to the orogenic belt. Many of these basins can be directly related to active wrench and extensional faults. The large southern North Sea Permian basin developed on the site of an earlier, Westphalian basin. By the Late Permian it became a broad feature, some 1500 km long and 500 km wide, parallel to and in front of the east-west Variscan mountain belt (Figure 5). Subsidence may have been a response to crustal thinning (Glennie, 1986), but its shape and the lack of extensional faults may be better explained by a foreland basin model. The crust subsided beneath the load of the Variscan orogenic wedge.

Sedimentation in the southern North Sea basin continued into the Late Cretaceous. During the Alpine orogeny, northwest Europe experienced north–south-directed compressional forces reactivating some of the Variscan structural elements. Thus the existing northwest–southeast-oriented normal faults moved in a right-lateral strike-slip direction. As a result, the Cimmerian–early Tertiary sedimentary basins were inverted, e.g., Sole Pit and West Netherlands basins (Ziegler, 1982; Glennie, 1986). The inversion caused erosion of up to 1525 m (5000 ft) of sediments from the area of the Leman field.

From the Eocene onwards, the compressional stress regime relaxed and the North Sea subsided. A continuous, undisturbed stratigraphic sequence was deposited from Miocene to present.

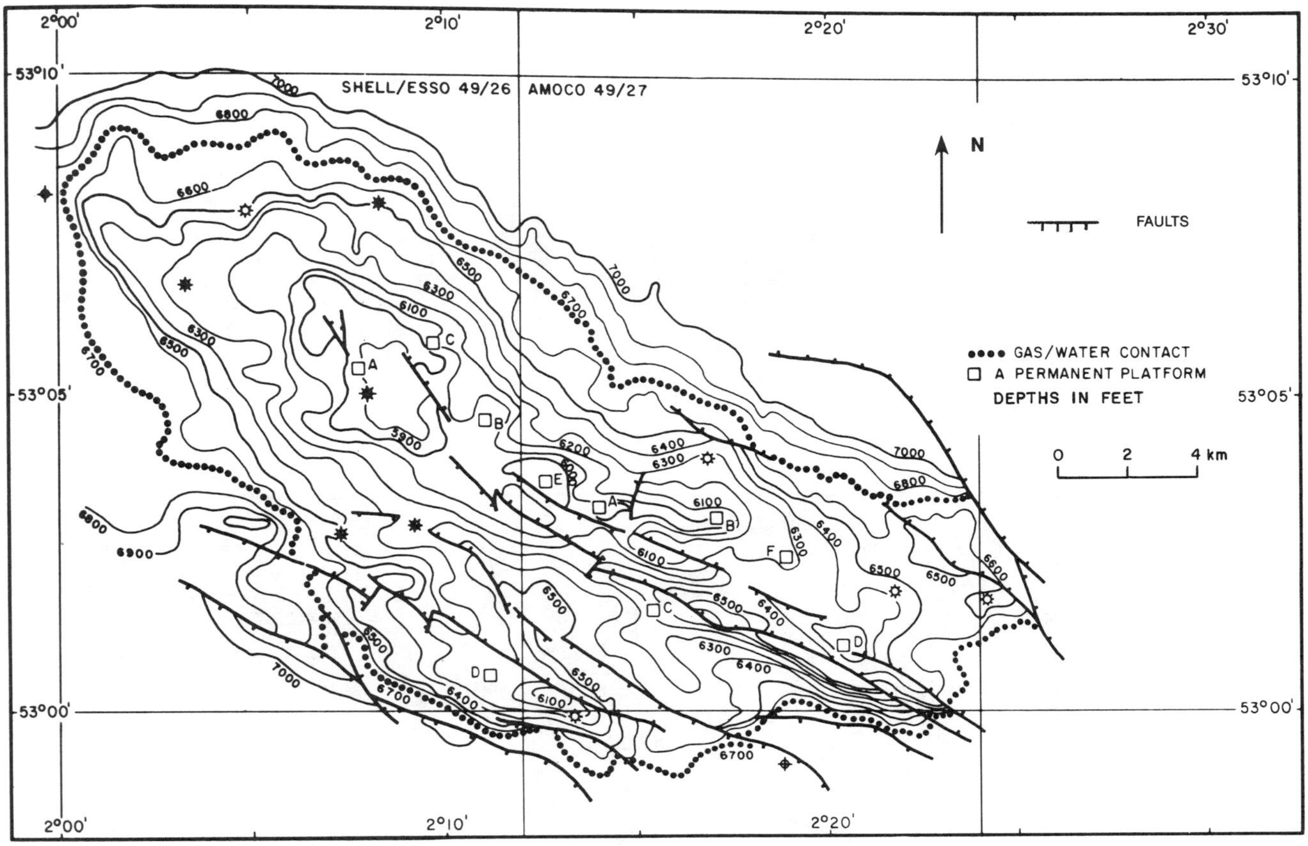
2°00'
2°10'
2°20'
2°30'
53°10'
53°05'
53°00'
SHELL/ESSO 49/26
AMOCO 49/27
N
FAULTS
GAS/WATER CONTACT
A PERMANENT PLATFORM
DEPTHS IN FEET
0 2 4 km

Figure 4. Top reservoir structure map of Leman field midfield development, 1974, based on map from 1 km seismic grid interpreted at BUS with isochore added to reach top reservoir.

Regional Structure

The Leman field is located at the southern end of the Sole Pit basin between two major fault zones: the Dowsing/South Hewett faults and the Swarte Bank Hinge (Figure 6). They are part of a northwest-southeast-trending set of steep normal faults created during the Hercynian orogeny. They were reactivated (with right-lateral strike-slip movements) during the Alpine orogeny and were most active during Late Cretaceous and early Tertiary (Glennie and Boegner, 1981). The Alpine compressive stress regime that was oriented in a NNE-SSW direction resulted in the right-lateral strike-slip movement.

Local Structure

The field is a densely faulted, broad, low relief anticline oriented northwest-southeast (Figures 7 and 8). The faults (Figure 18) show two main trends of strike directions: 150°/160° and 70°. A less pronounced third trend at 100°/110° is also present. The first (150°/160°) trend is roughly parallel to the Dowsing Fault Zone. All three trends seem to be related to the compressive Alpine stress regime that reactivated the Dowsing basement fault system.

The NNW-SSE-striking faults of the first trend could represent synthetic Riedel shears and the faults of the second trend antithetic Riedel shears.

STRATIGRAPHY

The stratigraphy of the field as shown in Figure 9 is as follows:

The oldest rocks penetrated in the field are the Carboniferous Coal Measures, a maximum of 382 m (1253 ft) of which were drilled by well 49/26-4. The coals in this formation form the source rock for the field.

Pre-Carboniferous stratigraphy is unknown from either well penetration or seismic interpretation; but continental Devonian redbeds have been encountered in wells to the north and south.

The Rotliegend sandstone was deposited on the eroded Carboniferous in an arid environment. The source was the London Brabant Massif, a part of the Variscan mountain chain that lay to the south of the field (Glennie et al., 1978). At the end of the Rotliegend deposition the sea inundated the area depositing four cycles of Zechstein evaporites.

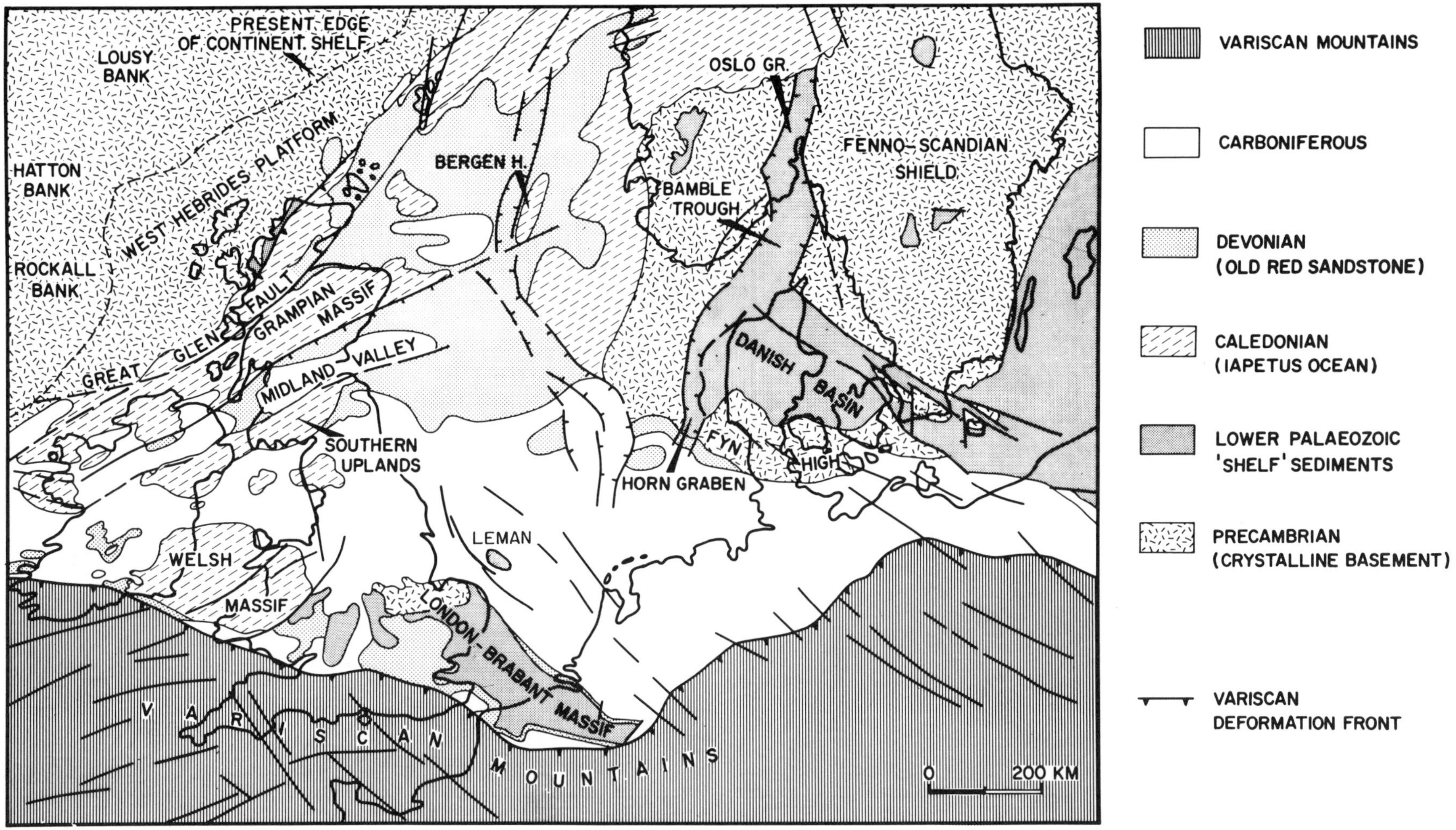
PRESENT EDGE OF CONTINENT. SHELF
LOUSY BANK
HATTON BANK
ROCKALL BANK
WEST HEBRIDES PLATFORM
GREAT GLEN FAULT
GRAMPIAN MASSIF
MIDLAND VALLEY
SOUTHERN UPLANDS
WELSH MASSIF
BERGEN H.
LEMAN
LONDON-BRABANT MASSIF
OSLO GR.
BAMBLE TROUGH
FENNO-SCANDIAN SHIELD
DANISH BASIN
FYN HIGH
HORN GRABEN
VARISCAN MOUNTAINS
0 200 KM
VARISCAN MOUNTAINS
CARBONIFEROUS
DEVONIAN (OLD RED SANDSTONE)
CALEDONIAN (IAPETUS OCEAN)
LOWER PALAEOZOIC 'SHELF' SEDIMENTS
PRECAMBRIAN (CRYSTALLINE BASEMENT)
VARISCAN DEFORMATION FRONT

Figure 5. Leman field location in relation to pre-Permian geology (simplified from Ziegler, 1982).

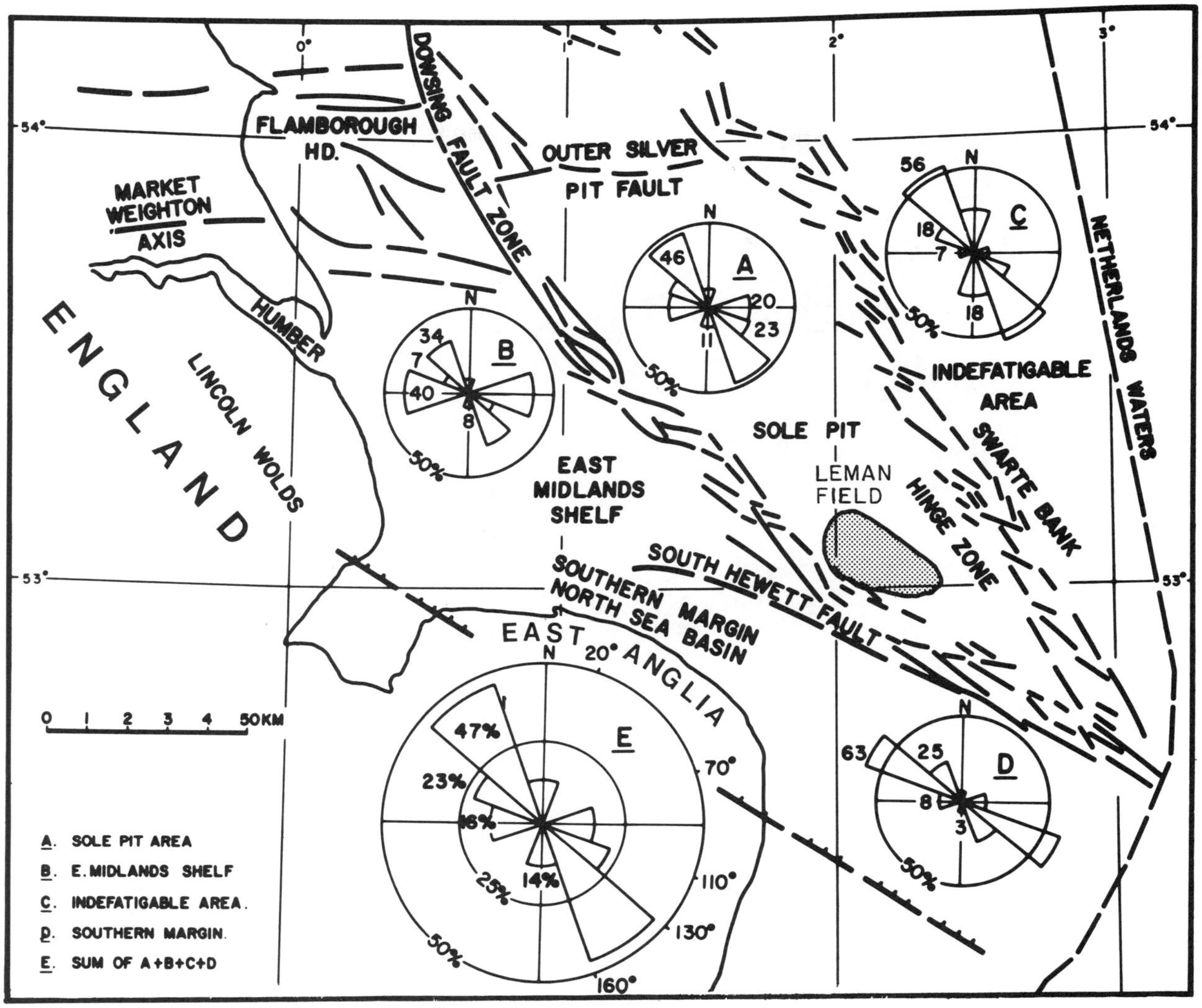

FLAMBOROUGH HD.
MARKET WEIGHTON AXIS
DOWSING FAULT ZONE
OUTER SILVER PIT FAULT
ENGLAND
HUMBER
LINCOLN WOLDS
EAST MIDLANDS SHELF
SOLE PIT
LEMAN FIELD
INDEFATIGABLE AREA
NETHERLANDS WATERS
SWARTE BANK HINGE ZONE
SOUTH HEWETT FAULT
SOUTHERN MARGIN NORTH SEA BASIN
EAST ANGLIA
0 1 2 3 4 50KM
A. SOLE PIT AREA
B. E.MIDLANDS SHELF
C. INDEFATIGABLE AREA.
D. SOUTHERN MARGIN.
E. SUM OF A+B+C+D

Figure 6. Regional faulting in southern North Sea (from Glennie and Boegner, 1981). Note insert rose diagrams showing strike direction of faulting in labeled areas.

The evaporite deposition was terminated by the onset of Triassic continental claystone and sandstone deposition. A sandstone deposited at this time, the Bunter Sandstone Formation, has good reservoir properties but is uncharged in the field.

Jurassic marine claystones followed by Cretaceous chalk were then deposited. These were subsequently eroded over the major part of the field.

In the Tertiary a series of marine clays and sands was deposited.

TRAP

The Leman field is a single reservoir, dip-closed anticlinal trap. The maximum closure is 335 m (1100 ft) with the top and lateral seals being formed by the Zechstein evaporite cap rock. The gas-water contact for the field is taken as 2042 m ss (6700 ft), some 60 m (200 ft) above the dip closure. The exact gas-water contact is difficult to define due to the capillary effects in the reservoir that result in long transition zones between gas and water. Apparent gas-water contacts ranging between 2018 and 2067 m ss (6620 and 6780 ft) have been interpreted in the wells. There are no unconformities within the reservoir. The trap is similar to several other fields in the southern North Sea, being unique only in terms of its large area and reserves. The original reservoir pressure was 22,160 kPa (3200 psi) at 1981 m ss (6500 ft) slightly above hydrostatic.

The trap was formed at the time of the Cimmerian inversion of the Sole Pit basin as evidenced by the occurrence of Jurassic and Cretaceous sediments on the flanks of the structure that are eroded from the crestal area (Figure 10).

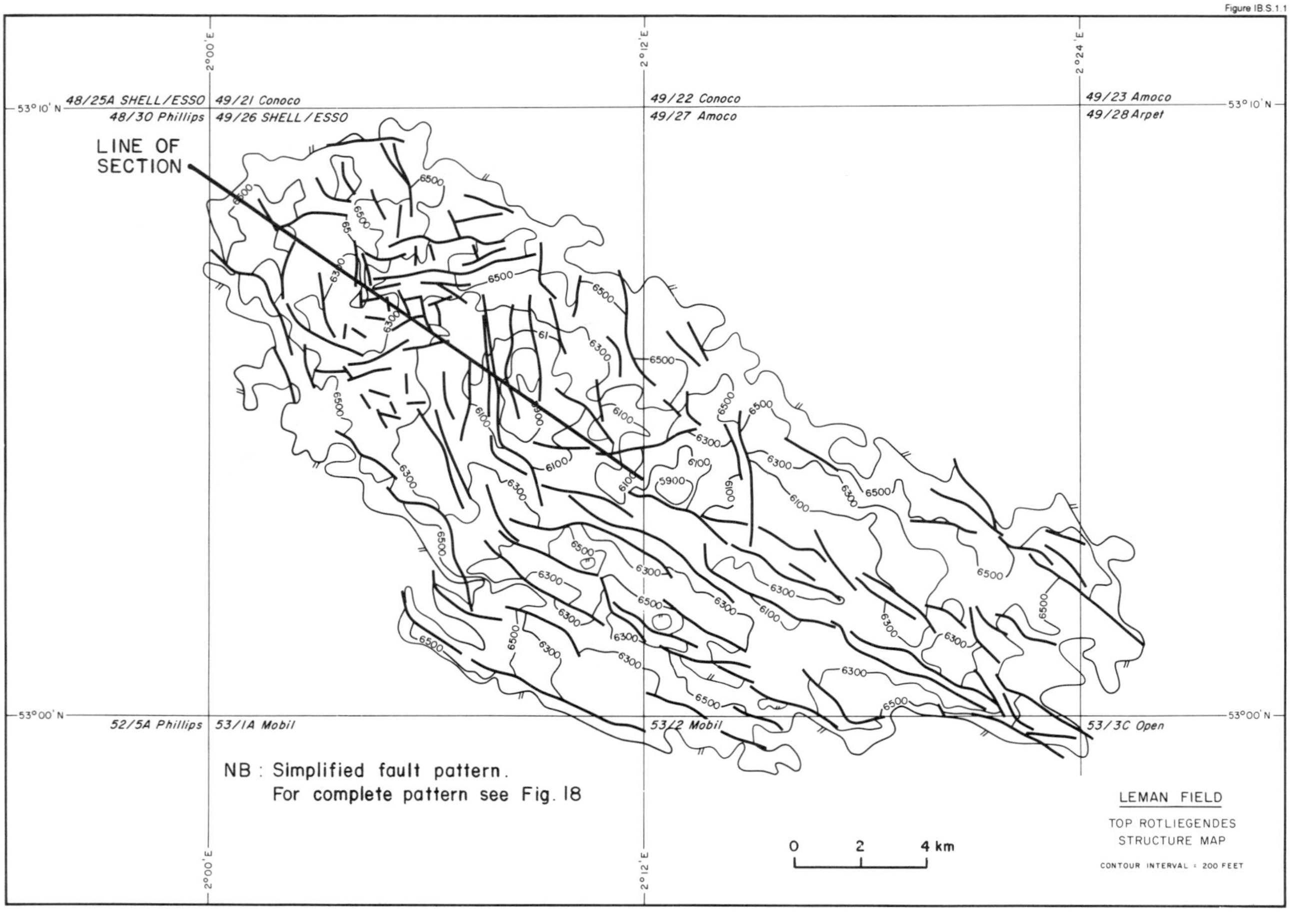
Figure IB.S.1.1
53°10'N
48/25A SHELL/ESSO
48/30 Phillips
49/21 Conoco
49/26 SHELL/ESSO
49/22 Conoco
49/27 Amoco
49/23 Amoco
49/28 Arpet
LINE OF SECTION
53°00'N
52/5A Phillips
53/1A Mobil
53/2 Mobil
53/3C Open
NB : Simplified fault pattern.
For complete pattern see Fig. 18
0 2 4 km
LEMAN FIELD
TOP ROTLIEGENDES STRUCTURE MAP
CONTOUR INTERVAL = 200 FEET

Figure 7. Top reservoir structure map of Leman field at present, 1988. Fault pattern much simplified.

RESERVOIR

In the Leman field, the Rotliegend has been divided into three units based on facies. From the base upward, these are wadi, dune, and reworked or waterlain (van Veen, 1975). These are equivalent to the fluvial, eolian, and "Weissliegendes" of Glennie (1986) (Figure 11).

Overlying the Carboniferous is the wadi facies zone. It consists of minor dune sandstones interbedded with homogeneous sands, conglomerates, and shales deposited by ephemeral rivers running off the London Brabant Massif. This zone is characterized by clay clasts derived from both the underlying Carboniferous by erosion and intraformationally indicating occasional subaerial exposure and desiccation. The thickness of this zone is controlled by the paleotopography of the underlying Carboniferous and varies between 8 m (25 ft) and 107 m (350 ft).

The eolian dune sands that overlie the wadi sands form the second facies zone. The sands were probably fed into the Southern Permian basin by wadis and subsequently transported and redeposited by winds that blew from the east or northeast. Stacked sequences of dunes can be recognized in cores and on dipmeters. The foreset beds have average dips of 25° while the bottomset beds are horizontal. Alternating laminae have different grain sizes, but within each lamina the sorting is good. The bottomset beds of the dunes have erosive contacts on the foresets of the underlying dune. Dune heights of up to 27 m (90 ft) (average 5 m, 15 ft) have been recognized in cores. The dunes were of transverse type with sinuous crests oriented perpendicular to the wind direction (Glennie, 1972). This zone is the thickest of the three facies zones and varies between 122 m (400 ft) and 198 m (650 ft).

The final zone is the "Weissliegendes" consisting of structureless, homogenized white sandstone ascribed to reworking of the top of the dune sands by escaping air during the rapid Zechstein transgression (Glennie and Buller, 1983). This zone varies from 10 m (30 ft) to 46 m (150 ft).

Reservoir Zonation

The whole of the Rotliegend contains reservoir-quality rock. The eolian dune sands can be divided into two zones on the basis of porosity/permeability relationships, and a study of the sands showed that

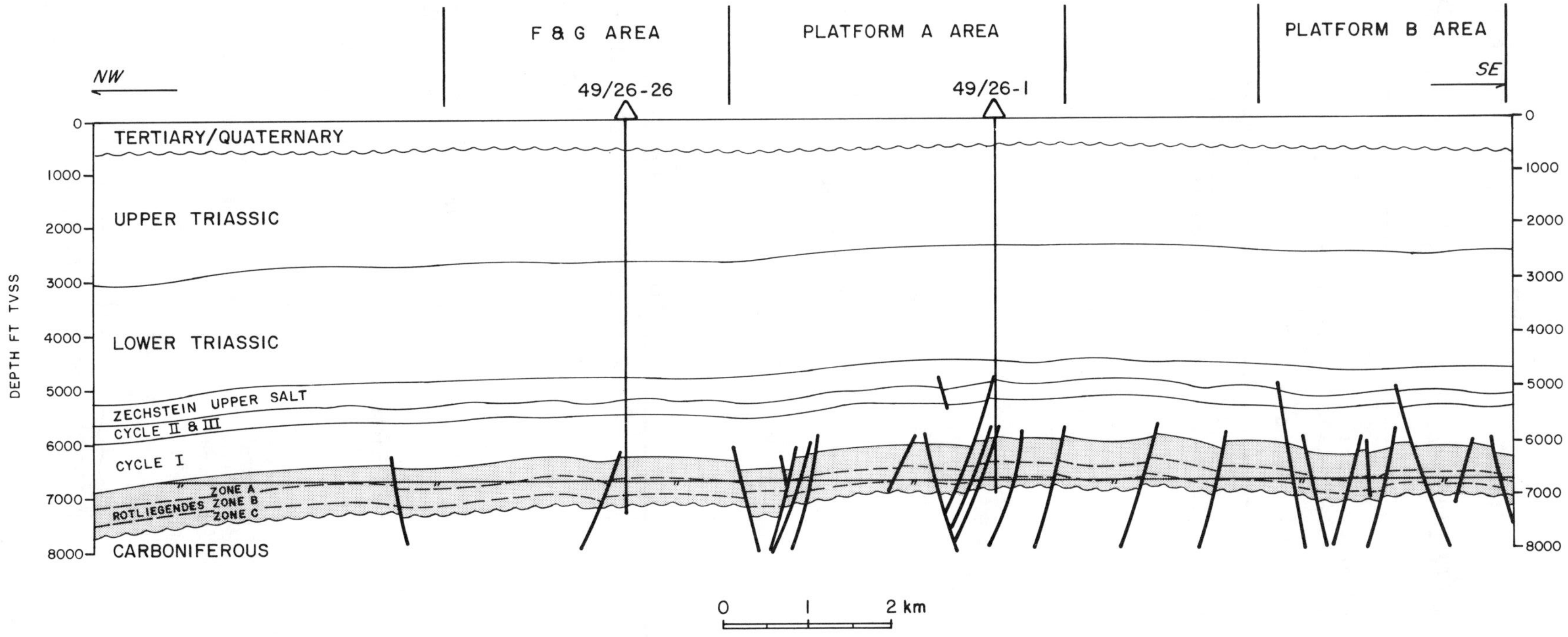
F & G AREA
PLATFORM A AREA
PLATFORM B AREA
NW
SE
49/26-26
49/26-1
DEPTH FT TVSS
0
1000
2000
3000
4000
5000
6000
7000
8000
TERTIARY/QUATERNARY
UPPER TRIASSIC
LOWER TRIASSIC
ZECHSTEIN UPPER SALT
CYCLE II & III
CYCLE I
ZONE A
ROTLIEGENDES ZONE B
ZONE C
CARBONIFEROUS
0
1
2 km

Figure 8. Structural cross-section of western half of Leman field. Note low relief and complex faulting.

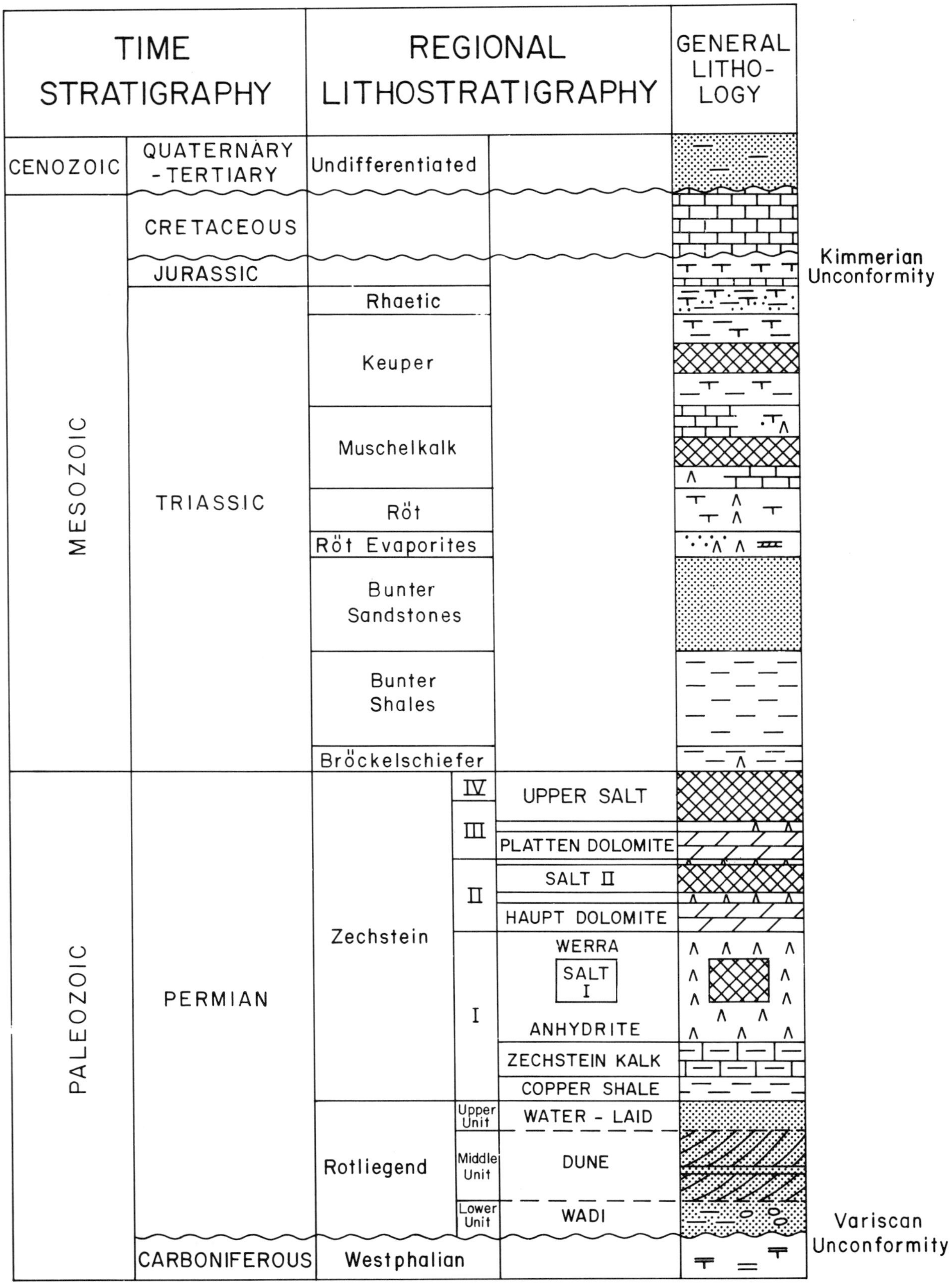

TIME STRATIGRAPHY
REGIONAL LITHOSTRATIGRAPHY
GENERAL LITHO-LOGY
CENOZOIC
QUATERNARY -TERTIARY
Undifferentiated
MESOZOIC
CRETACEOUS
JURASSIC
Kimmerian Unconformity
TRIASSIC
Rhaetic
Keuper
Muschelkalk
Röt
Röt Evaporites
Bunter Sandstones
Bunter Shales
Bröckelschiefer
PALEOZOIC
PERMIAN
Zechstein
IV
UPPER SALT
III
PLATTEN DOLOMITE
II
SALT II
HAUPT DOLOMITE
I
WERRA
SALT I
ANHYDRITE
ZECHSTEIN KALK
COPPER SHALE
Rotliegend
Upper Unit
WATER - LAID
Middle Unit
DUNE
Lower Unit
WADI
Variscan Unconformity
CARBONIFEROUS
Westphalian

Figure 9. Stratigraphic column of Leman field.

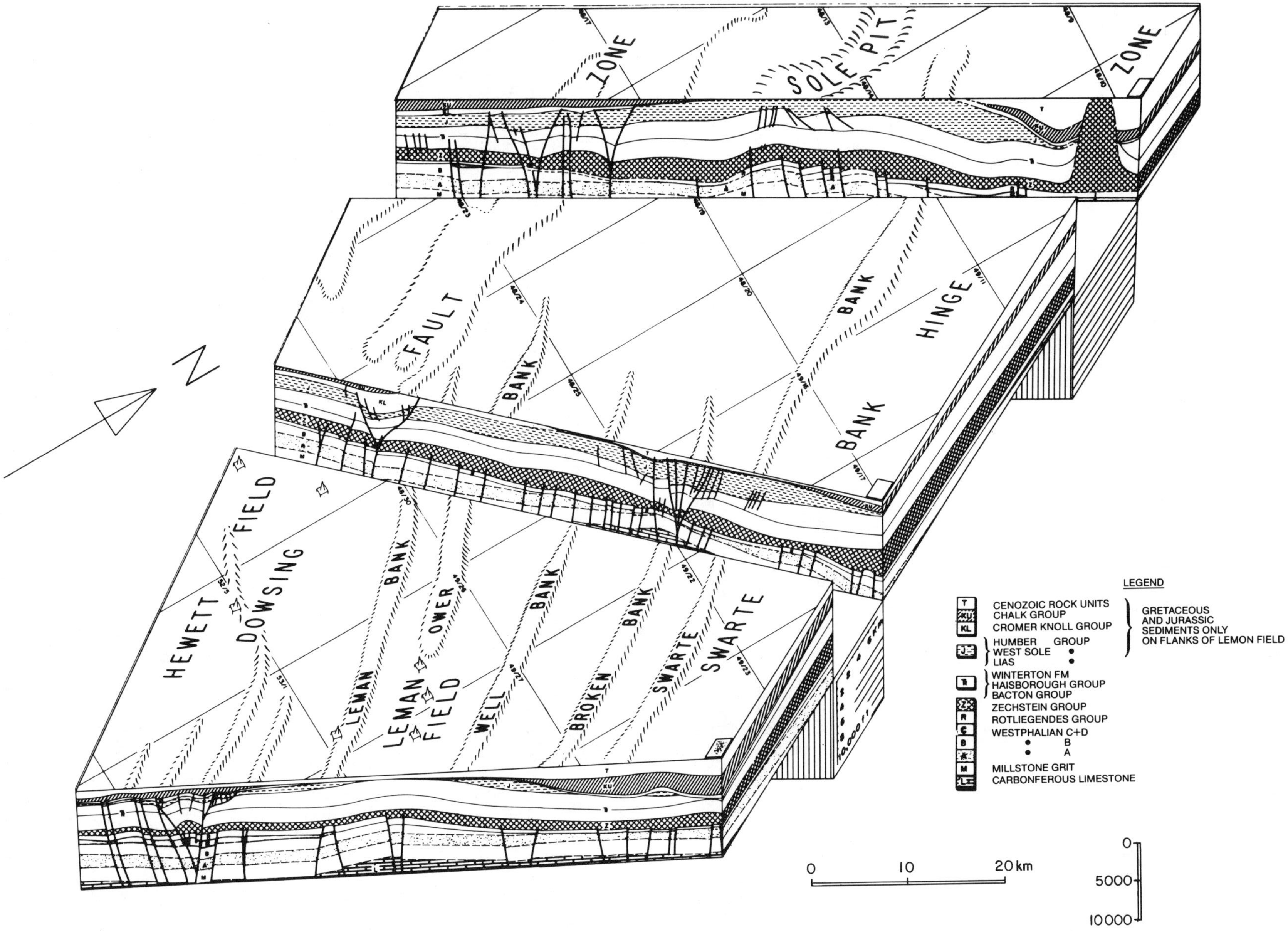
SOLE PIT
ZONE
ZONE
FAULT
BANK
HINGE
BANK
BANK
HEWETT
DOWSING
FIELD
LEMAN
BANK
OWER
LEMAN
FIELD
WELL
BANK
BROKEN
BANK
SWARTE
SWARTE
N
LEGEND
CENOZOIC ROCK UNITS
CHALK GROUP
CROMER KNOLL GROUP
HUMBER GROUP
WEST SOLE
LIAS
GRETACEOUS
AND JURASSIC
SEDIMENTS ONLY
ON FLANKS OF LEMON FIELD
WINTERTON FM
HAISBOROUGH GROUP
BACTON GROUP
ZECHSTEIN GROUP
ROTLIEGENDES GROUP
WESTPHALIAN C+D
B
A
MILLSTONE GRIT
CARBONFEROUS LIMESTONE
0
10
20 km
0
5000
10000

Figure 10. Block diagram of Leman field and southern North Sea (from Glennie and Boegner, 1981). Note eroded Jurassic and Cretaceous sediments on flanks of the Leman field.

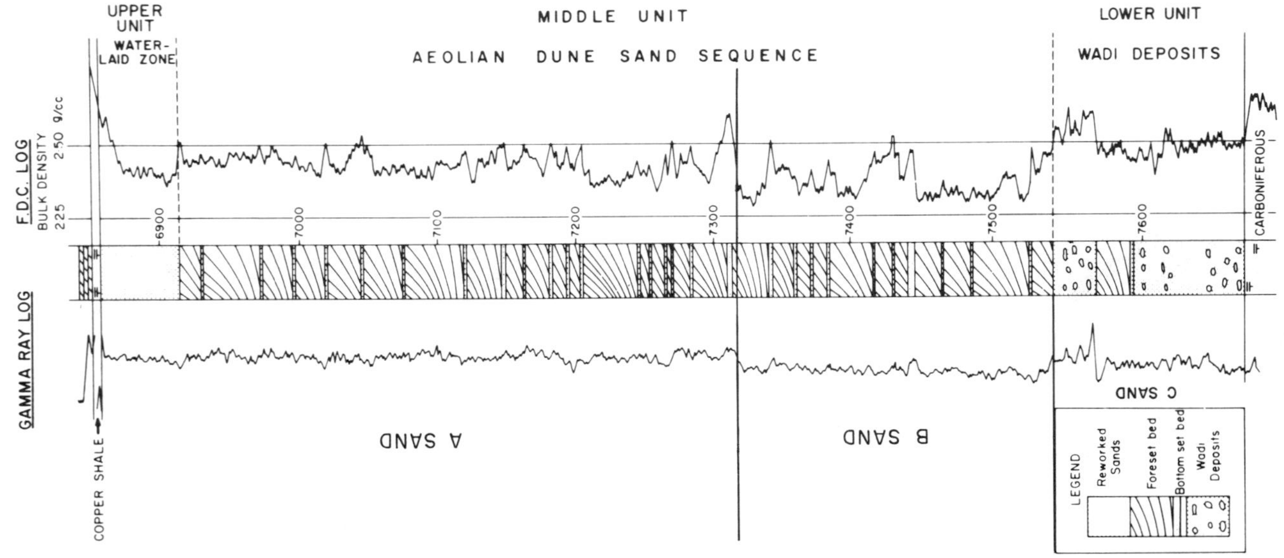
UPPER UNIT
WATER-LAID ZONE
MIDDLE UNIT
AEOLIAN DUNE SAND SEQUENCE
LOWER UNIT
WADI DEPOSITS
F.D.C. LOG
BULK DENSITY
2.25 2.50 g/cc
6900
7000
7100
7200
7300
7400
7500
7600
CARBONIFEROUS
GAMMA RAY LOG
COPPER SHALE
A SAND
B SAND
C SAND
LEGEND
Reworked Sands
Foreset bed
Bottom set bed
Wadi Deposits

Figure 11. Type log of Rotliegend sandstones reservoir Leman field.

Table 1. Average core porosity and permeability for the zones of the Rotliegend in Leman field. Note the increase in reservoir properties to the east in block 49/27.

Zone	49/26			49/27		
	Number of Samples	Average Por.	Average Perm.	Number of Samples	Average Por.	Average Perm.
A	4389	12.05	1.02	2638	13.70	2.45
B	1675	13.54	6.03	1264	15.06	15.60
C	693	9.30	0.55	340	12.61	4.67

the upper part, about 122 m (400 ft), was more cemented than the lower part. Hence for reservoir simulation work a threefold zonation was set up as follows (from the deepest upward). (See Table 1 for core average, porosity, and permeability data for the zones.)

The C zone is equivalent to the wadi facies zone described above. It is of little importance as a reservoir because it is below the gas-water contact over most of the field and has low porosity due to the poor sorting of the sandstone.

The B zone consists of stacked, gray-green or red-brown, fine- to medium-grained eolian dune sandstones. These sandstones are made up of foreset beds that have millimeter to centimeter laminae dipping at about 25°. Each lamina in itself is extremely well sorted, but there is great variation in median grainsize from one lamina to the next. This variation causes large changes in permeability. Each dune consists of a series of foreset beds separated from the preceding dune by poorly sorted, horizontally bedded bottomsets. The bottomset beds are usually of lower porosity and contain more clays than the foreset beds and can, thus, be readily identified on the density and gamma ray logs. The B zone sands are in general poorly cemented and are the best-quality reservoir rock in the field. The zone varies in thickness between 8 m (25 ft) and 122 m (400 ft).

The A zone consists of both dune sands, deposited under conditions similar to those of the B zone, and the Weissliegendes. The A zone is differentiated from the B zone by the greater amount of interstitial cement, particularly grain-coating hematite and illite. The illite, occurring in a fibrous form, tends to have a greater effect on the permeability of the rock than would be expected from its effect on the porosity (Stalder, 1973). The source of this additional cementation is not known for certain, but it could be derived from either a larger proportion of unstable silicates in the sediment source or a greater proportion of detrital clay derived from the sabkha facies that was present in the Sole Pit area to the north. Cementation within the Rotliegend is further complicated by the influence of the mineral-rich Zechstein pore water (Glennie et al., 1978). The thickness of the A zone is controlled by both the original Rotliegend sand composition and the depth of penetration of the Zechstein pore water. Thus, it is partly related to the original porosity and permeability of the dune sands. Synsedimentary faults within the dune sands would also provide fluid paths through the rock and enhance circulation of

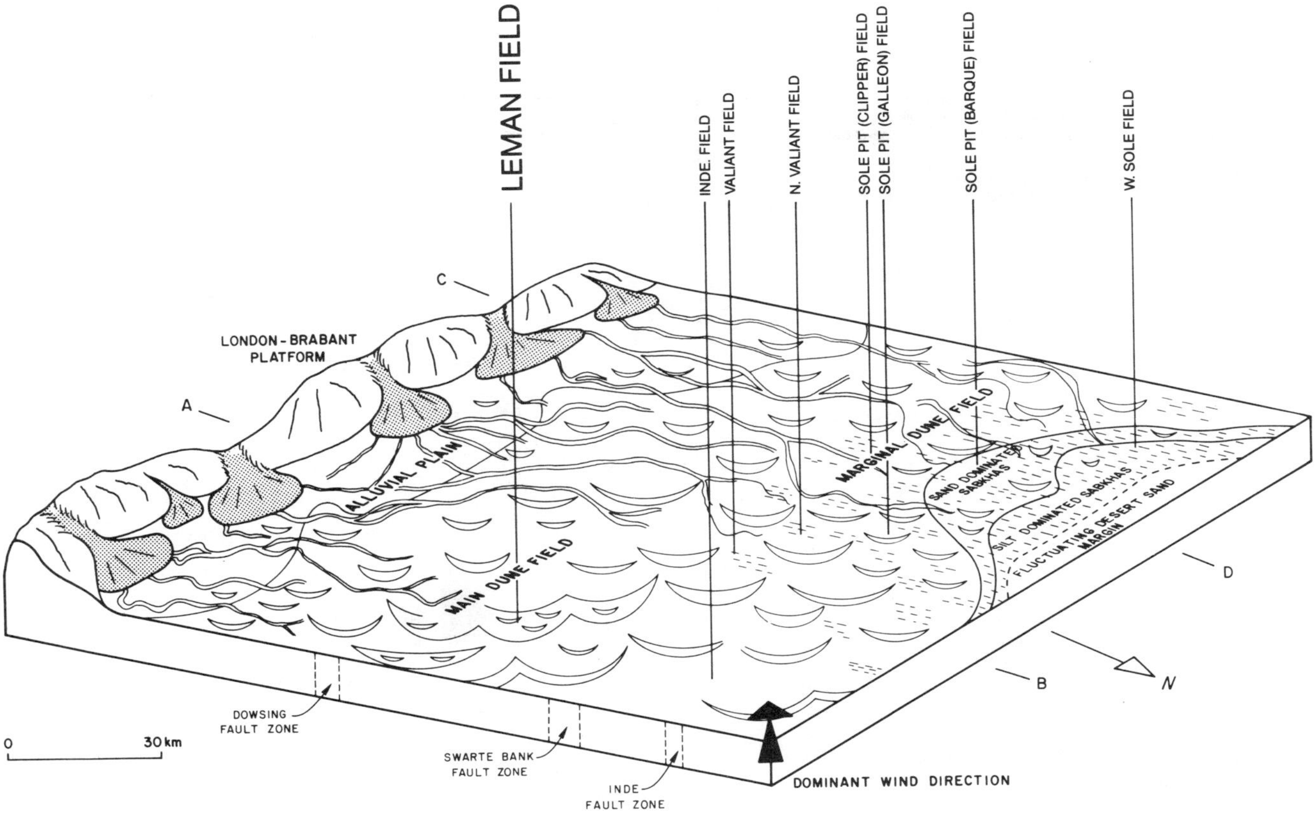
LEMAN FIELD
INDE. FIELD
VALIANT FIELD
N. VALIANT FIELD
SOLE PIT (CLIPPER) FIELD
SOLE PIT (GALLEON) FIELD
SOLE PIT (BARQUE) FIELD
W. SOLE FIELD
LONDON - BRABANT PLATFORM
ALLUVIAL PLAIN
MAIN DUNE FIELD
MARGINAL DUNE FIELD
SAND DOMINATED SABKHAS
SILT DOMINATED SABKHAS
FLUCTUATING DESERT SAND MARGIN
A
B
C
D
N
DOWSING FAULT ZONE
SWARTE BANK FAULT ZONE
INDE FAULT ZONE
0
30 km
DOMINANT WIND DIRECTION

Figure 12. Depositional setting of Leman field block diagram, showing relationship with other southern North Sea fields.

the cementing fluids. The sandstones of the A zone are easily recognized on cores because they are generally harder than those of the B zone due to the interstitial cement. The thickness of the A zone varies between 30 m (100 ft) and 200 m (650 ft) in the field. The A zone, although volumetrically greater than the other two zones, is a poorer reservoir than the B zone due to the increased cementation. It has been shown that the permeability for a given porosity in the B zone is three to seven times higher than that in the A zone.

The porosity in all zones is intergranular and modified by the diagenetic effects of paleoburial. As the field was buried unevenly (the northwest some 1220 m [4000 ft] and the southeast some 300 m [1000 ft] deeper than the present depth; see Table 1), the diagenetic effects are, accordingly, uneven across the field. Diagenesis consists of quartz overgrowth, dolomite, and fibrous illite growth in the pore throats (Glennie et al., 1978).

The depositional setting of the field is interpreted to be the main dune field between the eroding London Brabant platform and the desert lake to the north (Figures 12 and 13).

Fractures, both open and cemented, are visible in virtually all the cored wells in the field (Figure 14). Their contribution to the flow of the wells in the majority of the field is insignificant, but in the northwest, where the deeper paleoburial caused more severe permeability reduction through diagenesis, they have a measurable effect, raising the productivity of wells by up to 50%.

Porosity and permeability vary considerably across the field. Cross-plots of porosity and permeability show an areal variation in the relationship: For a given porosity, lower permeability is found in the northwest. Cross-plots for two wells are shown in Figure 15. The individual samples for a given porosity show a wide variation in permeability that is related to primary depositional characteristics such as median grain size and sorting. Figure 16 shows the very localized scale of this permeability variation.

Capillary pressure curves show that there is a very long transition zone above the free water level especially in the A and C zones.

The payzone is the entire Rotliegend in the most crestal parts of the field and is up to 275 m (900 ft) thick. However, due to the low general relief of the field some 80% of the gas is contained in the A zone.

The major barriers to flow within the reservoir are related to faulting and mode of sediment

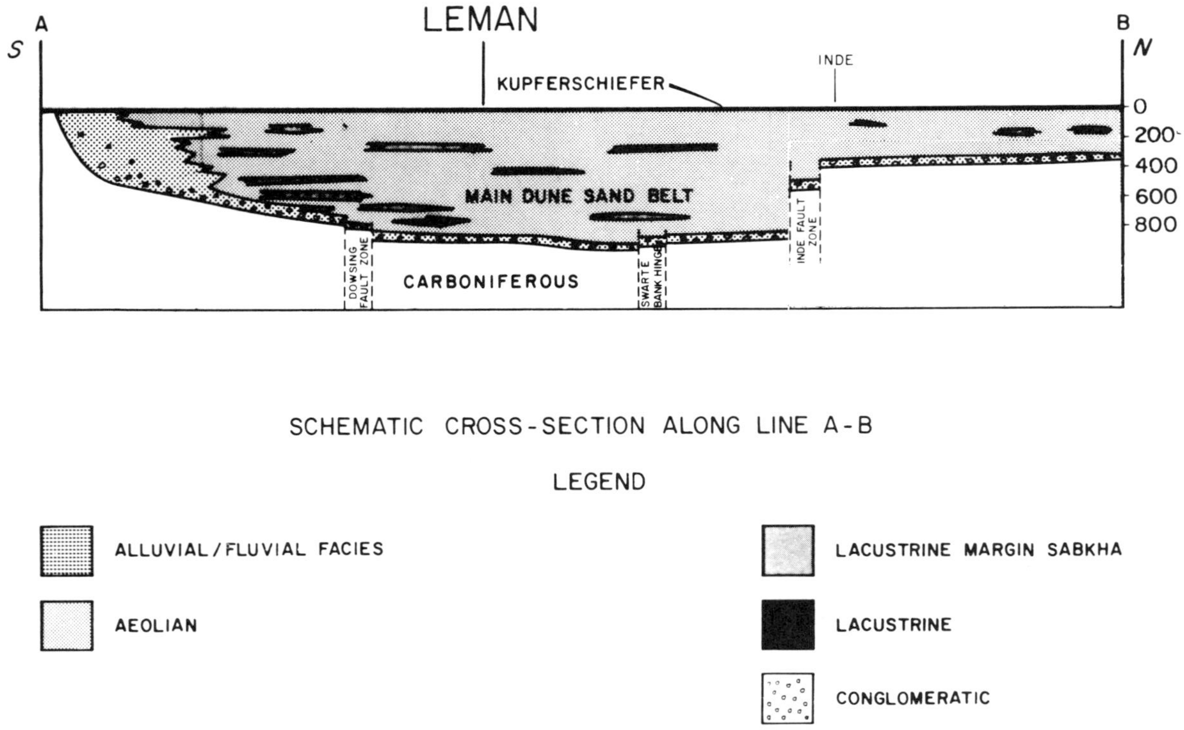
A
LEMAN
B
S
N
INDE
KUPFERSCHIEFER
0
200
400
600
800
MAIN DUNE SAND BELT
DOWSING FAULT ZONE
CARBONIFEROUS
SWARTE BANK HINGE
INDE FAULT ZONE
SCHEMATIC CROSS-SECTION ALONG LINE A-B
LEGEND
ALLUVIAL/FLUVIAL FACIES
AEOLIAN
LACUSTRINE MARGIN SABKHA
LACUSTRINE
CONGLOMERATIC

Figure 13. Depositional setting of Leman field cross-section across Figure 12.

deposition. Faulting, particularly in the southern and eastern parts of the field, created northwest-southeast-trending compartments that have very little pressure communication with each other. On the other hand, wind deposited dune bottomset beds separate successive cross-bedded dune foresets. The finely laminated, poorly sorted, carbonate cemented bottomset beds are relatively poor reservoir rocks. Their permeabilities are one or even two orders of magnitude lower than those of the better quality cross-bed dune foresets. The dune sands have strongly anisotropic permeability with the permeability along a dune slip face being much higher than that perpendicular to the slip face. Weber (1987) showed that, by analogy to the Permian De Chelly sandstone in northwestern Arizona, length to width to height ratios of around 200:75:1 can be expected. With an average cross-bed height of 4.5 m (15 ft) a dune cross-bed set in the Leman field could be 900 m (3000 ft) long and 343 m (1125 ft) wide. These dimensions are such that two wells from a cluster could penetrate a single cross-bed set.

The gas is sweet and lean with 95% methane, very little carbon dioxide, and a condensate gas ratio of only 1 bbl/mcf. An analysis is given in Table 2.

The initial development strategy of both operators of the field was to site drilling platforms close to culminations of the field in their respective leases and to drill a radial pattern of wells from these platforms. Further platforms were sited along the lease boundary to prevent, or at least minimize, crossflow over the border. At this time there was no direct seismic picture of the reservoir. With the availability of seismic maps of the reservoir, platforms were positioned to give a better areal drainage of the field and wells were sited to drain specific fault blocks. Figure 17 shows the current platform positions and well coverage. The field is developed as a straightforward depletion-type reservoir with an expected recovery factor of 83%. Drilling in both the northwest and east of the field has shown that the reservoir pressure in these areas has remained high and that a better spread of drainage points was required.

Faults

Faults are currently identified and mapped directly from seismic data. Faulting within the reservoir section has been recognized in some wells where fault zones were cored. However, due to the lack of character in the log response of the reservoir, it is not possible to identify fault cutouts from log correlation alone.

Faults that affect the reservoir generally die out within the salts of the Zechstein. Therefore, the majority of the faulting was not visible on the earlier surveys, hence the simplicity of the earlier maps. The direct seismic mapping of the top reservoir led to many more faults being identified. These fault indications from seismic are linked to form fault

WELL
UK 49/26-26
PLATE ABZ 005/14
CORE -5 TRAY -13
CORE -5 TRAY - 14
CORE -5 TRAY -15
CORE -5 TRAY -16
CORE -5 TRAY - 17
CORE -5 TRAY - 18
4" DIAMETER CORE
6596.00'
6599.00'
6602.00'
6605.00'
6608.00'
6611.00'
FEET
0
1
2
3
#204
#205
#206
#207
#208
#209
#210
#211
#212
#213
#214
#215
#216
#217
#218
#219 X
#220
221 S(V)
221 S(H)
#221
6599.00'
6602.00'
6605.00'
6608.00'
6611.00'
6614.00'

Figure 14. Core photograph Rotliegend sandstone well 49/26-26 Leman field. Note fractures both cemented and open, finely laminated sandstone, and truncated dips.

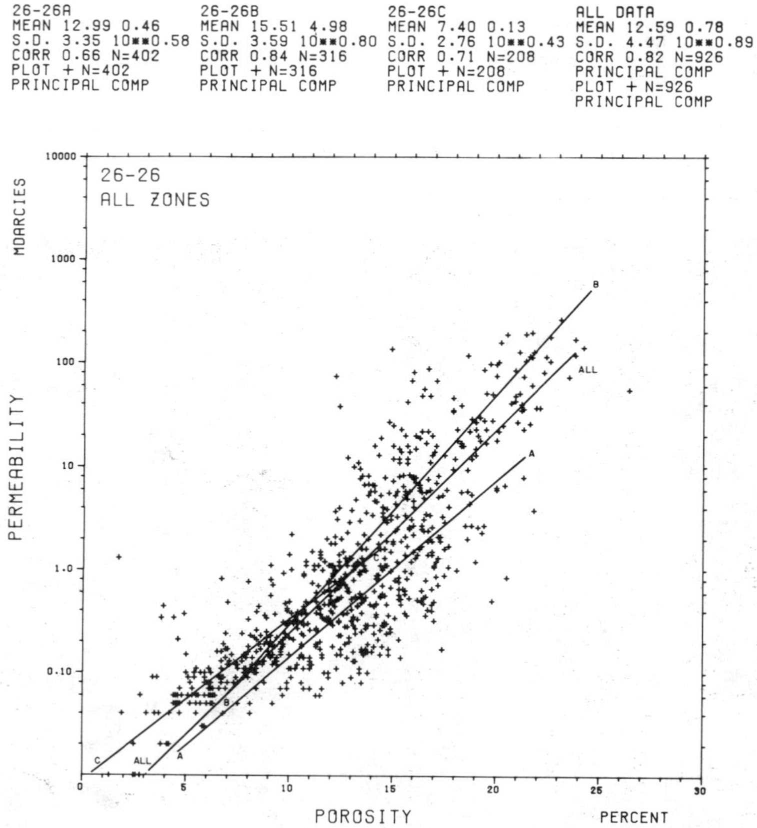
26-26A
MEAN 12.99 0.46
S.D. 3.35 10**0.58
CORR 0.66 N=402
PLOT + N=402
PRINCIPAL COMP
26-26B
MEAN 15.51 4.98
S.D. 3.59 10**0.80
CORR 0.84 N=316
PLOT + N=316
PRINCIPAL COMP
26-26C
MEAN 7.40 0.13
S.D. 2.76 10**0.43
CORR 0.71 N=208
PLOT + N=208
PRINCIPAL COMP
ALL DATA
MEAN 12.59 0.78
S.D. 4.47 10**0.89
CORR 0.82 N=926
PRINCIPAL COMP
PLOT + N=926
PRINCIPAL COMP
26-26
ALL ZONES
PERMEABILITY
MDARCIES
POROSITY
PERCENT

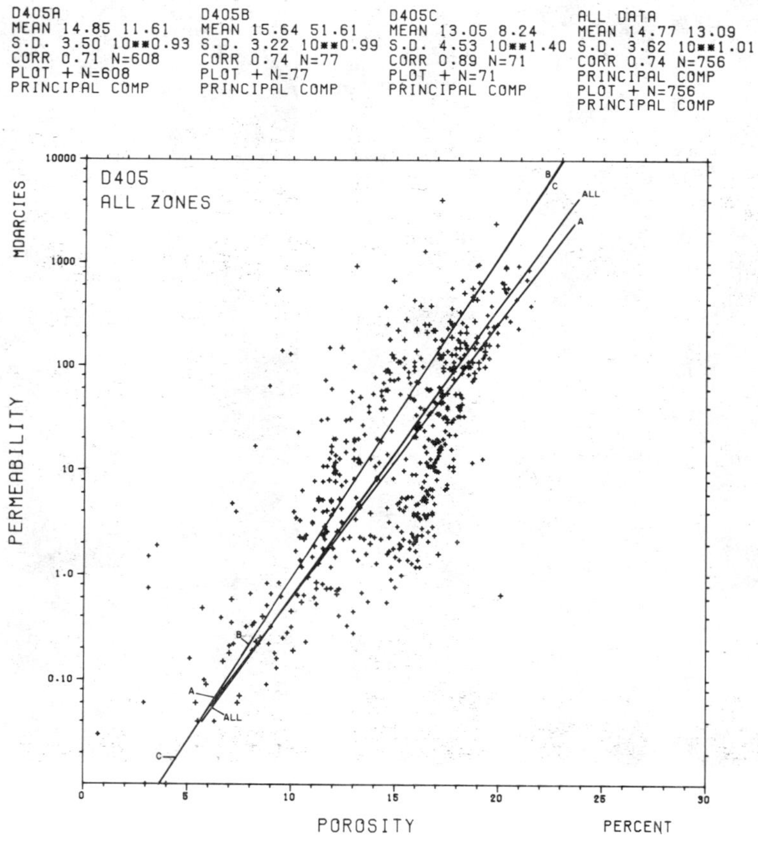
D405A
MEAN 14.85 11.61
S.D. 3.50 10**0.93
CORR 0.71 N=608
PLOT + N=608
PRINCIPAL COMP
D405B
MEAN 15.64 51.61
S.D. 3.22 10**0.99
CORR 0.74 N=77
PLOT + N=77
PRINCIPAL COMP
D405C
MEAN 13.05 8.24
S.D. 4.53 10**1.40
CORR 0.89 N=71
PLOT + N=71
PRINCIPAL COMP
ALL DATA
MEAN 14.77 13.09
S.D. 3.62 10**1.01
CORR 0.74 N=756
PRINCIPAL COMP
PLOT + N=756
PRINCIPAL COMP
D405
ALL ZONES
PERMEABILITY
MDARCIES
POROSITY
PERCENT

Figure 15. Porosity and permeability cross-plots of cores, Leman field. Note different trend between zones on each well and also difference between wells 49/26-26 in northeast and 49/26-D405 in center of field.

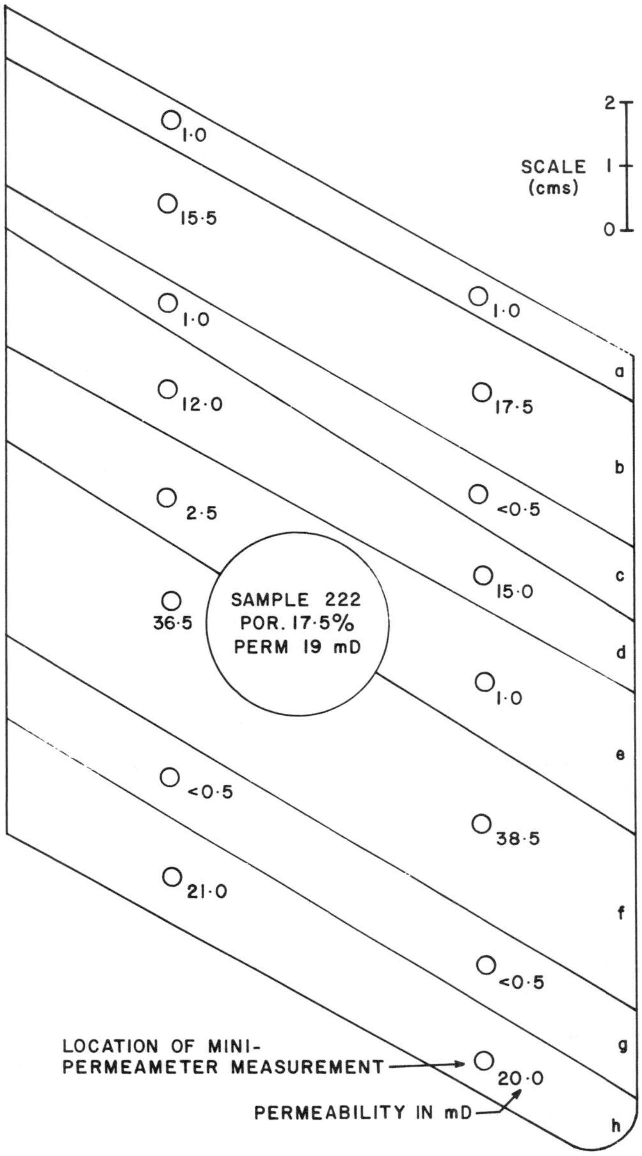

SCALE (cms)
2
1
0
1·0
15·5
1·0
12·0
2·5
36·5
SAMPLE 222
POR. 17·5%
PERM 19 mD
<0·5
21·0
1·0
17·5
<0·5
15·0
1·0
38·5
<0·5
20·0
a
b
c
d
e
f
g
h
LOCATION OF MINI-PERMEAMETER MEASUREMENT
PERMEABILITY IN mD

Figure 16. Permeability distribution in dune sand core, Leman field. Note higher permeability correlates with medium to fine grainsize, lower permeability with very fine sand to silt grainsize. a-h are alternating dune foreset laminae.

patterns in accordance with the structural model. However, due to the spacing of the seismic, aliasing of the faults is a problem. In the south and east of the field, there is a series of long en echelon faults that trend northwest-southeast. These have less vertical throw than the reservoir thickness but appear to be almost complete barriers to flow. These faults are strike-slip faults formed as a result of lateral movements along deep-seated basement faults. Their sealing capacity is due to grain crushing and cementation along the fault planes. Faulting in the north and west of the field does not form as regular a pattern as seen in the south and east, and the effect of these faults on flow is not yet apparent. Figure 18 shows the current fault pattern of the field with the field outline added.

Table 2. Composition of separator gas, separator oil (condensate), and reservoir fluid from well 49/26-1, Leman field. Based on composition of tank gas and tank oil obtained at 25°C (77°F).

Component	Separator Gas mol%	Separator Oil mol%	Reservoir Fluid mol%
Methane	95.05	14.00	94.94
Ethane	2.86	4.41	2.86
Propane	0.49	3.62	0.49
Iso-butane	0.08	1.68	0.08
N-butane	0.09	3.71	0.10
Iso-pentane	0.03	3.02	0.03
N-pentane	0.02	3.70	0.03
Hexanes	0.02	8.54	0.03
Heptanes plus	0.04	57.27	0.12
Helium	0.02	0.00	0.02
Nitrogen	1.26	0.02	1.26
Carbon dioxide	0.04	0.03	0.04
Total	100.00	100.00	100.00

Source

The source for the gas in Leman and indeed for all the other southern North Sea gas fields is the Westphalian Coal Measures that directly underlie the reservoir (Figure 19). Migration paths are supplied by the sands within the Westphalian that have huge areas of contact with both the coals and carbonaceous shales that are the actual sources. These sands form the conduits from the source at generation time to the reservoir. The reservoir unconformably overlies the Westphalian and hence has very good communication with it.

The time of migration coincides with the time of maximum depth of burial during the Jurassic and Cretaceous. The gas first migrated to the then structurally higher flanks of the Sole Pit basin. Later, during and after the structural inversion and formation of the trap in the Late Cretaceous, the gas remigrated back into the field (Figure 20). The fact that the gas remained trapped to the present demonstrates the extremely good seal formed by the Zechstein evaporites.

The source has two components, carbonaceous shales and coals with TOC values of 1% and 60%, respectively. The kerogen type is II/III–III. The potential yield for the shale is 0.14 mcf/ac-ft and for the coal 7.0 mcf/ac-ft (Cornford, 1986).

The source has a vitrinite reflectance maturation of 1.6 to 2.1 equivalent to a maximum depth of burial of between 3350 and 3960 m ss (11,000 and 13,000 ft).

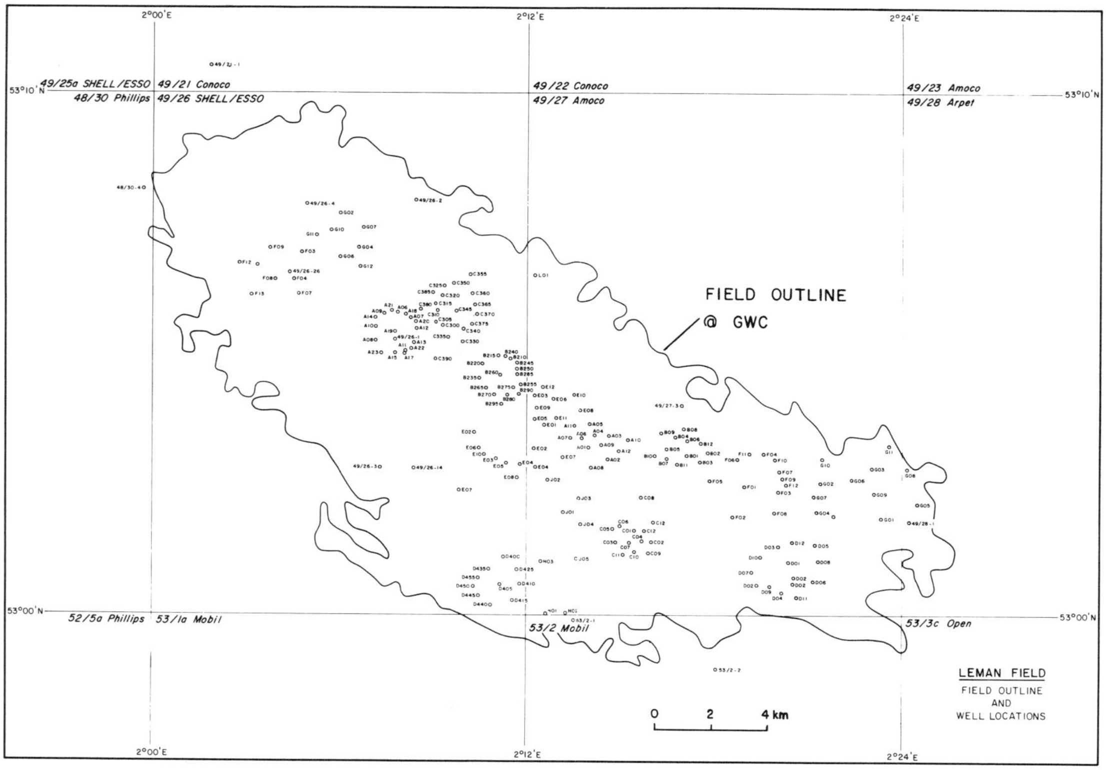
2°00'E
2°12'E
2°24'E
53°10'N
49/25a SHELL/ESSO
48/30 Phillips
49/21 Conoco
49/26 SHELL/ESSO
49/22 Conoco
49/27 Amoco
49/23 Amoco
49/28 Arpet
FIELD OUTLINE
@ GWC
53°00'N
52/5a Phillips
53/1a Mobil
53/2 Mobil
53/3c Open
LEMAN FIELD
FIELD OUTLINE
AND
WELL LOCATIONS
0
2
4 km

Figure 17. Development well location, Leman field. Note radial pattern of wells in early platforms.

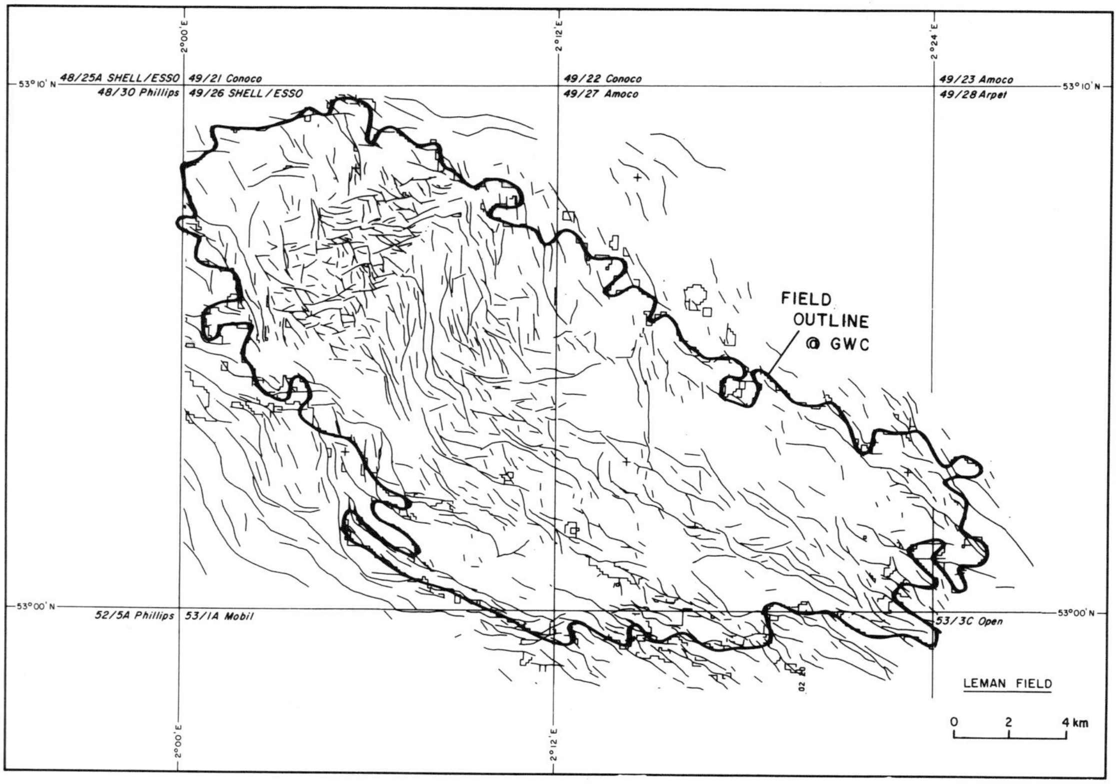
2°00'E
2°12'E
2°24'E
53°10'N
48/25A SHELL/ESSO
48/30 Phillips
49/21 Conoco
49/26 SHELL/ESSO
49/22 Conoco
49/27 Amoco
49/23 Amoco
49/28 Arpet
FIELD
OUTLINE
@ GWC
53°00'N
52/5A Phillips
53/1A Mobil
53/3C Open
LEMAN FIELD
0
2
4 km

Figure 18. Present-day fault pattern, Leman field.

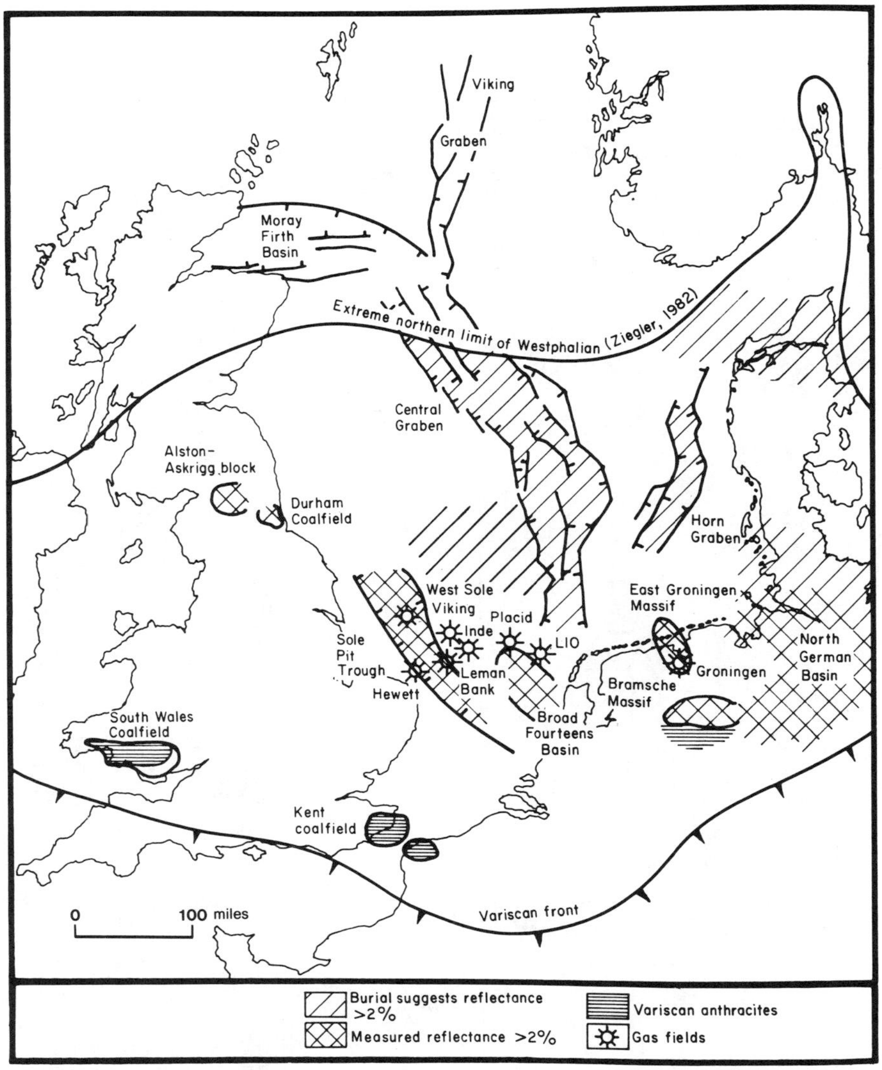
Viking
Graben
Moray
Firth
Basin
Extreme northern limit of Westphalian (Ziegler, 1982)
Central
Graben
Alston-
Askrigg block
Durham
Coalfield
Horn
Graben
West Sole
Viking
Inde
Placid
L10
East Groningen
Massif
Sole
Pit
Trough
Hewett
Leman
Bank
Groningen
North
German
Basin
Bramsche
Massif
Broad
Fourteens
Basin
South Wales
Coalfield
Kent
coalfield
0
100 miles
Variscan front
Burial suggests reflectance >2%
Measured reflectance >2%
Variscan anthracites
Gas fields

Figure 19. Distribution and maturity of Carboniferous source rock (from Cornford, 1986).

EXPLORATION CONCEPTS

Regional Play

The Rotliegend gas play in the southern North Sea is defined by the widespread presence of variable-quality Rotliegend sandstone reservoir, the near perfect Zechstein evaporite cap rock, and the coal-bearing Westphalian source rock (Figure 21).

These favorable conditions are offset by the complications of relative timing of generation and migration of hydrocarbon and trap formation, the deterioration of the reservoir quality of the Rotliegend due to diagenesis during the deep burial prior to the inversion and trap formation, and the possible absence of salt within the Zechstein cap rock allowing the charge to leak out.

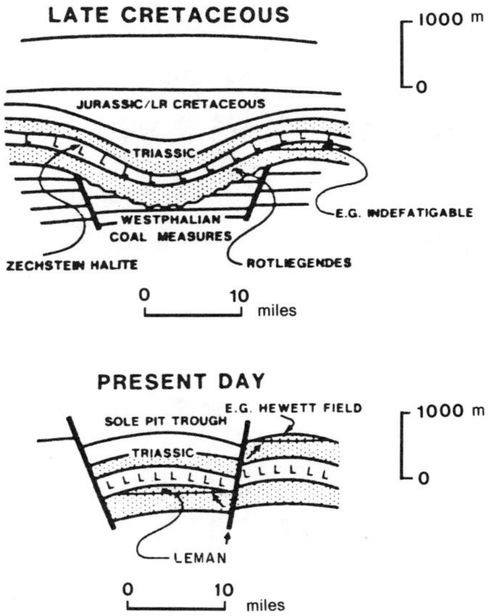
LATE CRETACEOUS
1000 m
0
JURASSIC/LR CRETACEOUS
TRIASSIC
E.G. INDEFATIGABLE
WESTPHALIAN
COAL MEASURES
ZECHSTEIN HALITE
ROTLIEGENDES
0
10
miles
PRESENT DAY
SOLE PIT TROUGH
E.G. HEWETT FIELD
1000 m
0
TRIASSIC
LEMAN
0
10
miles

Figure 20. Remigration on inversion. Schematic diagram showing remigration of gas (from Cornford, 1986).

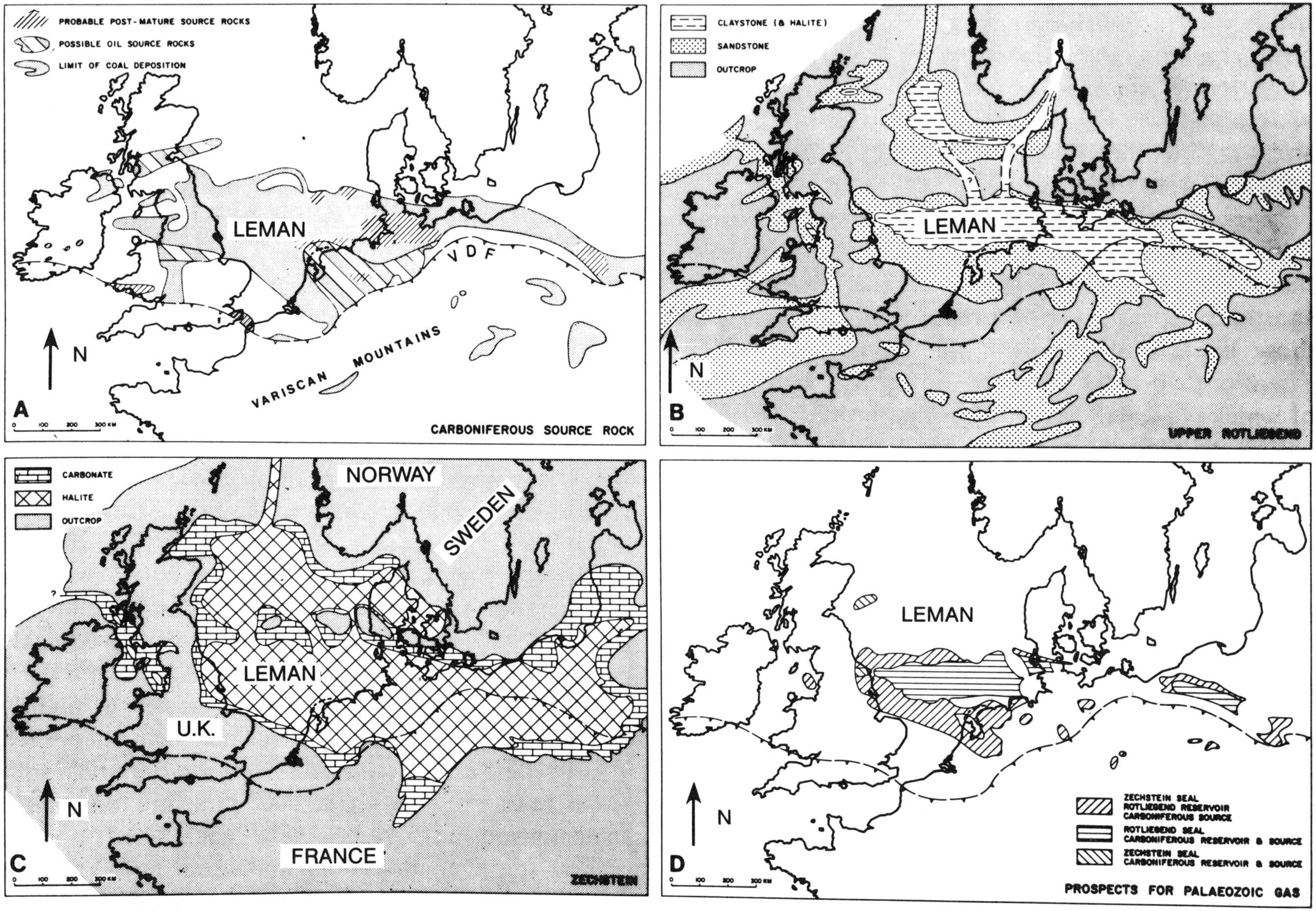
PROBABLE POST-MATURE SOURCE ROCKS
POSSIBLE OIL SOURCE ROCKS
LIMIT OF COAL DEPOSITION
LEMAN
V D F
VARISCAN MOUNTAINS
N
A
CARBONIFEROUS SOURCE ROCK
CLAYSTONE (& HALITE)
SANDSTONE
OUTCROP
LEMAN
N
B
UPPER ROTLIEGEND
CARBONATE
HALITE
OUTCROP
NORWAY
SWEDEN
LEMAN
U.K.
FRANCE
N
C
ZECHSTEIN
LEMAN
N
D
ZECHSTEIN SEAL
ROTLIEGEND RESERVOIR
CARBONIFEROUS SOURCE
ROTLIEGEND SEAL
CARBONIFEROUS RESERVOIR & SOURCE
ZECHSTEIN SEAL
CARBONIFEROUS RESERVOIR & SOURCE
PROSPECTS FOR PALAEOZOIC GAS

Figure 21. Facies maps defining the Rotliegend play North Sea (from Glennie, 1986). (A) Source. (B) Reservoir. (C) Cap. (D) Prospective area.

General Application of Geologic Parameters

Although no fault in the field has sufficient throw to offset the reservoir, there is strong evidence that the faults act as permeability barriers within the reservoir. This necessitates a wider spread of drainage points to fully develop the reserves. This aspect was unknown at the time of discovery of the field because the reservoir was not directly mappable from the available seismic data. Given today's seismic techniques, a full 3D seismic survey would have enabled the reservoir to be imaged in a much more detailed fashion before the development plan was made.

The reduced permeability of the A zone relative to the B zone also allowed large pressure gradients to develop in areas where the B zone was below the gas-water contact, thus effectively sealing off areas of the field from drainage. The original concept of the field behaving like a gas cylinder, and therefore needing only a limited number of centrally located drainage points to develop, has been found to be incorrect.

The use of RFT pressure measurements in recent wells has shown that the gas sands are being depleted, but due to much lower mobility of water relative to gas the pressures below the gas-water contact remain high.

ACKNOWLEDGMENTS

Shell U.K. Exploration and Production and Esso Exploration and Production U.K. Limited permitted publication of this paper. The advice and assistance of numerous colleagues within Shell and the use of much unpublished data on the field from internal reports has greatly eased the job of the author. Mr. Alan Gilbert prepared the figures.

REFERENCES CITED

Carmalt, S. W., and B. St. John, 1986, Giant oil and gas fields, *in* M. T. Halbouty, ed., Future petroleum provinces of the world: American Association of Petroleum Geologists Memoir 40, p. 11-54.

Cornford, C., 1986, Source rocks and hydrocarbons of the North Sea, *in* K. W. Glennie, ed., Introduction to petroleum geology of the North Sea: London, Blackwell Scientific Publications.

Glennie, K. W., 1972, Permian Rotliegendes of north west Europe interpreted in light of modern desert sedimentation studies: American Association of Petroleum Geologists Bulletin, v. 56, p. 1048-1071.

Glennie, K. W., 1986, Early Permian—Rotliegend, *in* K. W. Glennie, ed., Introduction to petroleum geology of the North Sea: London, Blackwell Scientific Publications, p. 63-85.

Glennie, K. W., and P.L.E. Boegner, 1981, Sole pit inversion tectonics, *in* L. V. Illing and G. D. Hobson, ed., Petroleum geology of the continental shelf of north-west Europe: London, Heyden & Son, p. 110-120.

Glennie, K. W., and A. T. Buller, 1983, The Permian Weissliegend of N.W. Europe: the partial deformation of aeolian dune sands caused by the Zechstein Transgression: Sedimentary Geology, v. 35, p. 43-81.

Glennie, K. W., G. C. Mudd, and P.J.C. Nagtegaal, 1978, Depositional environment and diagenesis of Permian Rotliegendes sandstone in Leman Bank and Sole Pit areas of the U.K. southern North Sea: Journal of the Geological Society of London, v. 135, p. 25-34.

Parsley, A. J., 1986, North Sea hydrocarbon plays, *in* K. W. Glennie, ed., Introduction to petroleum geology of the North Sea: Oxford, Blackwell Scientific Publications, p. 237-263.

Rhys, G. H., 1975, A proposed standard lithostratigraphic nomenclature for the southern North Sea, *in* A. W. Woodland, ed., Petroleum and the continental shelf of north-west Europe: London, Applied Science Publishers, v.1, p. 151-163.

Stalder, P. J., 1973, Influence of crystallographic habit and aggregate structure of authigenic clay minerals on sandstone permeability: Geologie en Mijnbouw, v. 52, p. 217-220.

Veen, F. R. van, 1975, Geology of the Leman gas field, *in* A. W. Woodland, ed., Petroleum and the continental shelf of north-west Europe: London, Applied Science Publishers, v. 1, p. 223-231.

Weber, K. J., 1987, Computation of initial well productivities in aeolian sandstone on the basis of a geological model, Leman gas field, U.K., *in* R. W. Tillman and K. J. Weber, eds., Reservoir sedimentology: Society of Economic Paleontologists and Mineralogists Special Publication No. 40, p. 333-354.

Ziegler, P. A., 1982, Geological atlas of western and central Europe: Amsterdam, Elsevier, 130 p.

SUGGESTED READING

Glennie, K. W., ed, 1986, Introduction to petroleum geology of the North Sea; Second Edition: Oxford, Blackwell Scientific Publications, contains an excellent overview of the geology of the whole of the North Sea including the Southern Permian basin.

Appendix 1. Field Description

Field name .... *Leman*

Ultimate recoverable reserves .... *11,523 MMMft^3 (326.3 MMMm^3)*

Field location:

- **Country** .... *United Kingdom*
- **State** .... *North Sea*
- **Basin/Province** .... *Southern North Sea Permian*

Field discovery:

- **Year field discovered** *1966*
- **Year second pay discovered** .... *NA*
- **Third pay** .... *NA*

Discovery well name and general location
(i.e., Jones No. 1, Sec. 2T12NR5E; or Smith No. 1, 5 mi west of Sheridan, Wyoming):

- **First pay** .... *49/26-1; 32 mi NE of Bacton on the north Norfolk coast*
- **Second pay** .... *NA*
- **Third pay** .... *NA*

Discovery well operator .... *Shell U.K. Exploration and Production*
(if more than one pay in field, list operators of discovery well in other pays)

- **Second pay** .... *NA*
- **Third pay** .... *NA*

IP in barrels per day and/or cubic feet or cubic meters per day:

- **First pay** .... *1087 10^6 ft^3/d*
- **Second pay** .... *NA*
- **Third pay** .... *NA*

All other zones with shows of oil and gas in the field:

Age	Formation	Type of Show

Permian age Zechstein dolomites with high pressure low volume gas in fractures and/or vugs.

Geologic concept leading to discovery and method or methods used to delineate prospect, e.g., surface geology, subsurface geology, seeps, magnetic data, gravity data, seismic data, seismic refraction, nontechnical:

Extrapolation of data from Rotliegend gas wells in the onshore area of Holland and Germany across the North Sea to similar sands in Durham, England, led to the belief that the Rotliegend sands should be present in the southern North Sea. Seismic data proved the existence of structures in the Zechstein, which overlies the Rotliegend. (The Zechstein was the deepest reflector visible.)

Structure:

Province/basin type (see St. John, Bally, and Klemme, 1984)

Inverted basin induced by basement wrench faulting

Tectonic history

Carboniferous gentle subsidence with deposition of coal measures. Hercynian uplift and consequent erosion at end Carboniferous. Permian to Cretaceous subsidence of the southern North Sea basin with continuous sedimentation. Cretaceous inversion of the Sole Pit basin and erosion of most of the Cretaceous and Jurassic sediments over the field. Paleocene to Recent very slow gentle subsidence with deposition of marine clays and sands.

Regional structure

Leman field is situated in the southeast of the Sole Pit inverted basin, which is bounded by the Dowsing fault zone, the Outer Silverpit Fault, and the Swarte Bank Hinge. The western end of the field being closer to the inversion center has therefore undergone greater paleo-burial than the east.

Local structure

An elongated anticlinal dome oriented NW-SE 18 mi long by 8 mi wide at the gas-water contact. The structure is highly faulted with three major fault trends being oriented parallel to the main structural trends of the area.

Trap

Trap type(s) .... *Single structural dome with a single pay*

Basin stratigraphy (major stratigraphic intervals from surface to deepest penetration in field):

Age	Formation	Depth to Top
Cenozoic	*North Sea Group*	*30-100 ft ss*
Cretaceous	*Chalk*	*400 ft ss only on flank*
Jurassic	*Lias Group*	*Eroded*
Triassic (Late)	*Haisborough Group*	*400-550 ft ss*
Triassic (Early)	*Bacton Group*	*2400-3000 ft ss*
Permian (Late)	*Zechstein Group*	*4500-5000 ft ss*
Permian (Early)	*Rotliegend Group*	*5900-6700 ft ss*
Carboniferous	*Westphalian*	*6500+ ft ss*

Location of well in field .... *NA*

Reservoir characteristics:

Number of reservoirs *1*

Formations .... *Leman Sandstone Formation of the Rotliegend Group*

Ages .... *Early Permian*

Depths to tops of reservoirs .... *5900 ft ss*

Gross thickness (top to bottom of producing interval) .... *800 ft maximum*

Net thickness—total thickness of producing zones

Average .... *340 ft*

Maximum .... *800 ft*

Average

Maximum

Lithology .... *Fine- to medium-grained cross-bedded dune sandstone with minor interbedded conglomerates and shales toward the base*

Porosity type .... *Intergranular modified by diagenetic cements of quartz, calcite, and illite*

Average porosity .... *NA*

Average permeability .... *NA*

Seals:

Upper

Formation, fault, or other feature

Lithology .... *Zechstein evaporites in four cycles of anhydrite, dolomite, and halite*

Lateral

Formation, fault, or other feature .... *Dip closure against the Zechstein evaporites*

Lithology

Source:

Formation and age .... *Westphalian Coal Measures of the Carboniferous*

Lithology .... *Coal and carbonaceous shale*

Average total organic carbon (TOC) .... *60% coal and 1% shale*

Maximum TOC .... *75% in coal*

Kerogen type (I, II, or III) .... *II/III-III*

Vitrinite reflectance (maturation) .... R_o *= 1.6- 2.1*

Time of hydrocarbon expulsion .... *Jurassic and Cretaceous*

Present depth to top of source .... *6500 ft ss*

Thickness .... *Base not seen in field but 3000-7500 ft in Germany*
Potential yield .... *0.14-7.0 mcf/ac-ft*

Appendix 2. Production Data

Field name .... *Leman*

Field size:

- **Proved acres** .... *69,456 ha. (28,109 ac)*
- **Number of wells all years** *175*
- **Current number of wells (as of year)** *144*
- **Well spacing** .... *482 ac/well overall, 125 ac/well in platform reach*
- **Ultimate recoverable** .... *11,523 bcf (326 bcm)*
- **Cumulative production** .... *8207 bcf (233 bcm)*
- **Annual production** .... *391 bcf (11.1 bcm)*
- **Present decline rate** *2.9%*
  - **Initial decline rate** *5%*
  - **Overall decline rate** *5%*
- **Annual water production** .... *Negligible*
- **In place, total reserves** .... *13,862 bcf (393 bcm)*
- **In place, per acre-foot** .... *585,523 ft^3*
- **Primary recovery** .... *11,523 bcf (326 bcm)*
- **Secondary recovery** .... *Nil*
- **Enhanced recovery** .... *Nil*
- **Cumulative water production** .... *Negligible*

Drilling and casing practices:

- **Amount of surface casing set** .... *400 ft*
- **Casing program** .... *18⅝ in. @1600 ft; 9⅝ in. @ top Zechstein; 7 in. @ base Zechstein; 4½ in. liner @ total depth*
- **Drilling mud** .... *Water-based or invert oil emulsion mud to top Zechstein, invert oil mud to total depth*
- **Bit program** .... *Rotary rock bits to top Zechstein, PCD or diamond bits on turbines to total depth*
- **High pressure zones** .... *Platten Dolomite and Haupt Dolomite Formations sometimes contain high-pressure gas*

Completion practices:

- **Interval(s) perforated** .... *Either whole reservoir to 50 ft above gas-water contact or limited interval close to top of reservoir for fracturing*
- **Well treatment** .... *Diverted acid squeeze or propped hydraulic fracture stimulation depending on logged porosity*

Formation evaluation:

- **Logging suites** .... *Usually limited to reservoir interval where formation density, neutron density, sonic, induction, gamma ray, and caliper logs are run*
- **Testing practices** .... *Repeat formation tester is run at final logging; a production test is carried out after completion before wells are put on stream*

Mud logging techniques

A contractor is employed on each rig to describe the drilled cuttings, gas chromatograph and H_2S detectors are used to continually monitor the mud returns.

Gas characteristics:

- **Type** .... *Lean*
- **Gas gravity** .... *0.585 with respect to air at standard conditions*

Initial condensate:gas ratio .... *1.0 bbl/mmft³*
Condensate density .... *47 lb/ft³*
Gas composition .... *95% methane, 3% ethane, 1% nitrogen, 1% C_3+, 0.04 carbon dioxide, 0.02% helium*

Field characteristics:

Average elevation .... *6500 ft ss*
Initial pressure .... *3022 psi*
Present pressure .... *800-2900 psi*
Pressure gradient .... *0.46 psi/ft*
Temperature .... *125°F*
Geothermal gradient .... *0.01°F/ft*
Drive .... *Depletion*
Gas column thickness .... *800 ft maximum*
Gas-water contact .... *6700 ft ss*
Connate water *25%*
Water salinity, TDS .... *240,000 ppm*
Resistivity of water .... *0.026 ohm-m @ 125°F*
Bulk volume water (%) .... *NA*

Transportation method and market for oil and gas:

Gas is gathered on three terminal platforms offshore and brought ashore to Bacton via three 30-in. pipelines; offshore compression has been installed to maintain deliverability. The gas is cleaned, dried, and the condensate removed at Bacton before the gas is sold to British Gas.

Drake Point Gas Field, Canadian Arctic Islands

D. C. WAYLETT
Panarctic Oils, Ltd.
Calgary, Alberta, Canada

FIELD CLASSIFICATION

BASIN: Sverdrup
BASIN TYPE: Cratonic Sag
RESERVOIR ROCK TYPE: Sandstone
RESERVOIR ENVIRONMENT OF DEPOSITION: Shelf; Beach (Barrier Bar)
RESERVOIR AGE: Jurassic
PETROLEUM TYPE: Gas
TRAP TYPE: Faulted Anticline

LOCATION

The first significant gas discovery in the Canadian Arctic Islands was made at Drake Point, which is on the east side of the Sabine Peninsula on Melville Island near lat. 76°25′N, long. 108°30′ W, in the District of Franklin, Northwest Territories, Canada. The geographic location of Melville Island and the general location of the Drake Point field on the Sabine Peninsula are shown in Figure 1. There are two additional gas fields in the Sabine Peninsula area: the Hecla and Roche Point fields (Figure 2). Other known oil and gas fields in the Canadian Arctic Islands are shown in Figure 3.

Marketable natural gas reserves in the Arctic Islands are estimated to be 17 trillion cubic feet (tcf) and "in-place" oil reserves are estimated to be 1600 million bbl. All reserve estimates used in this publication are those calculated by Panarctic Oils, Ltd.

Waylett (1979) described several of the natural gas fields in the Arctic Islands. Except for the Bent Horn oil field on Cameron Island, all the oil and gas reserves in the Arctic Islands are found in clastic rocks of the Mississippian to Tertiary Sverdrup basin (Figure 4), which is 1200 km (750 mi) long and up to 400 km (250 mi) wide.

The Drake Point field covers an area of 40,500 ha. (100,000 acres), with almost 22,275 ha. (55,000 acres) being offshore in water depths to 425 m (1400 ft). The onshore area is a low-lying, nearly vegetation-free tundra that overlies thick, continuous permafrost.

The estimated discovered, marketable gas reserves in the Drake Point field are 5.3 tcf. Additional small, undrilled fault blocks offshore could bring the total marketable gas reserves to 6 tcf. This field is ranked 237th in world size (Carmalt and St. John, 1986). The gas is found in sandstones of the Lower Jurassic Intrepid Inlet Member of the Jameson Bay Formation, and the Drake Point Member of the MacLean Strait Formation (Figure 5).

The Drake Point field still is in an early stage of exploration, with three offshore and six onshore wells having penetrated the gas reservoirs. Information on the field is relatively limited.

HISTORY

Pre-Discovery

Panarctic Oils, Ltd., selected the Sabine Peninsula of Melville Island for exploration because the stratigraphic sequence in the Mesozoic and upper Paleozoic rocks are discernible using surface geology (Tozer and Thorsteinsson, 1964). Available reconnaissance surface geological and photogeological maps indicated an anticlinal trend extending east to west across the Sabine Peninsula (Figure 6). Seismic work on this trend in 1968 outlined an area of closure on the eastern side of the peninsula, approximately 12.8 km (8 mi) long and 4 km (2.5 mi) wide. This interpretation further indicated that the closure was practically all onshore. The location of the Drake Point discovery well, N-67 (Figure 6), was selected on the basis of this control.

Discovery

Drake Point N-67 was the third well drilled in the Sverdrup basin and the sixth well in the Arctic Islands. The first two Sverdrup basin wells were drilled on western Melville Island near some outcropping Triassic tar sands. Drake Point N-67 was an obligatory deep test on Prairie Oil Royalty acreage, which was farmed out to Panarctic. This well was drilled to test a number of Jurassic and Triassic sandstones that outcrop south of the well. The well was spudded April 4, 1969.

On May 4, a drilling break occurred in what was later determined to be the Lower Jurassic Borden Island Formation (now the Drake Point Member of the MacLean Strait Formation) at 1100 m (3610 ft). On May 5, a partial blowout occurred and the well was in a precarious

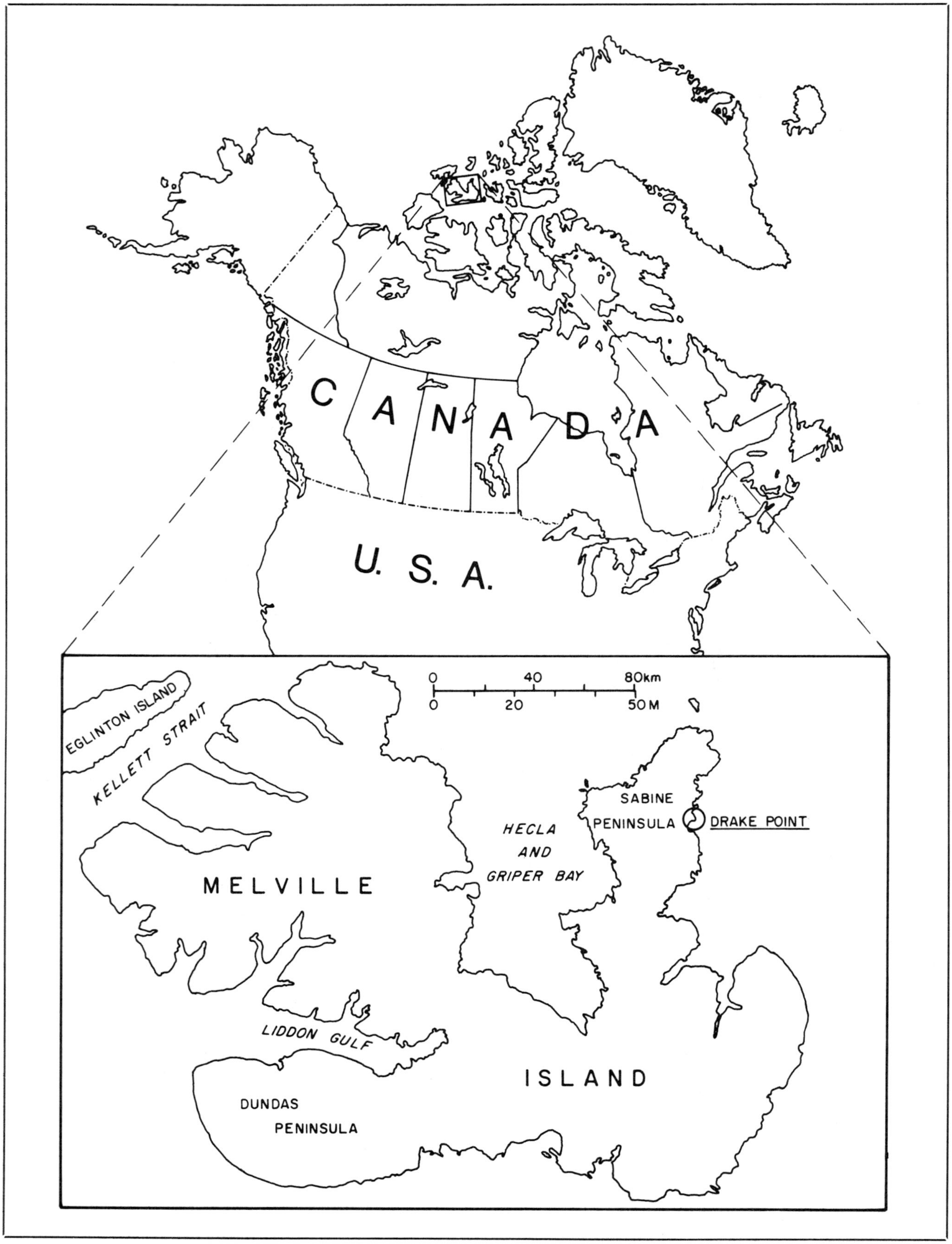
CANADA
U. S. A.
0 40 80km
0 20 50M
EGLINTON ISLAND
KELLETT STRAIT
SABINE
PENINSULA
DRAKE POINT
HECLA
AND
GRIPER BAY
MELVILLE
LIDDON GULF
ISLAND
DUNDAS
PENINSULA

Figure 1. Location map, showing location of the Drake Point field on the Sabine Peninsula.

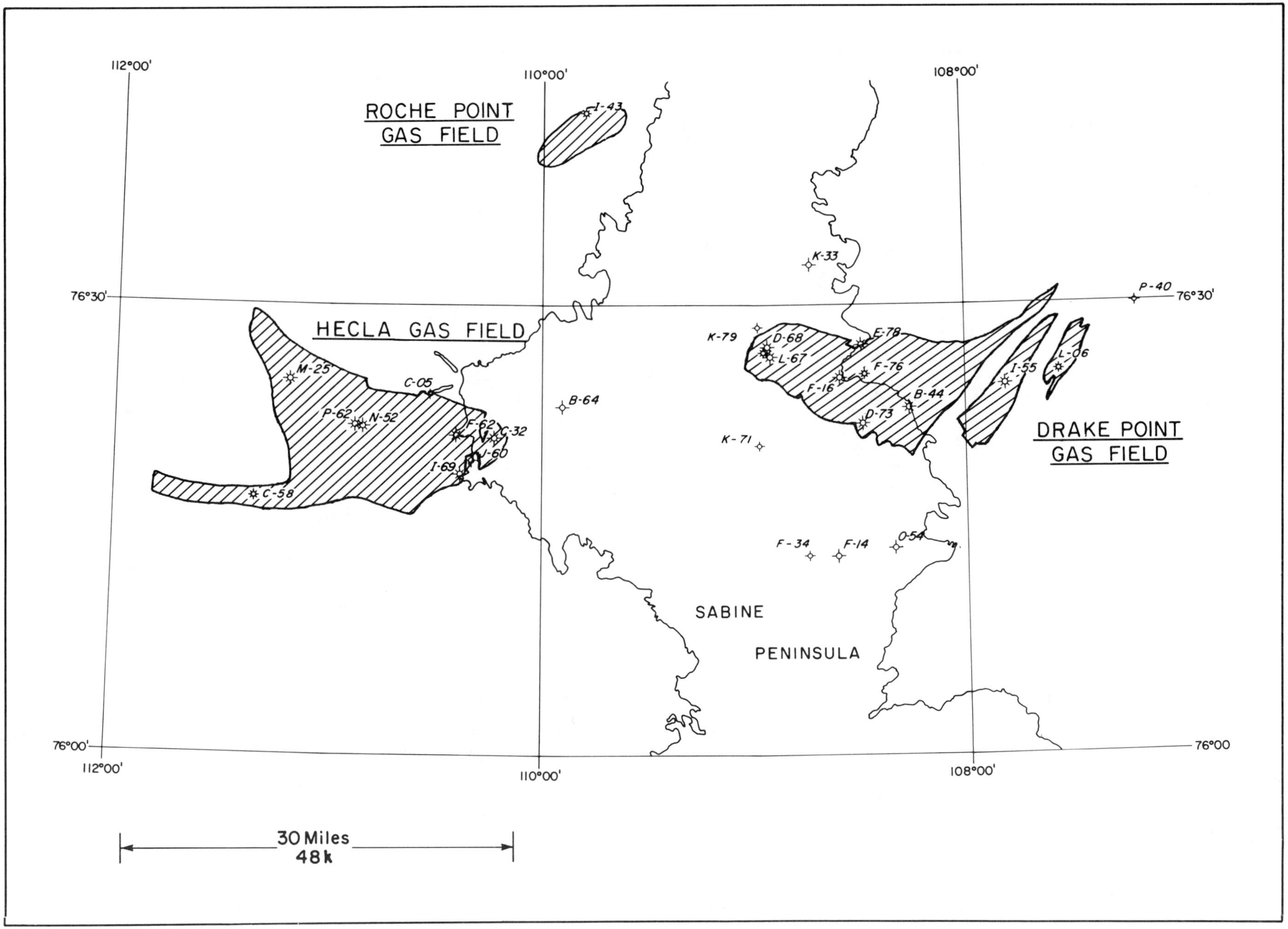

112°00'
110°00'
108°00'
76°30'
76°00'
76°00
ROCHE POINT GAS FIELD
HECLA GAS FIELD
DRAKE POINT GAS FIELD
SABINE PENINSULA
I-43
M-25
P-62
N-52
C-05
F-62
C-32
I-69
C-58
B-64
K-79
D-68
L-67
E-78
F-76
F-16
B-44
D-73
K-71
K-33
P-40
I-55
L-06
F-34
F-14
O-54
30 Miles
48k

Figure 2. Sabine Peninsula, Melville Island, Canada.

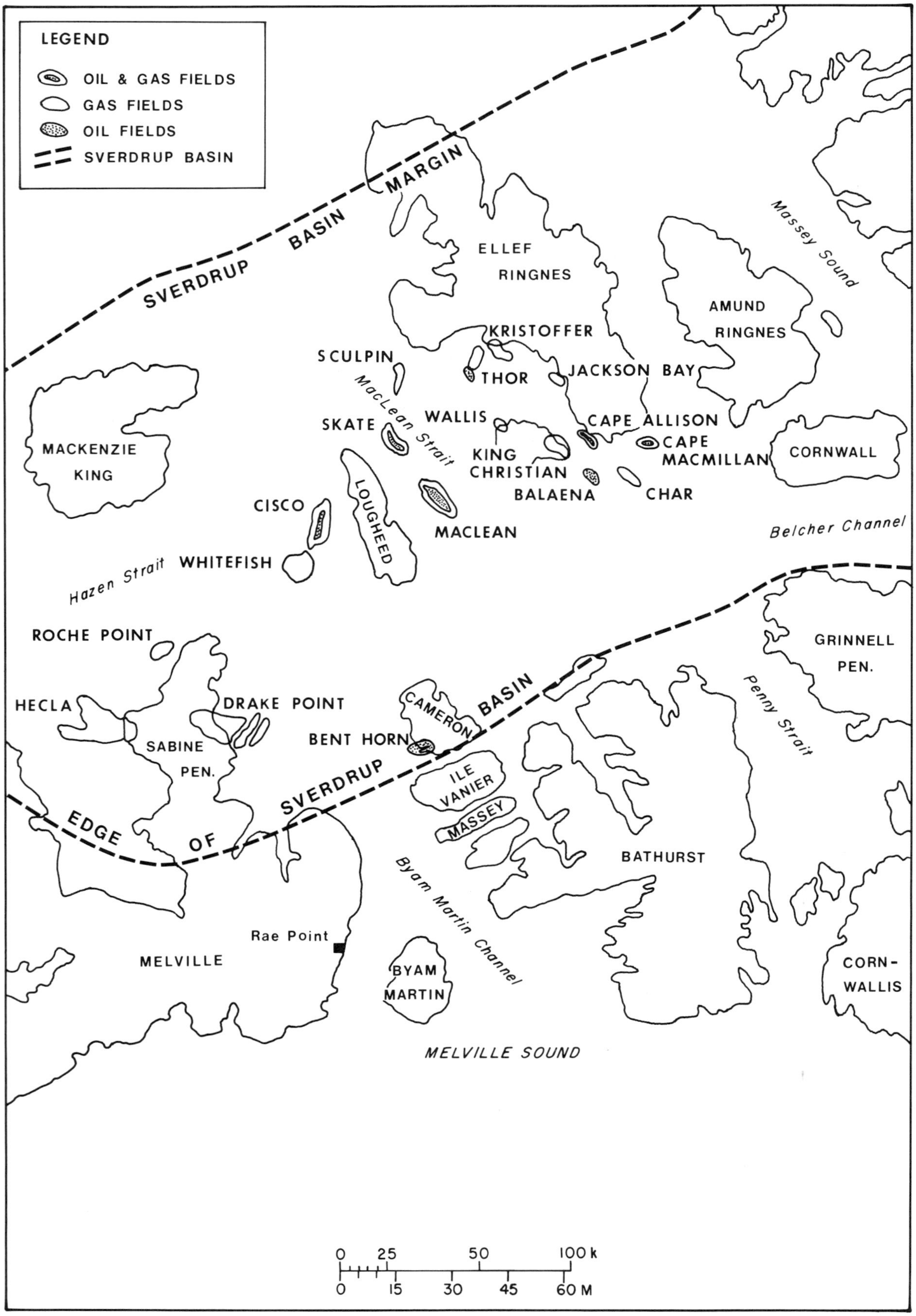
LEGEND
OIL & GAS FIELDS
GAS FIELDS
OIL FIELDS
SVERDRUP BASIN
SVERDRUP BASIN MARGIN
ELLEF RINGNES
AMUND RINGNES
Massey Sound
KRISTOFFER
SCULPIN
THOR
JACKSON BAY
MacLean Strait
WALLIS
SKATE
CAPE ALLISON
CAPE MACMILLAN
CORNWALL
MACKENZIE KING
KING CHRISTIAN
BALAENA
CHAR
CISCO
LOUGHEED
MACLEAN
Belcher Channel
Hazen Strait
WHITEFISH
ROCHE POINT
GRINNELL PEN.
HECLA
DRAKE POINT
CAMERON
BASIN
Penny Strait
SABINE PEN.
BENT HORN
ILE VANIER
MASSEY
EDGE OF SVERDRUP
BATHURST
Byam Martin Channel
Rae Point
MELVILLE
BYAM MARTIN
CORN-WALLIS
MELVILLE SOUND
0 25 50 100 k
0 15 30 45 60 M

Figure 3. Hydrocarbon fields, western Sverdrup basin. (Adapted from Leythaeuser and Stewart, 1986.)

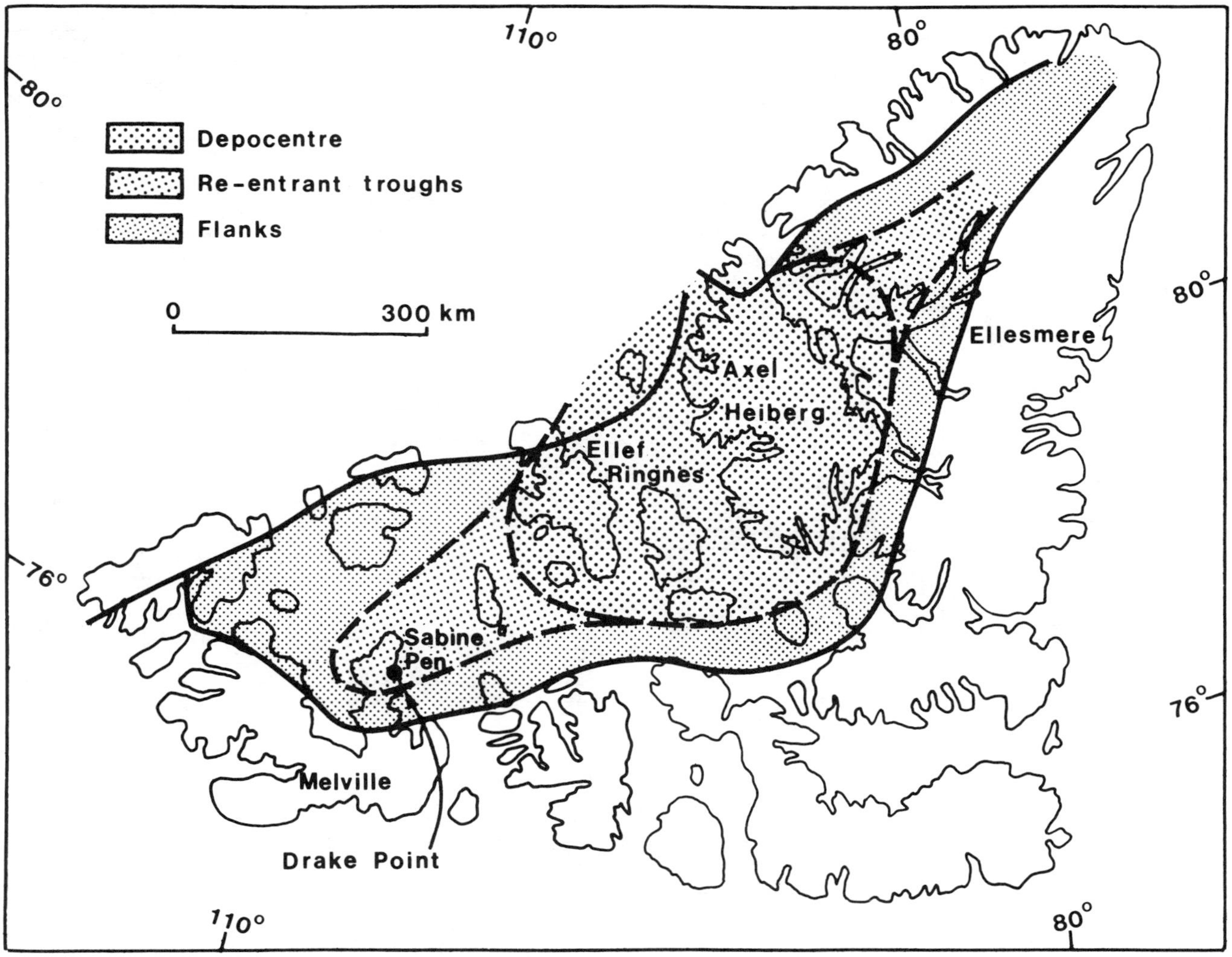
110°
80°
80°
Depocentre
Re-entrant troughs
Flanks
0
300 km
80°
Ellesmere
Axel
Heiberg
Ellef
Ringnes
76°
Sabine
Pen
76°
Melville
Drake Point
110°
80°
Figure 4. Sverdrup basin. (From Balkwill and Fox, 1982.)

condition for several days. On July 12, while a core was being recovered from the interval 2567 to 2576 m (8424 to 8454 ft), in the Lower Triassic Bjorne Formation, the well blew out completely and the blowout could not be contained. By July 30 the rig was moved from the location and the well was blowing gas and salt water that, by the following winter, built up a spectacular ice cone about 61 m (200 ft) high (Figure 7).

A new well, Drake L-67, located about 610 m (2000 ft) south-southwest of the wild well, was spudded on September 22, 1969, and completed on February 17, 1970, after reaching a depth of 3252 m (10,671 ft). The sand at 1134 m (3720 ft) tested dry gas at the rate of 10 MMCFGD. This sand, later identified as the Drake Point Member, proved to contain the main reserves in the field. A second gas sand was encountered at 1405 m (4610 ft) in the Lower Scheii Point Formation (now the Roche Point Formation), and tested dry gas at the rate of 13 MMCFGD with some condensate and oil. This reservoir subsequently was found to be of limited areal extent.

The L-67 well was plugged back to 1333 m (4375 ft), where a whipstock was set in order to start a directional hole to kill the wild well. The directional hole, L-67A, was unsuccessful and a new directional well, K-67, was spudded on July 19, 1970, closer to the wild well. The new well successfully completed the job on November 6, 1970, about 16 months after N-67 had blown wild.

Additional seismic work was carried out over the structure in 1969 in order to better define the productive area discovered by the first well on the structure. Further seismic surveys, mainly offshore, were done in 1973, 1974, and 1975 and the combination of additional drilling and the correct solution of the permafrost effects on the seismic structural picture made it evident that the Drake Point field extended some distance offshore.

The second exploratory well in the Drake Point field, Drake F-16, was spudded in May 1972, about 10 km (6 mi) southeast of the discovery well on the southeast side of a narrow northeasterly trending graben that cuts the structure. This well was drilled to determine whether the graben acted as a seal, and was successfully completed as a Drake Point gas producer. It confirmed that this graben did not break the continuity of the Drake Point reservoir. The Roche Point sandstone, productive

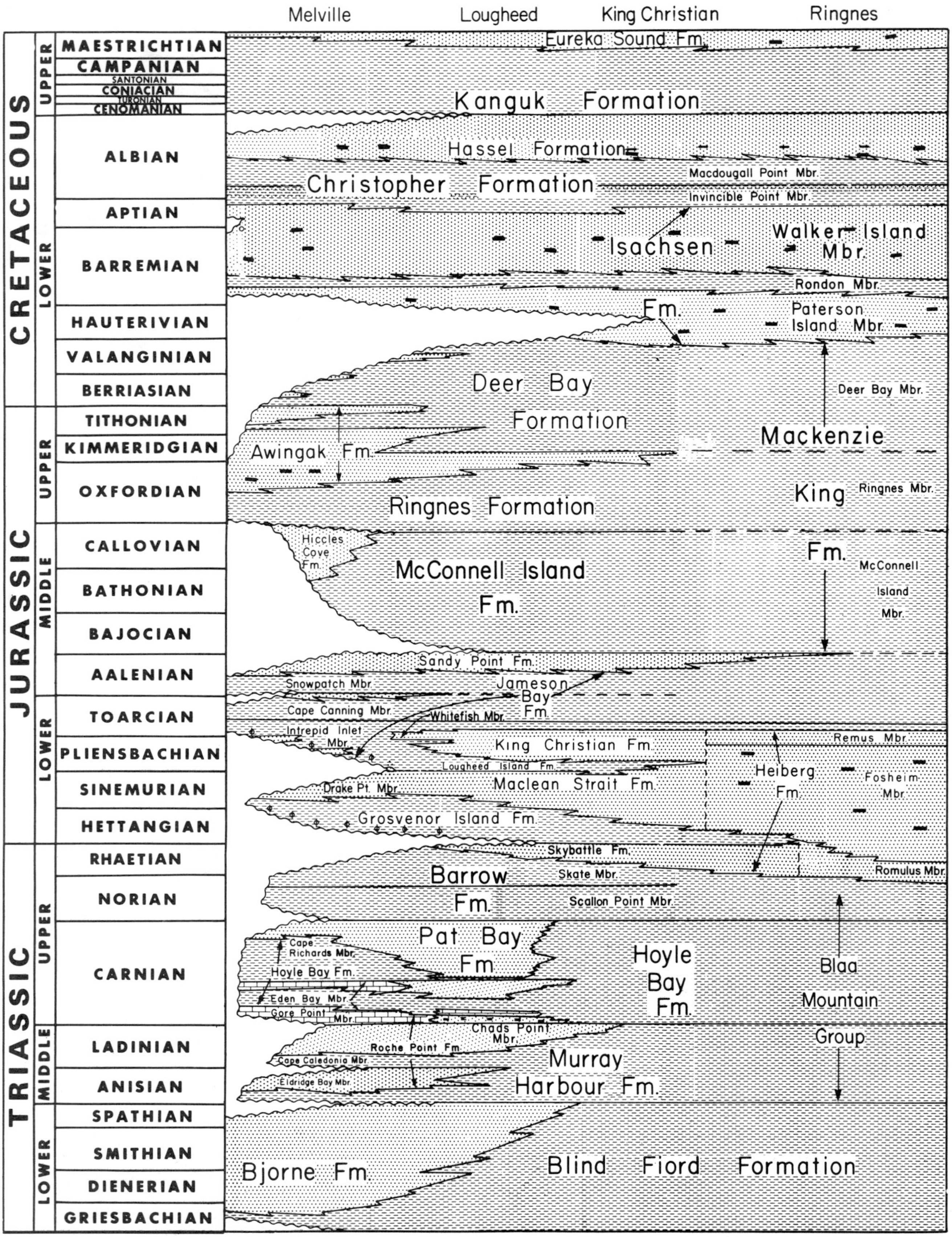
Melville
Lougheed
King Christian
Ringnes
CRETACEOUS
JURASSIC
TRIASSIC
UPPER
LOWER
MIDDLE
MAESTRICHTIAN
CAMPANIAN
SANTONIAN
CONIACIAN
TURONIAN
CENOMANIAN
ALBIAN
APTIAN
BARREMIAN
HAUTERIVIAN
VALANGINIAN
BERRIASIAN
TITHONIAN
KIMMERIDGIAN
OXFORDIAN
CALLOVIAN
BATHONIAN
BAJOCIAN
AALENIAN
TOARCIAN
PLIENSBACHIAN
SINEMURIAN
HETTANGIAN
RHAETIAN
NORIAN
CARNIAN
LADINIAN
ANISIAN
SPATHIAN
SMITHIAN
DIENERIAN
GRIESBACHIAN
Eureka Sound Fm.
Kanguk Formation
Hassel Formation
Christopher Formation
Macdougall Point Mbr.
Invincible Point Mbr.
Isachsen Fm.
Walker Island Mbr.
Rondon Mbr.
Paterson Island Mbr.
Deer Bay Formation
Deer Bay Mbr.
Awingak Fm.
Mackenzie King Fm.
Ringnes Formation
Ringnes Mbr.
Hiccles Cove Fm.
McConnell Island Fm.
McConnell Island Mbr.
Sandy Point Fm.
Snowpatch Mbr.
Jameson Bay Fm.
Cape Canning Mbr.
Whitefish Mbr.
Intrepid Inlet Mbr.
King Christian Fm.
Remus Mbr.
Lougheed Island Fm.
Heiberg Fm.
Fosheim Mbr.
Maclean Strait Fm.
Drake Pt. Mbr.
Grosvenor Island Fm.
Skybattle Fm.
Barrow Fm.
Skate Mbr.
Scallon Point Mbr.
Romulus Mbr.
Pat Bay Fm.
Cape Richards Mbr.
Hoyle Bay Fm.
Eden Bay Mbr.
Gore Point Mbr.
Hoyle Bay Fm.
Blaa Mountain Group
Chads Point Mbr.
Roche Point Fm.
Cape Caledonia Mbr.
Eldridge Bay Mbr.
Murray Harbour Fm.
Bjorne Fm.
Blind Fiord Formation

Figure 5. Mesozoic stratigraphic nomenclature of the western and central parts of the Sverdrup basin.

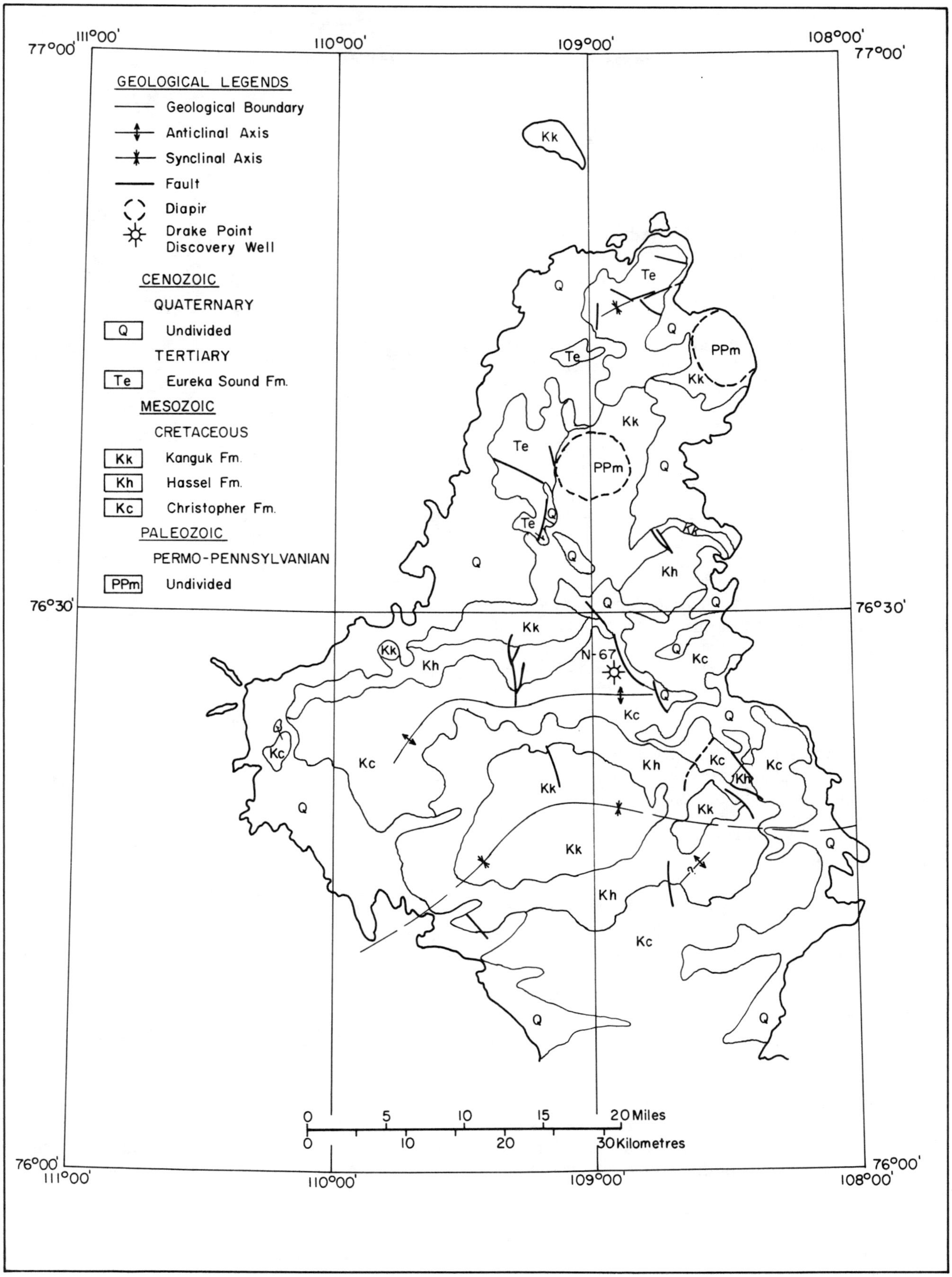
GEOLOGICAL LEGENDS
Geological Boundary
Anticlinal Axis
Synclinal Axis
Fault
Diapir
Drake Point Discovery Well
CENOZOIC
QUATERNARY
Q Undivided
TERTIARY
Te Eureka Sound Fm.
MESOZOIC
CRETACEOUS
Kk Kanguk Fm.
Kh Hassel Fm.
Kc Christopher Fm.
PALEOZOIC
PERMO-PENNSYLVANIAN
PPm Undivided
77°00'
111°00'
110°00'
109°00'
108°00'
76°30'
76°00'
N-67
0 5 10 15 20 Miles
0 10 20 30 Kilometres

Figure 6. Generalized geologic map of the Sabine Peninsula, Melville Island.

in L-67, was found structurally higher but the reservoir has low permeability and is ineffective.

By the end of 1974, two additional step-out wells, Drake B-44 and Drake E-78, had been drilled in the field to outline the productive limits and to assess reserves. A deep pool wildcat, Drake D-68, was spudded in June 1973, and completed in March 1974. Located near L-67, this well was drilled to test a large closure below 4877 m (16,000 ft), on what was originally interpreted from refraction data and geological considerations to be a Devonian carbonate buildup. The well bottomed in gas- and water-bearing carbonates of Middle Pennsylvanian age. Log analysis and drilling results indicated that no significant volume of gas was contained in the porous carbonate. The overlying Permian section, containing shales and thin sandstones, gave strong indications of gas but appeared to lack any effective reservoir. The seismic reflector on which the deep closure was mapped turned out to be an igneous sill above the carbonate. At a total depth of 5418 m (17,776 ft), this well remains one of the deepest tests in the Arctic Islands to date.

During 1975, two additional wells were drilled in the Drake Point field. The first well, East Drake I-55, was drilled 12.8 km (8 mi) offshore and was the second ice platform test drilled in the Arctic Islands using the technique that was developed by Panarctic Oils, Ltd., for a 1974 offshore test at the Hecla gas field. This well penetrated a separate pool in the Drake Point Member. The east and west boundaries of this pool were sealing faults and the gas–water contact was 35 m (116 ft) higher than the main pool contact. Drake D-73, the second 1975 test, was drilled on the south flank of the structure in the main pool. This well tested gas down to 1197 m (3928 ft) subsea, lending credence to the gas–water contact (– 1215 m/ – 3986 ft) using formation pressures.

In the spring of 1978, plans to complete an offshore gas well in the Arctic came to fruition with the drilling of Drake F-76. Located in the middle of the main pool, approximately 3.2 km (2 mi) east of Drake F-16, in 61 m (200 ft) of water and 914 m (3000 ft) from land, this well was the first offshore completion in the Arctic Islands. Although the well provided some additional geologic and reservoir data, its main purpose was to test various aspects of a subsea well completion and the transmission to shore of gas from a well so completed.

The spring of 1978 also saw a Drake Point main pool delineation well on the western edge of the pool. This well, Drake K-79, was located approximately 4.8 km (3 mi) west of D-68 and L-67, across a series of faults. It was felt that these faults could be sealing faults resulting in a separate pool with a structurally lower gas–water contact. A seismic amplitude anomaly at the level of the Drake Point Member enhanced this interpretation. This is the only well in the Arctic Islands that was drilled on the basis of a seismic amplitude anomaly. The well flowed gas at the rate of 526 MCFD during a test of the Intrepid Inlet Member of the Jameson Bay Formation. The Drake Point Member flowed only water on test. This well showed that the nearby faults were not sealing and confirmed that the gas–water contact was near 1219 m (4000 ft) subsea.

Figure 7. Relief drilling operations at Drake Point. The 200-ft ice cone, in the background, formed over the wild well as a result of sea water that froze after being blown out with the gas.

Drake L-06 was drilled in 1985 on a fault block between the I-55 and P-40 wells in the offshore. A drill-stem test of the Drake Point Member at L-06 flowed gas and provided the pressure data used to establish a third gas–water contact in the area at 999 m (3280 ft) subsea, some 180 m (590 ft) higher than the contact at the I-55 pool.

The Drake Point field now is waiting for gas markets to develop before further delineation and development wells will be drilled.

In December 1972, the Hecla field, which lies 32 km (20 mi) west of the Drake Point field, was discovered. This field is very similar to the Drake Point field except that it also has some stratigraphically trapped gas (Figure 2).

DISCOVERY METHOD

The Drake Point field discovery was the result of drilling a gentle surface anticline for Mesozoic reservoirs that outcropped to the south of the well location. It was believed that the reservoirs would thicken and improve in quality toward the well location, which was basinward from the outcrops. The sandstone that contained the gas, the Drake Point sandstone, was not the original objective because it did not outcrop. The major objectives were the sandstones of the Lower Triassic Bjorne Formation. Unfortunately, these sandstones generally are fresh-water flushed along the south margin of the basin.

The presence of a closed structure was confirmed by reflection seismic prior to drilling the discovery well.

It is unlikely that anything different would be done today if we were starting to explore the Sabine Peninsula, except that perhaps more seismic would be shot to define better the extent of the prospect.

There are no seeps associated with the Drake Point field, although oil-saturated sandstones in Carboniferous and Triassic rocks are known to outcrop along the south margin of the basin on Melville Island.

STRUCTURE

Tectonic History

The Drake Point field is located on the southwest margin of the Sverdrup basin. The Sverdrup basin (Figure 4) is a Mississippian to Tertiary regional depression superposed on deformed parts of the lower Paleozoic Ellesmerian Orogenic Belt. It is an episutural basin on continental crust (classification of Bally and Snelson, 1980). Subsurface and surface data show that the basin has locally sharp, chronologically persistent margins, where some thick wedges of coarse-grained, mature clastics along with some carbonates (mainly Carboniferous) accumulated. It contains minor but significant Paleozoic and Mesozoic basalt flows and gabbroic intrusions, and it has long-lasting, directionally consistent fracture patterns. These traits indicate the basin was generated by episodic rifting (Balkwill and Fox, 1982).

The basin was filled with clastics and some carbonates in three major megasequences: Carboniferous to Middle Jurassic, Oxfordian to late Neocomian, and Barremian to Maastrichtian. Many smaller sequences are present within these megasequences. The depocenter occupied the central part of the Sverdrup basin throughout its evolution (Figure 4). The depocenter was flanked by elongated re-entrants; the southwestern re-entrant (sometimes called the Hazen subbasin or the Western Sverdrup basin) is where the Drake Point field is located. The rocks in the depocenter probably are up to 12,200 m (40,000 ft) thick and consist predominantly of shale. In contrast, the succession in the re-entrants and on the margins is considerably thinner and sandier, and has many significant unconformities. A thick unit of Pennsylvanian evaporites, thought to be equivalent to the Otto Fiord Formation, underlies the depocenter and the re-entrant troughs.

During the Late Cretaceous to middle Tertiary, the Eurekan orogeny deformed the central and eastern parts of the Sverdrup basin. Compression of the central part of the basin resulted in large detached folds above upper Paleozoic evaporites. Eurekan folds in the western part of the Sverdrup basin consist of a number of broad folds involving rocks as young as Maastrichtian—the Drake Point structure being one of them. Seismic profiles and outcrop data show that salt movement played a significant role in the development of structures in this region as early as the middle Triassic. The northeast-trending structural grain in faults and salt structures (Figure 8) likely parallels an incipient upper Paleozoic rift system (Balkwill and Fox, 1982).

Regional Structure

Balkwill and Fox (1982) discuss in considerable detail the regional structural geology of the Western Sverdrup basin. Normal faults, linear magnetic anomalies, gabbroic dikes, aligned evaporite domes and salt walls, and modern earthquake epicenters define a broad tectonic belt in upper Paleozoic and Mesozoic rocks of the Western Sverdrup basin (Figure 8). From Melville Island the belt strikes northeastward toward northern Ellef Ringnes Island.

In the central part of the Western Sverdrup basin, evaporite diapirs are the preeminent structures. These northeast-trending domes and walls probably are cored with Carboniferous salt. Seismic data indicate episodic growth of these diapirs from middle Triassic to Tertiary time. Faults are less common within the region of salt diapirism.

Offshore seismic and drillhole data reveal an array of northeast-striking faults, with normal dip separation cutting Cretaceous and older rocks in a broad zone between Sabine Peninsula and Ellef Ringnes Island (Figure 8). The faults are straight and have throws as great as several thousand feet, although throws up to a few hundred feet are more the norm. In some places the normal faults are paired as narrow, straight keystone grabens. One particular graben system along the east side of the Drake Point field is in an area of modern earthquakes. Sixty-seven earthquakes occurred over a period of 2 months, indicating that episodic activity has continued to the present (Hasegawa, 1977).

Local Structure

The Drake Point gas field is located on a broad, low-relief, generally east-to-west-trending anticlinal structure (Figure 9) that is part of a trend of Eurekan folds extending from Prince Patrick Island on the west to Cameron Island on the east (Figure 8). This trend is a series of en echelon culminations, the two most significant being on the Hecla and Drake Point structures. The Drake Point fold is developed along a major hingeline that marks a sharp increase in dip and thickening of stratigraphic section northward into the Sverdrup basin. The anticline is broken up by a series of northeast-trending faults, shown in Figure 9. These faults become more numerous to the east.

STRATIGRAPHY

Mesozoic

The Mesozoic stratigraphy and stratigraphic terminology for the western part of the Sverdrup basin are described in detail by Embry (1984b, 1985; and in press).

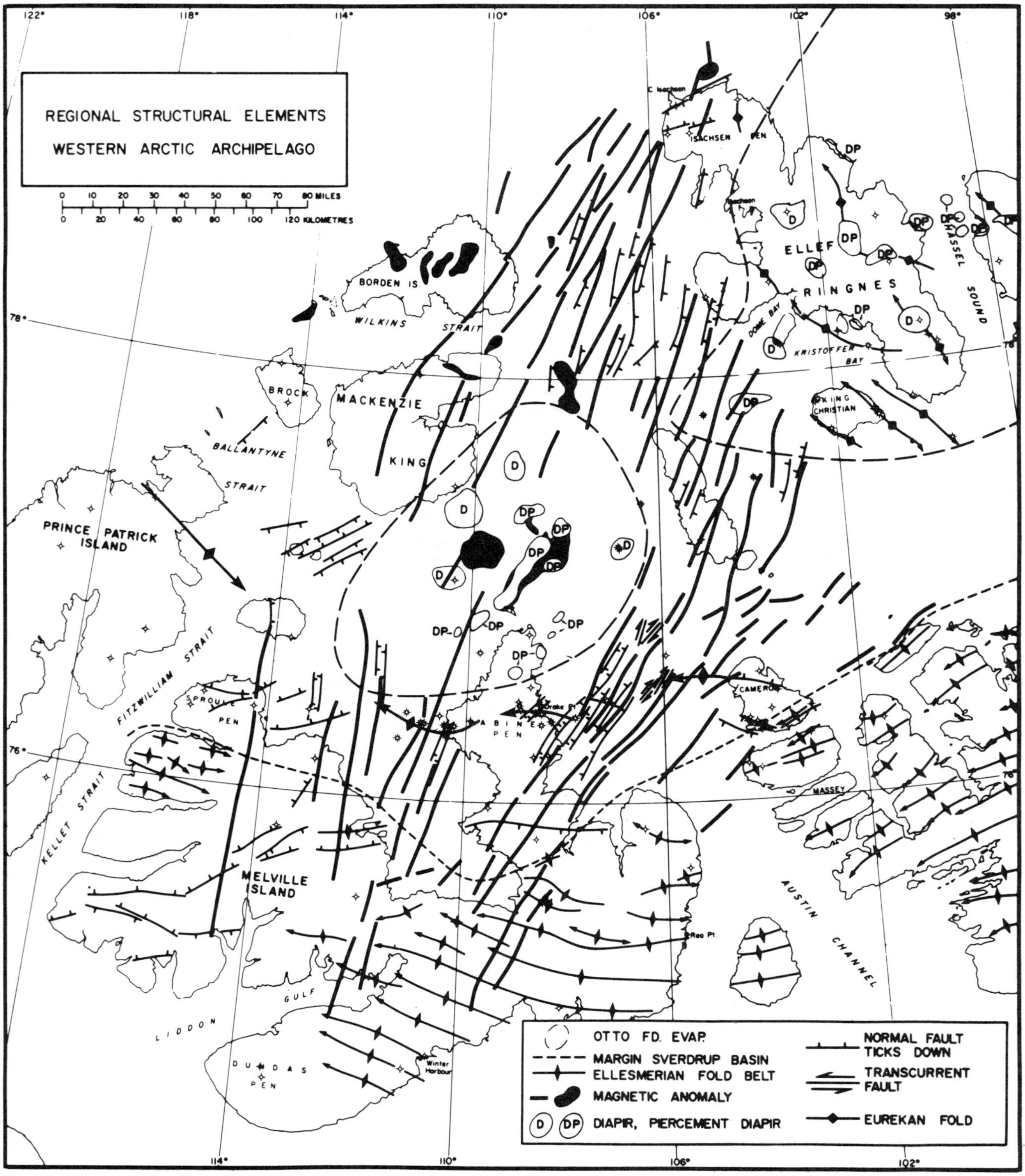
REGIONAL STRUCTURAL ELEMENTS
WESTERN ARCTIC ARCHIPELAGO
80 MILES
120 KILOMETRES
BORDEN IS
WILKINS STRAIT
BROCK
MACKENZIE KING
BALLANTYNE STRAIT
PRINCE PATRICK ISLAND
FITZWILLIAM STRAIT
SPROULE PEN
SABINE PEN
MELVILLE ISLAND
KELLET STRAIT
LIDDON GULF
DUNDAS PEN
Winter Harbour
ISACHSEN PEN
ELLEF RINGNES
HASSEL SOUND
DOME BAY
KRISTOFFER BAY
KING CHRISTIAN
CAMERON
MASSEY
AUSTIN CHANNEL
OTTO FD. EVAP.
MARGIN SVERDRUP BASIN
ELLESMERIAN FOLD BELT
MAGNETIC ANOMALY
DIAPIR, PIERCEMENT DIAPIR
NORMAL FAULT TICKS DOWN
TRANSCURRENT FAULT
EUREKAN FOLD

Figure 8. Regional structural elements, western Arctic Archipelago. (Modified from Balkwill and Fox, 1982.)

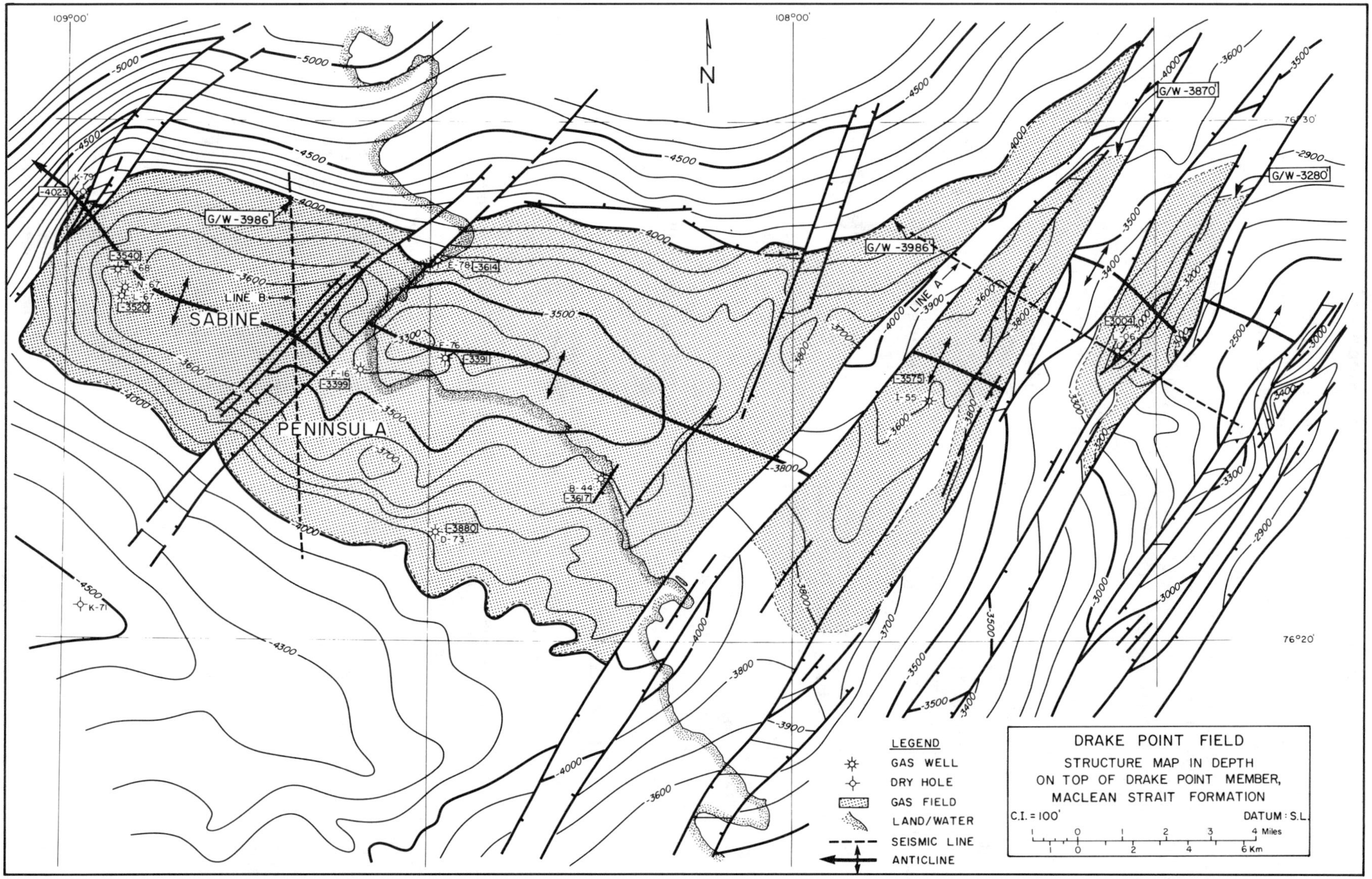
109°00'
108°00'
76°30'
76°20'
N
SABINE
PENINSULA
LINE A
LINE B
G/W -3986'
G/W -3986'
G/W -3870'
G/W -3280'
-4023
-3540
-3520
-3614
-3391
-3399
-3617
-3880
-3575
-3004
K-79
K-71
F-16
E-78
B-44
D-73
I-55
LEGEND
GAS WELL
DRY HOLE
GAS FIELD
LAND/WATER
SEISMIC LINE
ANTICLINE
DRAKE POINT FIELD
STRUCTURE MAP IN DEPTH
ON TOP OF DRAKE POINT MEMBER,
MACLEAN STRAIT FORMATION
C.I. = 100'
DATUM: S.L.
4 Miles
6 Km

Figure 9. Structure map in depth on top of Drake Point member, MacLean Strait Formation, Drake Point field.

Figure 5 shows the generalized lithology, stratigraphic nomenclature, conformable/unconformable relationships and correlations for the Western and Central Sverdrup basin. Figures 10 and 11 show the lithology and thickness of these units in the vicinity of the Drake Point field.

The Mesozoic part of the Western Sverdrup basin consists of a thick series of sandstones and shales along with some Middle Triassic limestones, all deposited in relatively shallow water.

The youngest rocks are a thick sequence of alternating sandstone and shale units from early Tertiary to Early Cretaceous in age. This succession outcrops extensively over much of the area and is of little economic interest. An unconformity separates this succession from the underlying Lower Cretaceous to Upper Jurassic sequence. These thick sandstones and shales include the deltaic sandstones of the Awingak Formation, which form important oil and gas reservoirs in the basin. These sandstones and shales unconformably overlie a relatively thin sequence of shallow-water sandstones, shales, and limestones of Middle Jurassic to Middle Triassic age. Lower Jurassic sandstones contain over 75% of the discovered gas reserves in the Arctic Islands, including the reserves in the Drake Point field. Fair to excellent source rocks are found in the shales of this sequence. These rocks unconformably overlie the oldest Mesozoic succession, which consists of a very thick sequence of deltaic sandstones and shales of Early Triassic age. These sandstones contain only minor reserves of gas in the Sverdrup basin. The Lower Triassic rocks unconformably overlie Permian rocks.

Permian-Carboniferous

The Permian-Carboniferous consists of a thick sequence of clastics, carbonates, and evaporites. These rocks outcrop around much of the basin margin and have been extensively studied at the eastern end of the Sverdrup basin (Thorsteinsson, 1974). Little work is published on the Permian-Carboniferous of the western part of the Sverdrup basin. Only four wells on the Sabine Peninsula have penetrated a significant portion of this section. Figure 12 shows the stratigraphic relationship of the Permian-Carboniferous strata in the vicinity of the Drake Point field. The Permian-Carboniferous section thickens rapidly in the vicinity of the field, from a thickness of 2621 m (8600 ft) just south of the field to probably in excess of 4572 m (15,000 ft) just north of the field.

The Upper Permian ranges in thickness from 457 to 762 m (1500 to 2500 ft); it consists of limestones of the Degerbols Formation and siliceous siltstones and cherts of an (as yet) unnamed unit.

The Lower Permian ranges in thickness from 1158 m (3800 ft) to (probably) in excess of 2438 m (8000 ft), and consists of shelfal sandstones, shales and limestones of the Belcher Channel Formation. These rocks change facies, in a northward direction, to the medium- and dark-gray shales of the Hare Fiord Formation. These rocks unconformably overlie a thick sequence of Middle Pennsylvanian and older rocks.

The Middle and Lower Pennsylvanian strata consist of sandstone (Canyon Fiord Formation) and limestones (Nansen Formation) in the vicinity of the Drake Point field. These rocks thicken from 823 m (2700 ft), just south of the field, to possibly in excess of 3048 m (10,000 ft) north of the field. North of the field, the rocks change facies to the basinal Hare Fiord shales and a thick evaporite sequence equivalent in age to the Lower Pennsylvanian Otto Fiord Formation found in the eastern and central part of the Sverdrup basin. An area of salt domes and salt walls is present 16 km (10 mi) north of Drake Point field (Figures 6 and 8).

The Permian-Carboniferous rocks are cut by several thick gabbroic sills.

Unconformably underlying the Permian-Carboniferous sequence of rocks are Devonian and older rocks of the Franklinian basin that outcrop south of the Drake Point area. These rocks are well out of reach of the drill except on the basin margin.

TRAP

The gas in the Drake Point field is trapped structurally. The structural configuration of the field, an elongate, faulted anticline, resulted from post-Maastrichtian (probably Early Tertiary) folding. The faults cutting the field probably are of the same age but are parallel to older Carboniferous structural trends.

The faulting has resulted in at least three separate pools with the gas–water contacts rising to the east (Figures 9 and 13). Each pool appears to be filled to the spill point. The structural configuration of the field is shown in the structure map on the top of the Drake Point Member of the MacLean Strait Formation (Figure 9), a north to south cross section across the field (Figure 11), and the seismic cross sections in Figure 14. The locations of the two seismic sections are shown in Figure 9.

The depth to the top of the main reservoir, the MacLean Strait Formation, varies due to the anticlinal nature of the structure and the associated faulting (Figure 9). The top of the pay zone occurs at a depth of just above −914 m (−3000 ft) subsea at the east end of the field, and deepens to −1215 m (−3986 ft), the main pool gas–water contact, along most of the north and south boundaries of the field.

RESERVOIR

Stratigraphy

Detailed stratigraphic relationships in and around the Drake Point field are described by Douglas and Oliver (1979) and by Low (1983) (Panarctic Oils, Ltd. report). Figure 15 summarizes, in some detail, the characteristics of the reservoir formations of the Drake Point field. The two major gas-bearing units are the

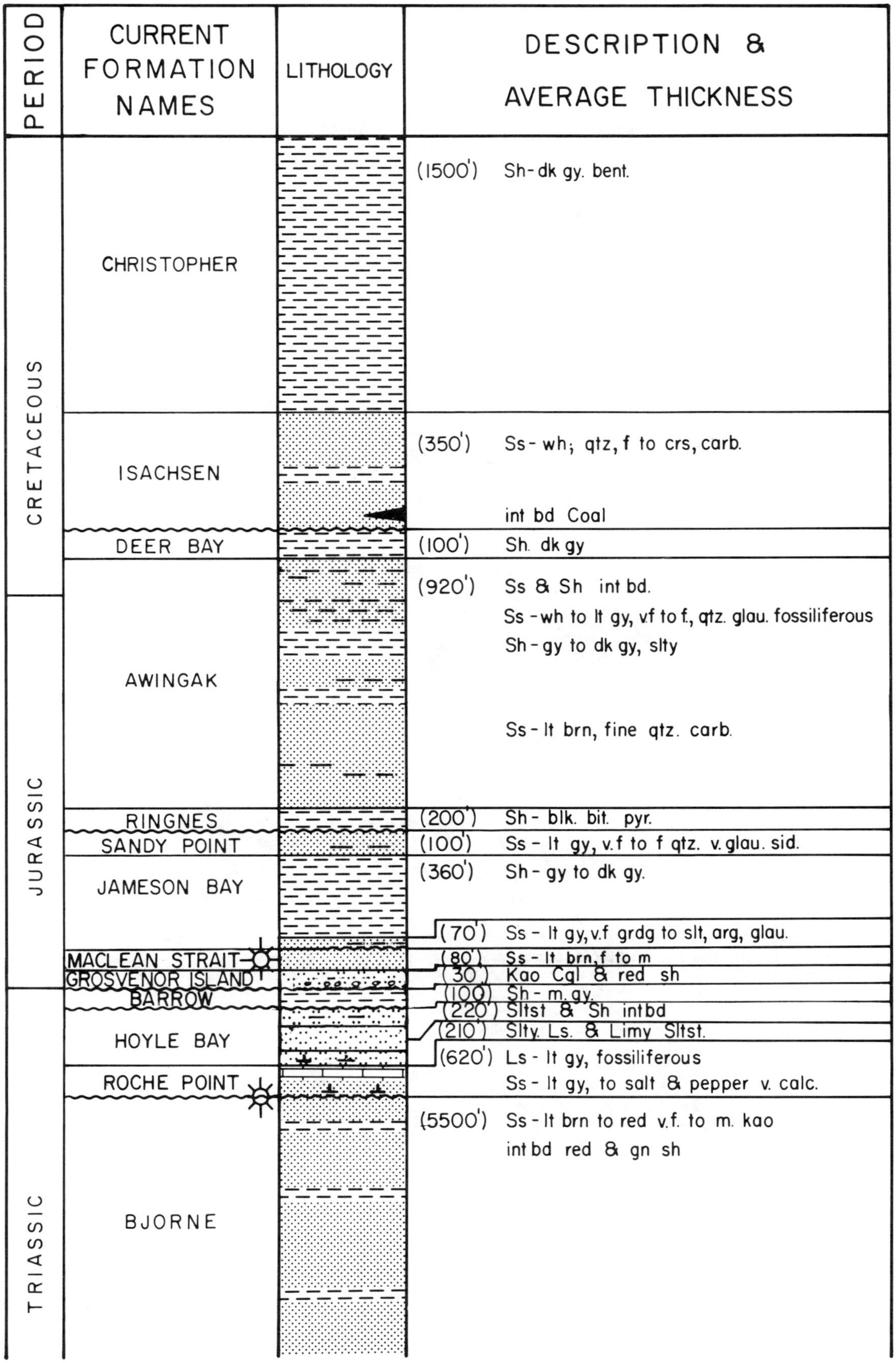
PERIOD
CURRENT FORMATION NAMES
LITHOLOGY
DESCRIPTION & AVERAGE THICKNESS
CRETACEOUS
CHRISTOPHER
(1500') Sh-dk gy. bent.
ISACHSEN
(350') Ss-wh; qtz, f to crs, carb.
int bd Coal
DEER BAY
(100') Sh. dk gy
JURASSIC
AWINGAK
(920') Ss & Sh int bd.
Ss-wh to lt gy, v.f to f., qtz. glau. fossiliferous
Sh-gy to dk gy, slty
Ss-lt brn, fine qtz. carb.
RINGNES
(200') Sh-blk. bit. pyr.
SANDY POINT
(100') Ss-lt gy, v.f to f qtz. v. glau. sid.
JAMESON BAY
(360') Sh-gy to dk gy.
(70') Ss-lt gy, v.f grdg to slt, arg, glau.
MACLEAN STRAIT
(80') Ss-lt brn, f to m
GROSVENOR ISLAND
(30') Kao Cgl & red sh
BARROW
(100') Sh-m. gy.
(220') Sltst & Sh intbd
TRIASSIC
HOYLE BAY
(210') Slty. Ls. & Limy Sltst.
(620') Ls-lt gy, fossiliferous
ROCHE POINT
Ss-lt gy, to salt & pepper v. calc.
BJORNE
(5500') Ss-lt brn to red v.f. to m. kao
int bd red & gn sh

Figure 10. Generalized stratigraphic column, Drake Point field.

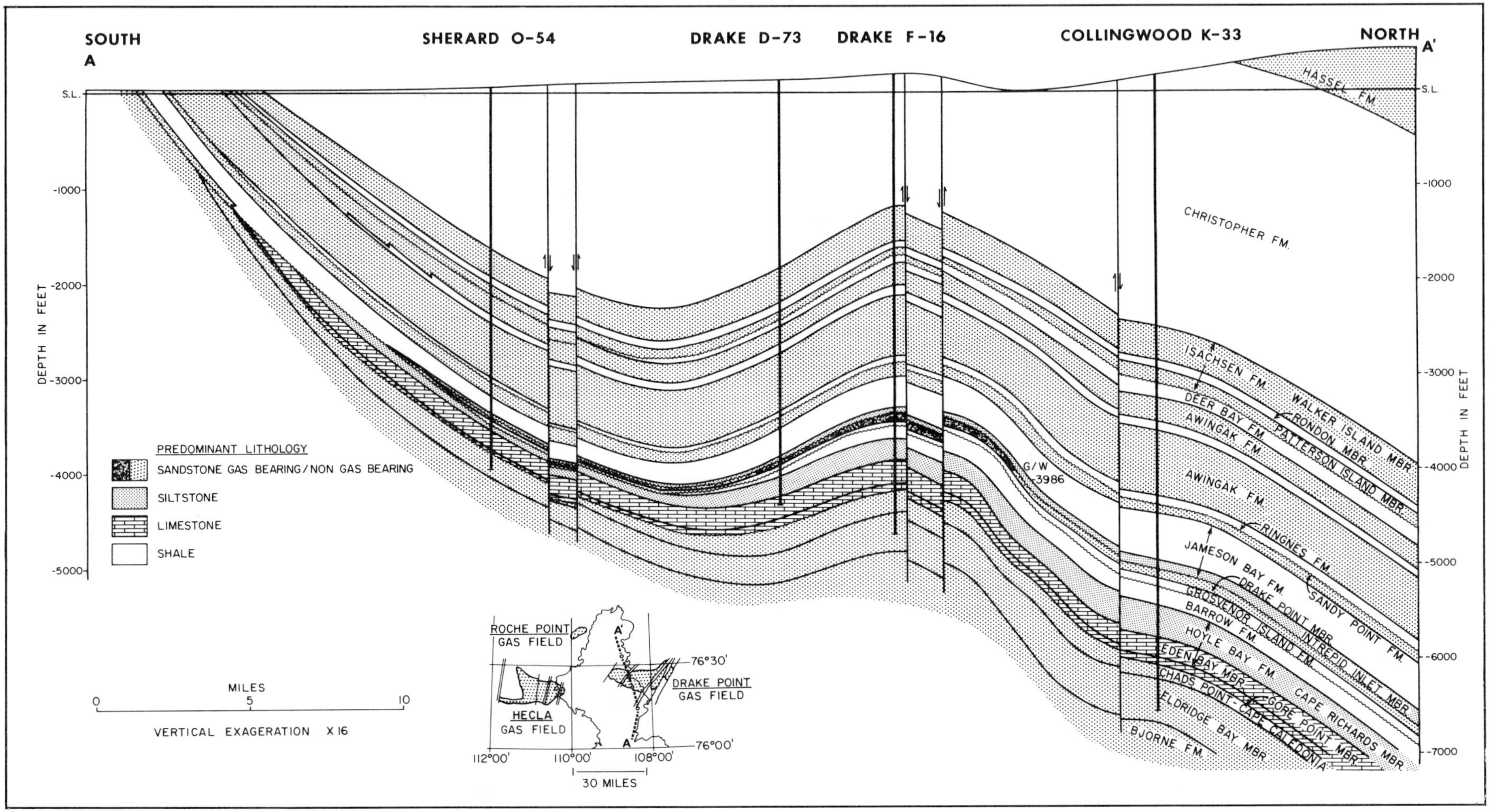
SOUTH
A
SHERARD O-54
DRAKE D-73
DRAKE F-16
COLLINGWOOD K-33
NORTH
A'
S.L.
-1000
-2000
-3000
-4000
-5000
-6000
-7000
DEPTH IN FEET
HASSEL FM.
CHRISTOPHER FM.
ISACHSEN FM.
WALKER ISLAND MBR.
RONDON MBR.
PATTERSON ISLAND MBR.
DEER BAY FM.
AWINGAK FM.
AWINGAK FM.
RINGNES FM.
JAMESON BAY FM.
SANDY POINT FM.
DRAKE POINT MBR.
INTREPID INLET MBR.
GROSVENOR ISLAND FM.
BARROW FM.
HOYLE BAY FM.
CAPE RICHARDS MBR.
EDEN BAY MBR.
GORE POINT MBR.
CHADS POINT-CAPE CALEDONIA
ELDRIDGE BAY MBR.
BJORNE FM.
G/W
-3986
PREDOMINANT LITHOLOGY
SANDSTONE GAS BEARING/NON GAS BEARING
SILTSTONE
LIMESTONE
SHALE
MILES
0
5
10
VERTICAL EXAGERATION X16
ROCHE POINT
GAS FIELD
HECLA
GAS FIELD
DRAKE POINT
GAS FIELD
76°30'
76°00'
112°00'
110°00'
108°00'
30 MILES

Figure 11. North-south cross section A-A′, Drake Point field.

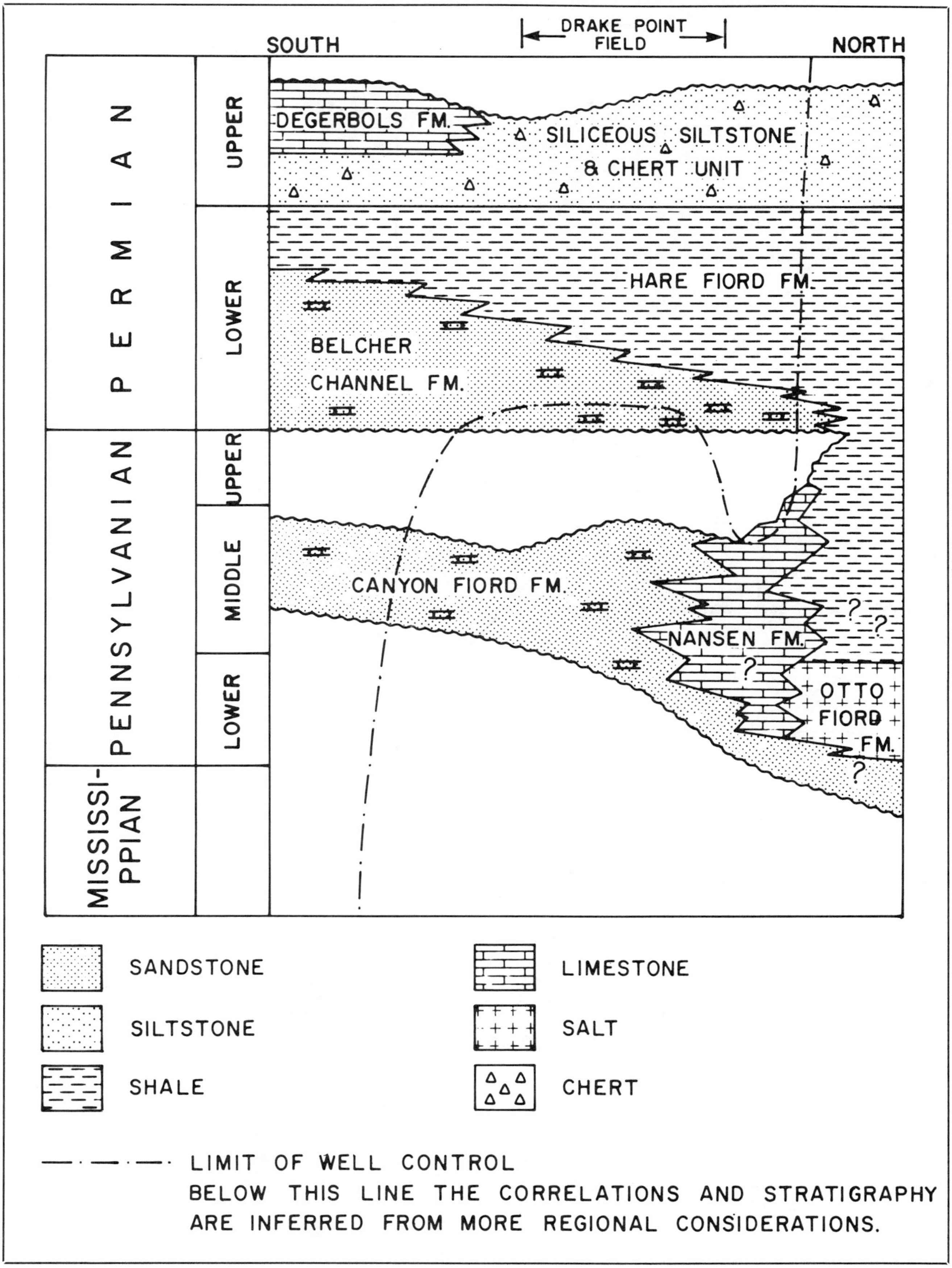
DRAKE POINT FIELD
SOUTH
NORTH
PERMIAN
UPPER
LOWER
PENNSYLVANIAN
UPPER
MIDDLE
LOWER
MISSISSIPPIAN
DEGERBOLS FM.
SILICEOUS SILTSTONE & CHERT UNIT
HARE FIORD FM.
BELCHER CHANNEL FM.
CANYON FIORD FM.
NANSEN FM.
OTTO FIORD FM.
SANDSTONE
SILTSTONE
SHALE
LIMESTONE
SALT
CHERT
LIMIT OF WELL CONTROL
BELOW THIS LINE THE CORRELATIONS AND STRATIGRAPHY ARE INFERRED FROM MORE REGIONAL CONSIDERATIONS.

Figure 12. Permian-Carboniferous correlation chart, Sabine Peninsula, Melville Island.

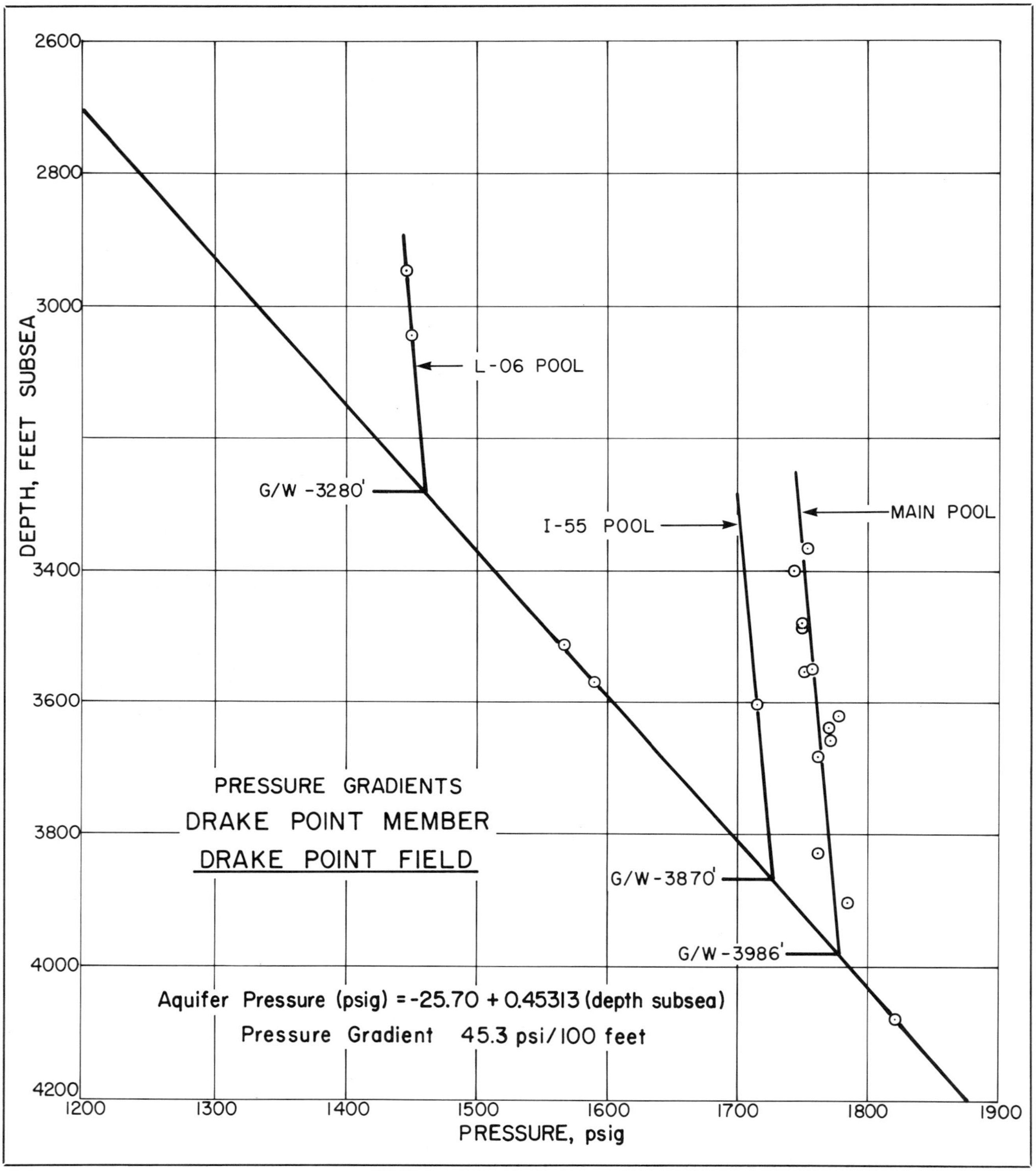

PRESSURE GRADIENTS
DRAKE POINT MEMBER
DRAKE POINT FIELD
L-06 POOL
I-55 POOL
MAIN POOL
G/W -3280'
G/W-3870'
G/W-3986'
Aquifer Pressure (psig) = -25.70 + 0.45313 (depth subsea)
Pressure Gradient 45.3 psi/100 feet
DEPTH, FEET SUBSEA
PRESSURE, psig
2600
2800
3000
3400
3600
3800
4000
4200
1200
1300
1400
1500
1600
1700
1800
1900

Figure 13. Pressure gradients, Drake Point member, Drake Point field.

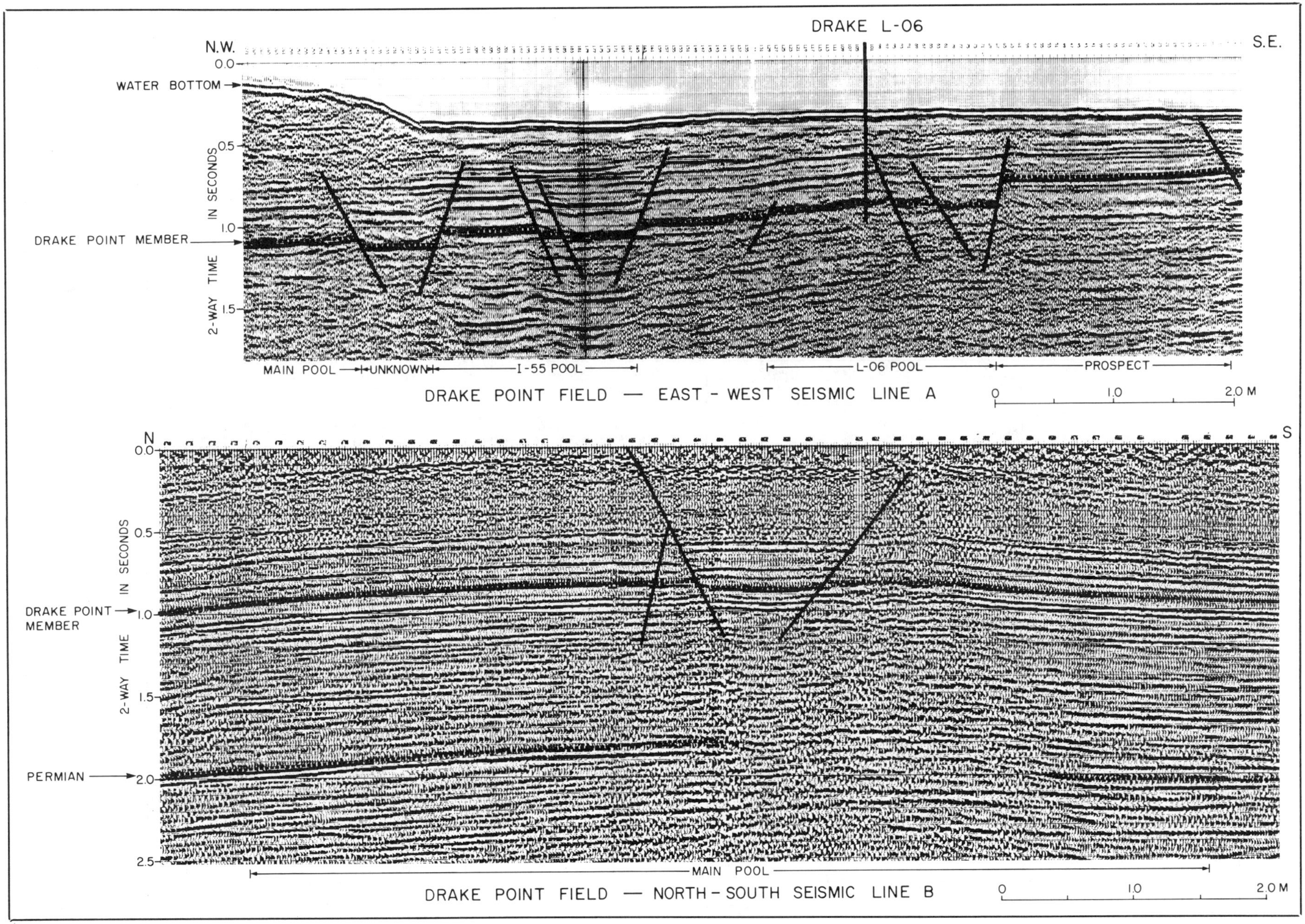
DRAKE L-06
N.W.
S.E.
WATER BOTTOM
DRAKE POINT MEMBER
2-WAY TIME IN SECONDS
0.0
0.5
1.0
1.5
MAIN POOL
UNKNOWN
I-55 POOL
L-06 POOL
PROSPECT
DRAKE POINT FIELD — EAST - WEST SEISMIC LINE A
0
1.0
2.0 M
N
S
DRAKE POINT MEMBER
PERMIAN
2.0
2.5
MAIN POOL
DRAKE POINT FIELD — NORTH - SOUTH SEISMIC LINE B

Figure 14. East-to-west seismic line A, and north-to-south seismic line B, Drake Point field. See Figure 9 for locations.

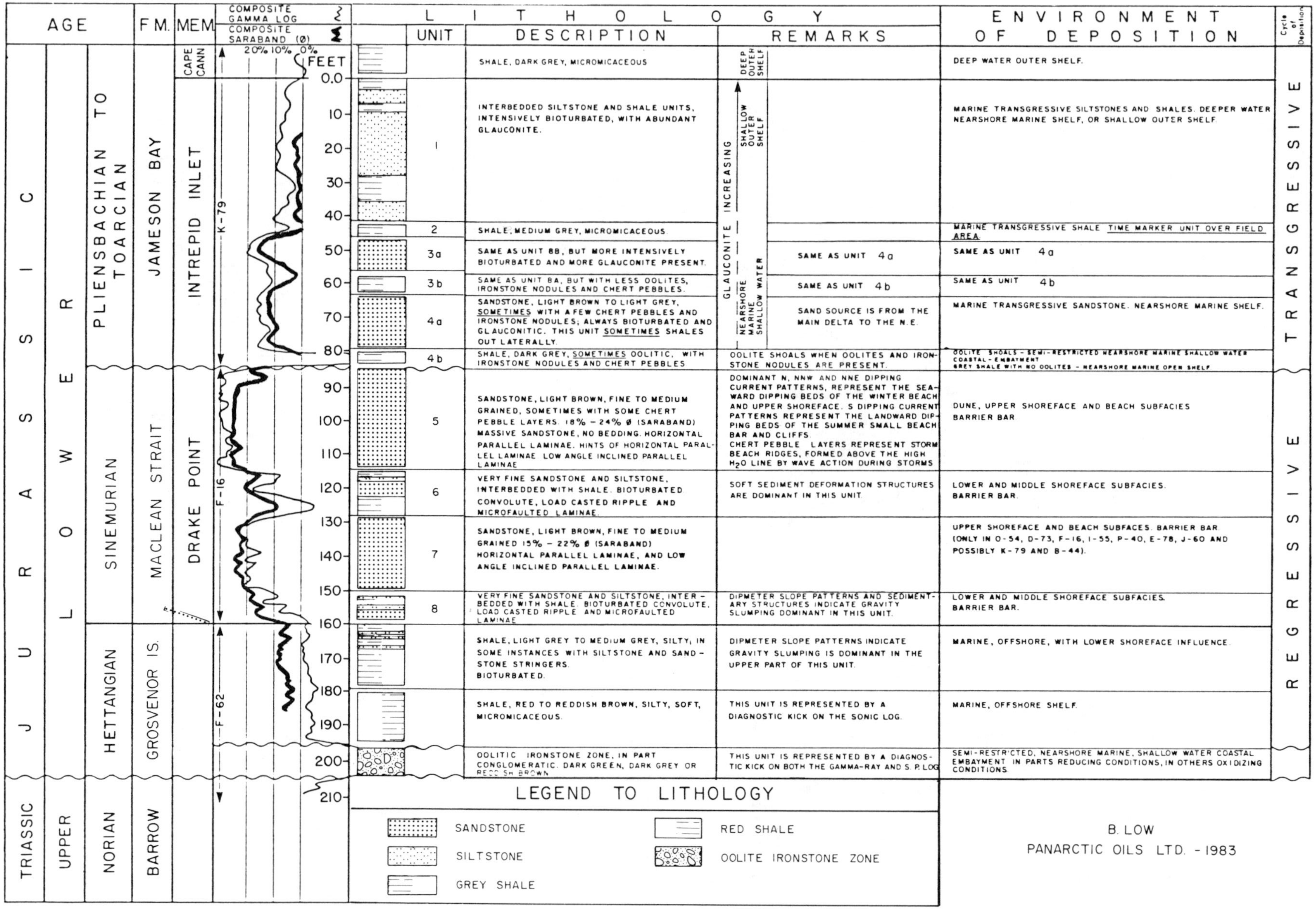
AGE
F.M.
MEM.
COMPOSITE GAMMA LOG
COMPOSITE SARABAND (Ø)
20% 10% 0%
FEET
L I T H O L O G Y
UNIT
DESCRIPTION
REMARKS
ENVIRONMENT OF DEPOSITION
Cycle of Deposition
JURASSIC
LOWER
PLIENSBACHIAN TO TOARCIAN
JAMESON BAY
CAPE CANN
INTREPID INLET
K-79
SINEMURIAN
MACLEAN STRAIT
DRAKE POINT
F-16
HETTANGIAN
GROSVENOR IS.
F-62
TRIASSIC
UPPER
NORIAN
BARROW
0.0
10
20
30
40
50
60
70
80
90
100
110
120
130
140
150
160
170
180
190
200
210
SHALE, DARK GREY, MICROMICACEOUS
DEEP OUTER SHELF
DEEP WATER OUTER SHELF.
1
INTERBEDDED SILTSTONE AND SHALE UNITS, INTENSIVELY BIOTURBATED, WITH ABUNDANT GLAUCONITE.
SHALLOW OUTER SHELF
MARINE TRANSGRESSIVE SILTSTONES AND SHALES. DEEPER WATER NEARSHORE MARINE SHELF, OR SHALLOW OUTER SHELF.
GLAUCONITE INCREASING
NEARSHORE MARINE SHALLOW WATER
2
SHALE, MEDIUM GREY, MICROMICACEOUS
MARINE TRANSGRESSIVE SHALE TIME MARKER UNIT OVER FIELD AREA.
3a
SAME AS UNIT 8B, BUT MORE INTENSIVELY BIOTURBATED AND MORE GLAUCONITE PRESENT.
SAME AS UNIT 4a
SAME AS UNIT 4a
3b
SAME AS UNIT 8A, BUT WITH LESS OOLITES, IRONSTONE NODULES AND CHERT PEBBLES.
SAME AS UNIT 4b
SAME AS UNIT 4b
4a
SANDSTONE, LIGHT BROWN TO LIGHT GREY, SOMETIMES WITH A FEW CHERT PEBBLES AND IRONSTONE NODULES; ALWAYS BIOTURBATED AND GLAUCONITIC. THIS UNIT SOMETIMES SHALES OUT LATERALLY.
SAND SOURCE IS FROM THE MAIN DELTA TO THE N.E.
MARINE TRANSGRESSIVE SANDSTONE. NEARSHORE MARINE SHELF.
4b
SHALE, DARK GREY, SOMETIMES OOLITIC, WITH IRONSTONE NODULES AND CHERT PEBBLES
OOLITE SHOALS WHEN OOLITES AND IRONSTONE NODULES ARE PRESENT.
OOLITE SHOALS - SEMI-RESTRICTED NEARSHORE MARINE SHALLOW WATER COASTAL - EMBAYMENT
GREY SHALE WITH NO OOLITES - NEARSHORE MARINE OPEN SHELF
5
SANDSTONE, LIGHT BROWN, FINE TO MEDIUM GRAINED, SOMETIMES WITH SOME CHERT PEBBLE LAYERS. 18% – 24% Ø (SARABAND) MASSIVE SANDSTONE, NO BEDDING. HORIZONTAL PARALLEL LAMINAE. HINTS OF HORIZONTAL PARALLEL LAMINAE. LOW ANGLE INCLINED PARALLEL LAMINAE
DOMINANT N, NNW AND NNE DIPPING CURRENT PATTERNS, REPRESENT THE SEAWARD DIPPING BEDS OF THE WINTER BEACH AND UPPER SHOREFACE. S DIPPING CURRENT PATTERNS REPRESENT THE LANDWARD DIPPING BEDS OF THE SUMMER SMALL BEACH BAR AND CLIFFS. CHERT PEBBLE LAYERS REPRESENT STORM BEACH RIDGES, FORMED ABOVE THE HIGH H2O LINE BY WAVE ACTION DURING STORMS
DUNE, UPPER SHOREFACE AND BEACH SUBFACIES BARRIER BAR
6
VERY FINE SANDSTONE AND SILTSTONE, INTERBEDDED WITH SHALE. BIOTURBATED CONVOLUTE, LOAD CASTED RIPPLE AND MICROFAULTED LAMINAE.
SOFT SEDIMENT DEFORMATION STRUCTURES ARE DOMINANT IN THIS UNIT.
LOWER AND MIDDLE SHOREFACE SUBFACIES. BARRIER BAR.
7
SANDSTONE, LIGHT BROWN, FINE TO MEDIUM GRAINED 15% – 22% Ø (SARABAND) HORIZONTAL PARALLEL LAMINAE, AND LOW ANGLE INCLINED PARALLEL LAMINAE.
UPPER SHOREFACE AND BEACH SUBFACES. BARRIER BAR. (ONLY IN O-54, D-73, F-16, I-55, P-40, E-78, J-60 AND POSSIBLY K-79 AND B-44).
8
VERY FINE SANDSTONE AND SILTSTONE, INTERBEDDED WITH SHALE. BIOTURBATED CONVOLUTE, LOAD CASTED RIPPLE AND MICROFAULTED LAMINAE
DIPMETER SLOPE PATTERNS AND SEDIMENTARY STRUCTURES INDICATE GRAVITY SLUMPING DOMINANT IN THIS UNIT.
LOWER AND MIDDLE SHOREFACE SUBFACIES. BARRIER BAR.
SHALE, LIGHT GREY TO MEDIUM GREY, SILTY, IN SOME INSTANCES WITH SILTSTONE AND SANDSTONE STRINGERS. BIOTURBATED.
DIPMETER SLOPE PATTERNS INDICATE GRAVITY SLUMPING IS DOMINANT IN THE UPPER PART OF THIS UNIT.
MARINE, OFFSHORE, WITH LOWER SHOREFACE INFLUENCE.
SHALE, RED TO REDDISH BROWN, SILTY, SOFT, MICROMICACEOUS.
THIS UNIT IS REPRESENTED BY A DIAGNOSTIC KICK ON THE SONIC LOG.
MARINE, OFFSHORE SHELF.
OOLITIC IRONSTONE ZONE, IN PART CONGLOMERATIC. DARK GREEN, DARK GREY OR REDDISH BROWN
THIS UNIT IS REPRESENTED BY A DIAGNOSTIC KICK ON BOTH THE GAMMA-RAY AND S. P. LOG
SEMI-RESTRICTED, NEARSHORE MARINE, SHALLOW WATER COASTAL EMBAYMENT IN PARTS REDUCING CONDITIONS, IN OTHERS OXIDIZING CONDITIONS
TRANSGRESSIVE
REGRESSIVE
LEGEND TO LITHOLOGY
SANDSTONE
SILTSTONE
GREY SHALE
RED SHALE
OOLITE IRONSTONE ZONE
B. LOW
PANARCTIC OILS LTD. - 1983

Figure 15. Composite stratigraphic column of reservoir formations in the Drake Point and Hecla gas fields. (From Low, 1983).

A
.5 MM
B
.2 MM

Figure 16. Drake Point member thin sections A and B; very fine- to fine-grained quartz arenite/sublitharenite.

Intrepid Inlet Member of the Jameson Bay Formation and the Drake Point Member of the MacLean Strait Formation.

Jameson Bay Formation, Intrepid Inlet Member

The Intrepid Inlet Member basically is a transgressive, glauconitic, mud-rich, very fine-grained, argillaceous sandstone that has shale at its base. It grades upward into a transgressive, glauconitic siltstone that is shaly at the top. Quartz, glauconite, and rock fragments are the most common grains. The main cement in these sandstones is siderite with minor amounts of calcite, dolomite, and pyrite. The member is made up of six units. From youngest to oldest they are: 1, 2, 3a, 3b, 4a, and 4b (Figure 15). Units 3a and 4a host gas in sandstone reservoirs that have porosities ranging from 8 to 15%. Younger units of the Intrepid Inlet Member progressively onlap units of the Drake Point Member, southward, and overlie them unconformably.

MacLean Strait Formation, Drake Point Member

The Drake Point Member consists of four units: 5, 6, 7, and 8 (Figure 15). The upper two units (5 and 6) consist of one barrier bar cycle, and the lower two units (7 and 8) constitute a second barrier bar cycle. Over the field the member ranges in thickness from 12.5 to 25 m (41 to 82 ft). The dominant lithology is sandstone, ranging from quartzwackes/lithic graywackes to quartz arenites/sublitharenites (Figure 16). The major framework grains are subangular to subrounded quartz and well-rounded chert. Silica, carbonate, and pyrite constitute the cements found in this member. Most of the silica cement occurs as diagenetic overgrowth on reworked grains. The major authigenic constituent is carbonate cement, the bulk of which is calcite. Most of the sandstones in this member have some very light residual oil staining. Porosities in units 5 and 7 range from 15 to 24%, and in units 6 and 8 range from 8 to 19%.

Gas and Field Characteristics

The porosity and permeability characteristics of the field are excellent. While there is a great range of porosities for the main sandstone reservoirs, the field average is near 20%. Similarly, while there is a wide range in permeability from a few millidarcys to several darcys, the field averages 500 to 1000 md. Net pays range from 13 to 29 m (42 to 96 ft). Because of the high porosity and permeability, individual well deliverabilities are very high. Although all wells in the field have been drill-stem tested, three wells were tested more extensively and flowed natural gas at rates up to 78 MMCFGD on restricted chokes. The absolute open flow (AOF) rates calculated for these wells range from 250 MMCFGD to nearly 400 MMCFGD.

A reservoir simulation study of the Drake Point Member reservoir completed for Panarctic Oils, Ltd., suggested that the recovery efficiency would be between 91 and 93%, and that abandonment pressure would be in the range of 1379 kPa (200 psia).

All the gas found in this field is dry, averaging 98% methane. It is free of hydrogen sulfide, low in carbon dioxide content, and has a gross heating value of approximately 37,260 kJ/m^3 (1000 Btu/ft^3) at 101 kPa (14.65 psia) and 60 °F.

Geochemistry

Several authors, including Powell (1978) and Leythaeuser and Stewart (1986), have done extensive work on the geochemistry of the Sverdrup basin. There still are some perplexing problems that have not been satisfactorily resolved relating to the Drake Point field. For

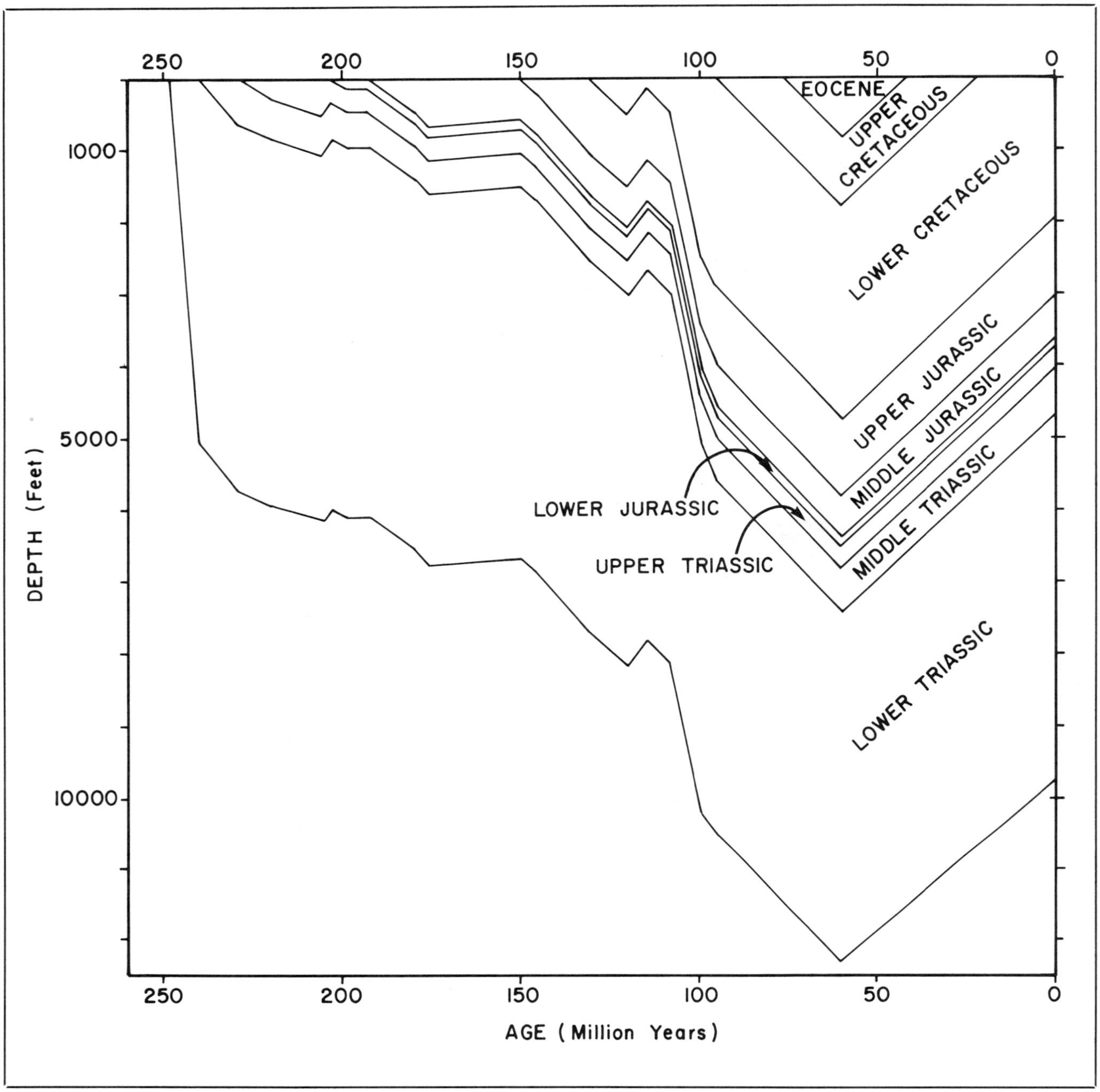
250
200
150
100
50
0
EOCENE
UPPER CRETACEOUS
LOWER CRETACEOUS
UPPER JURASSIC
MIDDLE JURASSIC
MIDDLE TRIASSIC
LOWER JURASSIC
UPPER TRIASSIC
LOWER TRIASSIC
1000
5000
10000
DEPTH (Feet)
AGE (Million Years)

Figure 17. Geohistory plot for Drake Point field.

example, the fact that dry gas (97.6 to 98.1% methane)—with an isotopic composition normally indicative of wet gas associated with oil generation—is the only hydrocarbon product recovered (to date) from the Drake Point Member is an enigma difficult to explain based on the geochemical information currently available. The methane fractions of gases recovered have a carbon isotopic composition ranging from −45.1 to −47.2 ‰ PDB. This range in $\delta^{13}C$ values is indicative of gas generated from marine source rocks in the peak oil generating stage (Leythaeuser and Stewart, 1986). Most of the Drake Point wells have encountered abundant to trace amounts of light-brown oil staining in the Drake Point Member sandstones. Analyses of oil-stained cores indicate that the residual oil in this sandstone was extensively biodegraded. The occurrence of residual oil indicates that the oil was initially reservoired in the Drake Point Member sandstones in the Drake Point structure. It is probable that most of this oil accumulation is lost, probably through leakage along faults and fractures. It also is probable that fresh water subsequently flowed hydrodynamically upward from the Bjorne Formation through these faults and fractures to cause residual oil to be degraded; likely by a combination of water washing and bacterial activity. Following the loss of the original oil accumulation and degradation of the residual oil, faults and fractures presumably resealed, allowing a second migration of late-stage thermogenic gas to become trapped. As previously mentioned, however, the isotopic composition of the methane presently in the Drake Point field reservoirs is not indicative of the late-stage thermogenic gas. This hypothesis is not, therefore, valid unless the isotopic composition of the gas has been modified by the addition of isotopically light biogenic methane or early thermogenic methane and/or a combination of both. The anomalous relationship between the chemical and isotopic compositions of gases in Drake Point reservoir structure implies that the factors controlling their generation, migration, and accumulation probably are unique.

The Hoyle Bay and Roche Point formations in the Drake Point area contain good to excellent quality oil-prone source beds that have an average weighted TOC of 3 to 5%, and a reflectance R_0 of 0.5 to 0.6. In the axial part of the basin these rocks are in a very mature to overmature stage, with R_0 up to 1.4 (Leythaeuser and Stewart, 1986). Most of the kerogen in the Middle Triassic is of type II. It is reasonable to assume that much of the natural gases accumulated in the Mesozoic reservoirs derived from oil-prone source beds in these two formations. Several of the Jurassic shales probably also contributed gas to the reservoirs. Some of these shales, while immature in the Drake Point field area, are mature to the north.

Another possible source of gas for the Drake Point field is the thick (+3048-m; +10,000-ft) sequence of shales, cherts, sandstones, and limestones of Permian and Pennsylvanian age that underlie the Drake Point field and are in a mature to overmature facies. Powell (1978) feels that the gas potential of this section may be considerable in view of the maturation level and the high recorded gas contents.

Figure 17, a geohistory plot for the Drake Point field, shows the two rapid periods of burial for this area. The first, in the Lower Triassic, would probably bury some of the Permian and Pennsylvanian rocks to a sufficient depth to initiate generation of hydrocarbons. The second period of rapid deposition, in the Lower Cretaceous, probably marked the time of initial generation of hydrocarbons from the Middle Triassic to Middle Jurassic sequence. Since the Drake Point trap did not form until Late Cretaceous to Early Tertiary time, it is felt that the main contributor to the gas reserves at the Drake Point field are Mesozoic rocks of the Middle Triassic to Middle Jurassic sequence.

Faults

Several of the northeast-trending faults that cut the field have throws that exceed the gross thickness of the Drake Point and Intrepid Inlet members. These faults displace the field gas-water contacts, resulting in at least three separate fault-bounded pools having a common aquifer (Figures 9 and 13). Whereas these faults effectively separate the individual pools, they may leak. Even though the pools appear to be filled to spill point, it may be because natural gas migration into the pools is continuing to the present, and probably exceeds the losses through leaky seals.

Where the reservoirs are juxtaposed across the faults in the Drake Point field there does not appear to be any barrier to fluid movement.

EXPLORATION CONCEPTS

Exploration for Lower Jurassic sandstones on gentle anticlines at the south margin of the Sverdrup basin resulted in the discovery of one more gas field, the Hecla gas field, which is similar to the Drake Point field and lies about 32 km (20 mi) to the west (Figure 2). Hecla contains 3.5 tcf of marketable natural gas.

One important and significant difference between the two fields is that there is a stratigraphic component to the Hecla trap. The field gas–water contact lies outside of structural closure, so the south edge of the field is defined by the southern pinchout of the Intrepid Inlet and Drake Point Member sandstones. Figure 11 shows this pinchout south of the Drake Point field. This stratigraphically trapped gas opened up a new play concept along the south margin of the basin. Although several dry holes were drilled because of the stratigraphic trap concept, many prospects remain to be drilled. For example, in the Drake Point field itself, several fault blocks at the east end of the field, and the graben between the main and I-55 pools, remain to be drilled. These are evident in Figures 9 and 14. It is estimated that additional gas reserves in these fault blocks could bring the ultimate marketable reserves in the Drake Point field up to 6 tcf.

Because the Drake Point field is not fully delineated, several parameters relative to the field may change. This is particularly true for net pay data in the northeast offshore part of the field, where additional delineation wells are required.

No further activity in this field is expected until markets for Arctic Islands natural gas are developed.

ACKNOWLEDGMENTS

The author wants to express his appreciation to Panarctic Oils, Ltd., for permission to publish this paper and for drafting services. Much of the data in this paper were originally collected for Panarctic Oils, Ltd., by former Panarctic employees, particularly S. Schmaltz, B. Low, and R. Stewart. A. Embry of the Geological Survey of Canada provided much of the regional stratigraphic information.

The manuscript benefited from reviews by A. Embry (Geological Survey of Canada) and H. Balkwill (PetroCanada Exploration). D. Powley of Amoco Production Co. was the American Association of Petroleum Geologists reviewer; his comments and suggestions for improving the manuscript were much appreciated.

REFERENCES CITED

Balkwill, H. R., and F. G. Fox, 1982, Incipient rift zone, Western Sverdrup basin, arctic Canada, *in* A. F. Embry and H. R. Balkwill, eds., Arctic geology and geophysics: Canadian Society of Petroleum Geologists Memoir 8, p. 171-187.

Bally, A. W., and S. Snelson, 1980, Realms of subsidence, *in* Facts and principles of world petroleum occurrence: Canadian Society of Petroleum Geologists Memoir 6, p. 9-94.

Carmalt, S. W., and B. St. John, 1986, Giant oil and gas fields, *in* M. T. Halbouty, ed., Future petroleum provinces of the world: American Association of Petroleum Geologists Memoir 40, p. 11-54.

Douglas, T. R., and T. A. Oliver, 1979, Environments of deposition of the Borden Island gas zone in the subsurface of the Sabine Peninsula area, Melville Island, arctic archipelago: Bulletin of Canadian Petroleum Geologists, v. 27, p. 273-313.

Embry, A. F., 1982, The Upper Triassic/Lower Jurassic Hieberg deltaic complex of the Sverdrup basin, *in* A. F. Embry and H. R. Balkwill, eds., Arctic geology and geophysics: Canadian Society of Petroleum Geologists Memoir 8, p. 189-217.

Embry, A. F., 1984a, The Scheii Point and Blaa Mountain groups (Middle-Upper Triassic), Sverdrup basin, Canadian arctic archipelago, *in* Current research, part B: Geological Survey of Canada, Paper 84-1B, p. 327-336.

Embry, A. F., 1984b, Stratigraphic subdivision of the Roche Point, Hoyle Bay, and Barrow formations (Scheii Point Group), western Sverdrup basin, Arctic Islands, *in* Current research, part B: Geological Survey of Canada, Paper 84-1B, p. 275-283.

Embry, A. F., 1985, Stratigraphic subdivision of the Isachsen and Christopher formations (Lower Cretaceous), Arctic Islands, *in* Current research, part B: Geological Survey of Canada, Paper 85-1B, p. 239-246.

Embry, A. F., in press, Jurassic-lowermost Cretaceous, Melville Island area: Geological Survey of Canada.

Hasegawa, H. S., 1977, Focal parameters of four Sverdrup basin, arctic Canada earthquakes in November and December of 1972: Canadian Journal of Earth Science, v. 14, p. 2481-2494.

Leythaeuser, D., and K. R. Stewart, 1986, Generation, migration and accumulation of hydrocarbons in the Sverdrup basin, Canada: unpublished Panarctic Oils, Ltd., report.

Low, B., 1983, The reservoir geology of the Intrepid Inlet and Drake Point members in the Drake and Hecla gas field areas, Melville Island, arctic archipelago: unpublished Panarctic Oils, Ltd., report.

Powell, T. G., 1978, An assessment of the hydrocarbon source rock potential of the Canadian Arctic Islands: Geological Survey of Canada Paper 78-12.

Thorsteinsson, R., 1974, Carboniferous and Permian stratigraphy of Axel Heiberg Island and western Ellesmere Island, Canadian arctic archipelago: Geological Survey of Canada Bulletin 224.

Tozer, E. T., and R. Thorsteinsson, 1964, Western Queen Elizabeth Islands, arctic archipelago: Geological Survey of Canada Memoir 332, p. 177-185.

Waylett, D. C., 1979, Natural gas in the Arctic Islands—discovered reserves and future potential: Journal of Petroleum Geology, v. 1, no. 3, p. 21-34.

Appendix 1. Field Description

Field name .... *Drake Point field*

Ultimate recoverable reserves

Field location:

Country .... *Canada*
State .... *Northwest Territory*
Basin/Province .... *Sverdrup*

Field discovery:

Year field discovered *1969*
Year second pay discovered
Third pay

Discovery well name and general location
(i.e., Jones No. 1, Sec. 2T12NR5E; or Smith No. 1, 5 mi west of Sheridan, Wyoming):

First pay .... *Panarctic Drake Point N-67 76°30'N lat., 108°30'W long.*
Second pay
Third pay

Discovery well operator .... *Panarctic Oils Limited*
(if more than one pay in field, list operators of discovery well in other pays)

Second pay
Third pay

IP in barrels per day and/or cubic feet or cubic meters per day:

First pay
Second pay
Third pay

All other zones with shows of oil and gas in the field:

Age	Formation	Type of Show
Middle Triassic	*Roche Point*	*Minor gas accumulation*

Geologic concept leading to discovery and method or methods used to delineate prospect, e.g., surface geology, subsurface geology, seeps, magnetic data, gravity data, seismic data, seismic refraction, nontechnical:

Initial indicator was a broad, gentle surface anticline with closure being confirmed prior to drilling by some seismic reflection data.

Structure:

Province/basin type (see St. John, Bally, and Klemme, 1984)

Bally, 1143; Kelmme, IICc

Tectonic history

The region is underlain by a thick miogeosynclinal sequence of Lower Paleozoic sediments that were uplifted and folded during Late Devonian and Mississippian time. A major rift basin developed during the Pennsylvanian and Permian time which resulted in the deposition of a very thick sequence of carbonates, clastics, and evaporites. During Mesozoic and early Tertiary time a thick sequence of shallow-water clastics were deposited along the axis of the older rifted basin. No major tectonic activity occurred in the western part of the Sverdrup basin during this time.

Regional structure

The field is on the south flank of the Sverdrup basin.

Local structure

Drake Point is located on the eastern culmination of a large, gentle east-west-trending anticline that is cut by numerous northeast-trending normal faults many of which have some strike-slip movement.

Trap

Trap type(s)

The Drake Point field has one anticlinal trap that has been cut by numerous faults yielding at least three separate pools having different gas-water contacts.

Basin stratigraphy (major stratigraphic intervals from surface to deepest penetration in field):

Age	Formation	Depth to Top in m (ft)
Lower Cretaceous (Albian)	*Christopher*	*Surface*
Lower Cretaceous (Aptian-Barremian)	*Isachsen*	*427 (1400)*
Upper Jurassic	*Awingak*	*518 (1700)*
Middle-Lower Jurassic	*Jameson Bay*	*975 (3200)*
Lower Jurassic	*MacLean Strait*	*1128 (3700)*
Upper Triassic	*Hoyle Bay*	*1219 (4000)*
Middle Triassic	*Roche Point*	*1295 (4250)*
Lower Triassic	*Bjorne*	*1555 (5100)*

Location of well in field

Reservoir characteristics:

Number of reservoirs *1*

Formations .... *Intrepid Inlet Member, Jameson Bay Formation, Drake Point Member, MacLean Strait Formation*

Ages .... *Lower Jurassic*

Depths to tops of reservoirs .... *1097 m (3600 ft); 1128 m (3700 ft)*

Gross thickness (top to bottom of producing interval) .... *20-33.5 m (65-110 ft); average 30.4 m (100 ft)*

Net thickness—total thickness of producing zones

Average .... *21.6 m (71 ft)*

Maximum .... *29.2 m (96 ft)*

Average

Maximum

Lithology .... *Fine- to very fine grained quartzose sandstone*

Porosity type .... *Intergranular porosity*

Average porosity *19-20%*

Average permeability .... *566 md*

Seals:

Upper

Formation, fault, or other feature .... *Cape Canning Member, Jameson Bay Formation*

Lithology .... *Shale*

Lateral

Formation, fault, or other feature

Lithology

Source:

Formation and age .... *Hoyle Bay and Roche Point formations (Middle Jurassic)*

Lithology .... *Sandstone, limestone, siltstone, shale*

Average total organic carbon (TOC) *3-5%*

Maximum TOC *13%*

Kerogen type (I, II, or III) .... *II*

Vitrinite reflectance (maturation) .... R_o *= 0.4-0.6 in vicinity of field to 1.4 north of field*

Time of hydrocarbon expulsion .... *Beginning in Albian and probably continuing to present*

Present depth to top of source .... *1220-1524 m (4000-5000 ft)*
Thickness .... *Gross Zone: 274-305 m (900-1000 ft);*
Net Source Shale: 30-46 m (100-150 ft)

Potential yield

Appendix 2. Production Data

Field name .... *Drake Point field*

Field size:

Proved acres .... *40,470 ha (100,000 ac)±*
Number of wells all years
Current number of wells (as of year)
Well spacing
Ultimate recoverable .... *5305 bcf*
Cumulative production .... *Nil; shut in*
Annual production .... *Nil; shut in*
Present decline rate
Initial decline rate
Overall decline rate
Annual water production
In place, total reserves .... *6070 bcf*
In place, per acre-foot
Primary recovery
Secondary recovery
Enhanced recovery
Cumulative water production

Drilling and casing practices:

Amount of surface casing set
Casing program
Drilling mud
Bit program
High pressure zones

Completion practices:

Interval(s) perforated
Well treatment

Formation evaluation:

Logging suites
Testing practices
Mud logging techniques

Oil characteristics:

Type
(Tissot and Welte Classification in "Petroleum Formation and Occurrence," 1984, Springer-Verlag, p. 419)
API gravity
Base
Initial GOR
Sulfur, wt%
Viscosity, SUS

Pour point
Gas-oil distillate

Field characteristics (Main Pool):

Average elevation .... *Sea level*
Initial pressure .... *12,259 kPa (1778 psi) at main pool g/w*
Present pressure
Pressure gradient .... *3.12 kPa (0.453 psi)*
Temperature .... *82°F*
Geothermal gradient
Drive .... *Water*
Gas column thickness .... *188 m (616 ft)*
Gas-water contact .... *−1215 m (−3986 ft) subsea*
Connate water *25%*
Water salinity, TDS .... *51,700 mg/L*
Resistivity of water
Bulk volume water (%)

Transportation method and market for oil and gas:

Greater Burgan Field

P. BRENNAN
Consultant
Longboat Key, Florida

FIELD CLASSIFICATION

BASIN: Arabian
BASIN TYPE: Foredeep Ramp
RESERVOIR ROCK TYPE: Sandstone
RESERVOIR ENVIRONMENT OF DEPOSITION: Littoral (Deltaic to Lagoonal)
RESERVOIR AGE: Cretaceous to Jurassic
PETROLEUM TYPE: Oil
TRAP TYPE: Anticline

LOCATION

The Greater Burgan field lies within the Arabian basin and in the State of Kuwait, an independent emirate situated in the northeastern part of the Arabian peninsula at the western side of the head of the Persian (Arabian) Gulf. Kuwait is bounded on the east by the waters of the Gulf; on the north and west by the Republic of Iraq; and on the south by the Kingdom of Sa'udi Arabia (Figure 1).

Although small in size, with a surface area of only 17,820 km^2 (6880 mi^2), Kuwait contains a number of important oil fields, including the Greater Burgan, Raudhatain, Sabriyah, and Minagish fields. The Greater Burgan field, located at lat. 29° N, long. 48° E, is by far the largest of these. The Greater Burgan field lies in southeastern Kuwait in typical desert terrain a few miles inland from the Gulf shoreline, at elevations ranging from 75 to 115 m (250 to 385 ft) (Figure 2). The field encompasses a surface area of 780 km^2 (300 mi^2) and, with recoverable reserves of at least 75 billion bbl, is certainly among the very largest producing oil fields of the world. The Greater Burgan field was ranked as the second largest known by Halbouty et al. (1970) and by Carmalt and St. John (1986).

HISTORY

Pre-Discovery

Reports of bitumen deposits on the desert plain of Burgan 45 km (28 mi) south of Kuwait Bay led to the visit in 1912 of a British Admiralty Commission that included four geologists. One of these, Pascoe of the Indian Geological Survey, submitted a report (Pascoe, 1913) that postulated the existence at Burgan of a gentle surface dome associated with the deposits of bituminous earth.

Anglo-Persian Oil Company (later Anglo-Iranian and later still British Petroleum) carried out reconnaissance surveys in Kuwait during 1914 and again during 1917, the orderly progress of the work being interrupted by World War I. The reports of these surveys confirmed the presence of bitumen at Burgan but made no mention of surface structure in the area. Subsequent surveys by Heim (1924) and Anglo-Persian (1925–1926 and 1931) yielded only ambiguous results.

During 1931–1932, Anglo-Persian undertook a more comprehensive survey of the Kuwait area. This included both pitting and shallow drilling at Burgan, confirming the presence there of extensive impregnations of bitumen in the surficial deposits, together with indications of gas seepage. The report of this survey (Cox, 1932) concluded that surface structure was present in the area south of Kuwait Bay. While accepting that no direct evidence was available regarding geological conditions beneath the widespread blanket of Neogene deposits, Cox (1932) recommended wildcat drilling at three localities where oil and/or gas seepages had been identified. One of the recommended drilling sites was the area of the Burgan bitumen deposits.

Gulf Oil Corporation had been seeking a petroleum concession in Kuwait in open competition with Anglo-Persian. Interest in the area intensified with the discovery of commercial oil on Bahrain Island in 1932 and with the grant of a petroleum concession in Sa'udi Arabia to Standard Oil of California in 1933. Gulf and Anglo-Persian agreed to cooperate in obtaining the Kuwait concession and formed the jointly owned Kuwait Oil Company as the vehicle for this. The joint negotiations were successful, and, in December 1934, the Kuwait Oil Company was granted a 75 year petroleum concession covering the entire territory of Kuwait. The concession included

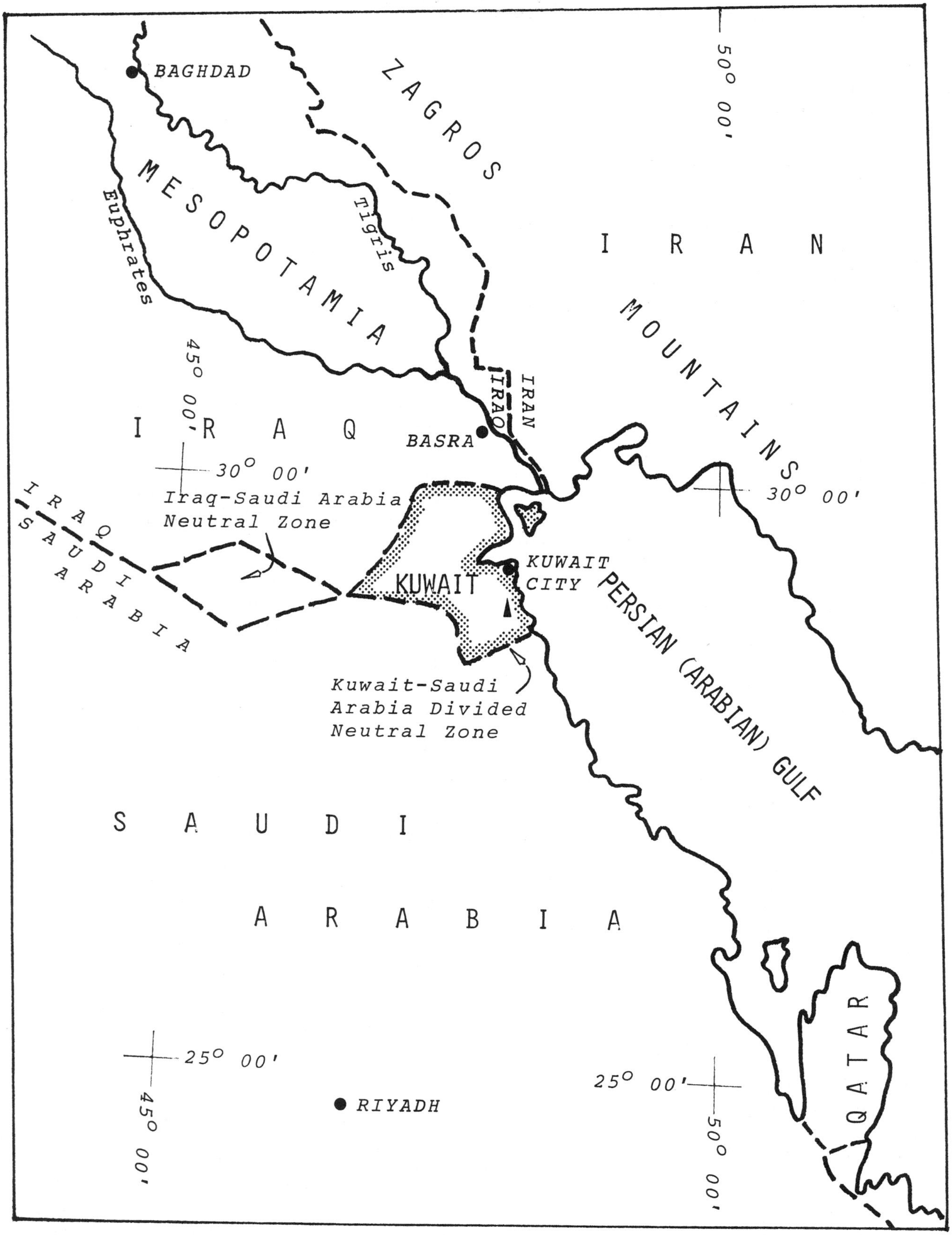
BAGHDAD
ZAGROS
MESOPOTAMIA
Euphrates
Tigris
IRAN
MOUNTAINS
IRAQ
IRAN
IRAQ
BASRA
45° 00'
30° 00'
Iraq-Saudi Arabia
Neutral Zone
IRAQ
SAUDI
ARABIA
KUWAIT
KUWAIT
CITY
PERSIAN (ARABIAN) GULF
Kuwait-Saudi
Arabia Divided
Neutral Zone
SAUDI
ARABIA
QATAR
25° 00'
RIYADH
50° 00'

Figure 1. Regional location map of Kuwait.

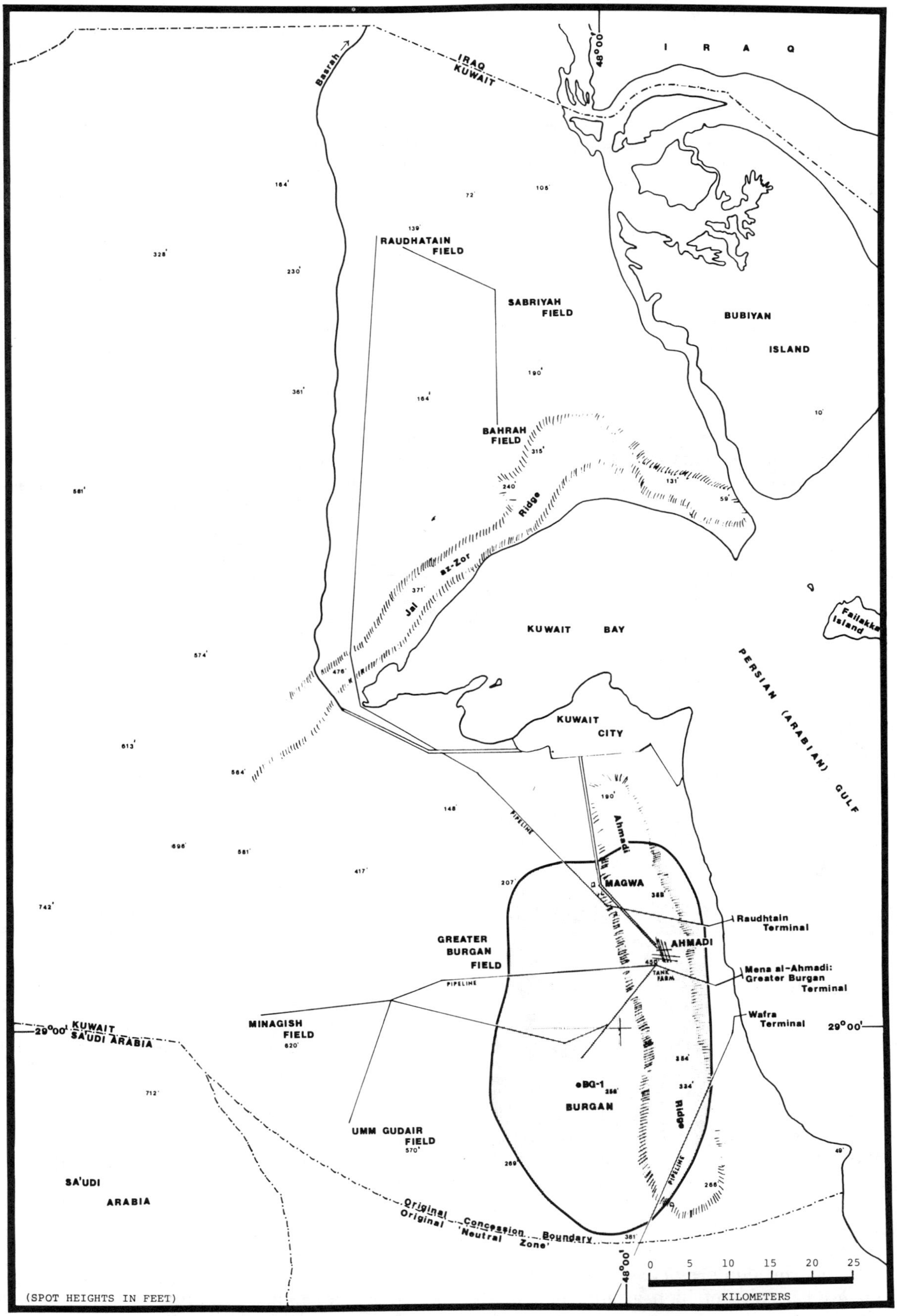

IRAQ
KUWAIT
Basrah
48°00'
I R A Q
RAUDHATAIN FIELD
SABRIYAH FIELD
BUBIYAN ISLAND
BAHRAH FIELD
Jal az-Zor Ridge
KUWAIT BAY
Failakka Island
PERSIAN (ARABIAN) GULF
KUWAIT CITY
Ahmadi
MAGWA
Raudhtain Terminal
AHMADI
GREATER BURGAN FIELD
TANK FARM
Mena al-Ahmadi: Greater Burgan Terminal
PIPELINE
Wafra Terminal
29°00'
KUWAIT
SA'UDI ARABIA
MINAGISH FIELD
BQ-1
BURGAN
Ridge
UMM GUDAIR FIELD
SA'UDI ARABIA
Original Concession Boundary
Original 'Neutral Zone'
0 5 10 15 20 25
KILOMETERS
(SPOT HEIGHTS IN FEET)

Figure 2. Greater Burgan field: location map.

the territorial waters and islands but excluded Kuwait's share of the jointly administered Kuwait–Sa'udi Arabia Neutral Zone to the south.

Discovery

With the concession in hand, the Kuwait Oil Company fielded a joint Gulf/Anglo-Persian geological party during 1935. Three wildcat locations were recommended for drilling, corresponding to Cox's seepage areas and including the area of the Burgan bitumen deposits. The first location to be drilled was at Bahrah, on the northern shore of Kuwait Bay 70 km (45 mi) north of Burgan, where Cox had mapped structure in association with a small, active seepage of oil and gas. While Bahrah-1 was being drilled, Kuwait Oil Company carried out a gravity meter-magnetometer survey of the entire concession area. A structural lead was outlined in the area of the Burgan bitumen deposits and was confirmed by a limited program of reflection seismic.

Bahrah-1 was completed dry and abandoned during April 1937, at total depth 2420 m (7950 ft) in strata of inferred Albian age. Although a dry hole, the well did confirm the presence of adequate sedimentary section below the Neogene Kuwait Series and even had some slight shows of oil. The drilling rig was moved south and a new wildcat was spudded on the plain of Burgan; this well, Burgan-1, encountered oil-bearing sandstone at a drilling depth of 1120 m (3672 ft) on 23 February 1938. Drilling was terminated for mechanical reasons at 1125 m (3692 ft), only 6 m (20 ft) into the indicated pay zone; the well was cased and perforated and flowed 31.8° API crude oil to surface. Because of a lack of testing equipment, no definite production rate was obtained at the time of completion (Milton, 1967); however, the well was properly tested in December 1938 and established an initial potential of 4386 BOPD on a 1-in. choke. It was established that the reservoir was a fine-grained quartz sandstone and that the shales immediately overlying it carried fossils of early Late Cretaceous (Cenomanian) age (Figure 6).

Post-Discovery

During the period 1938–1942, eight additional wells were drilled and completed at Burgan, in the assumed crestal area of the subsurface structure; all established oil production from the Cretaceous sandstone reservoir first tested by the discovery well. In 1942, drilling operations were suspended for the balance of the World War II years. They were resumed in earnest in 1945, and the first shipment of Burgan crude oil left the new Mena al-Ahmadi export terminal, located on the Gulf littoral 24 km (15 mi) east of the new field, in June 1946. The oil was fed down to the sea through gravity lines running from a tank farm on a ridge of high ground just to the east of the plain of Burgan. This was the Dhahar or Ahmadi ridge, suspected to be of structural origin but yielding negative results to seismic investigation because of near-surface weathering effects. Cox had suggested in his 1932 report that this ridge might prove to represent the east flank of a regional uplift lying to the south of Kuwait Bay.

By 1950, approximately 100 wells had been drilled in the Burgan field, all capable of producing oil, and the field production level was approaching 350,000 BOPD. Three major and one minor producing sand intervals had been penetrated, containing a common oil column and transected by a common oil-water contact. The shale section overlying the highest producing sand ("Cap Rock shales," later Ahmadi Formation) had been confirmed as Cenomanian in age, while below the first major sand section ("First sand," later Wara Formation) lay a thin but persistent unit of limestone carrying a late Albian or early Cenomanian microfauna. This "Orbitolina concava limestone," later Mauddud Formation, was in turn underlain by two thick sections of sand and sandstone with minor shales, initially termed the "Third and Fourth Burgan sands" and later included in the Burgan Formation. Lacking diagnostic fossils, these clastics were assigned a late Early Cretaceous, probably Albian, age, since they appeared to behave conformably with the overlying limestone (Figure 7). Up to 1950, the base of the pre-Mauddud clastic section had not been penetrated; most wells were terminated just below the field oil-water contact, an essentially plane surface at a subsea elevation of 1360 m (4466 ft).

During 1950–1951, attempts were made to drill through the entire Albian–Cenomanian clastic section, to investigate the potential of Early Cretaceous formations known to be oil-productive in southern Iraq, to the north, and of Late Jurassic formations known to be oil-productive in Sa'udi Arabia, to the south. The first attempt was unsuccessful. Loss of circulation into an underlying carbonate zone resulted in a blowout from the uncased Albian–Cenomanian oil reservoirs and, after an extensive fishing job, the well was cased and completed in the Albian Burgan Formation. The second deep test well, Burgan 113-HNC, was drilled to a depth of 4220 m (13,853 ft) in strata of believed Triassic age, the most complete stratigraphic penetration of the sedimentary section in the area up to that time. No commercial oil production was encountered below the Burgan Formation, but the well did establish the presence of a limited column of heavy oil in carbonate section of early Early Cretaceous age, and of oil staining in strata of Early to Middle Jurassic age. While the evaluation of the lower part of the borehole was not comprehensive, the identification of porosity and oil shows in the early Early Cretaceous and Early to Middle Jurassic intervals could be regarded as the first indication of the future potential of these older and deeper zones.

By 1951, the Burgan structure had been confirmed as a gentle dome, oriented almost north-south, measuring at least 24 by 13 km (15 by 8 mi) and with flank dips in the range of 2–3°. A program of reflection seismic and shallow drilling had already outlined a second, smaller subsurface feature 10 km (7 mi) north of the most northerly Burgan well drilled to that date. This new prospect lay close to the known gas seepage of Ma'adanyat, one of the areas recommended for drilling by Cox in his 1932 report. The northerly feature was now named Magwa, after a nearby company installation; it was drilled during 1951 and proved productive in the Wara and Burgan Formations. The new Magwa field contained crude oil of a slightly better grade than Burgan and possessed a small primary gas cap, but it shared a common oil-water contact and was evidently connected to the main Burgan accumulation across the intervening structural saddle. Development of the Magwa producing area began immediately.

East of Burgan-Magwa, and roughly paralleling the Gulf shore, lay the topographic feature of Ahmadi ridge, rising to a height of 120 m (400 ft) and the site of both the Burgan field tank farm and the newly built company town of Ahmadi. The surface ridge was suspected to be of structural origin, but reflection seismic across it had yielded inconclusive results. It was known from the drilling at Burgan that the Neogene Kuwait Series was underlain by a weathered and silicified erosional surface of Eocene age and that, because of structural rejuvenation, this unconformity reflected deeper structure in modified form. At Ahmadi ridge, shallow drilling to the Eocene unconformity confirmed the presence of an anticlinal stucture, oriented north-south and narrower and less domal in shape than either Burgan or Magwa.

The Ahmadi structure was drilled during 1952 and proved productive in the now-familiar Burgan-Wara clastic section. A gas cap was present, and the underlying oil column contained crude of a slightly higher grade than at Burgan or Magwa, but the oil-water contact was the same as in the Burgan and Magwa domes, once again indicating a connection across the intervening structural saddles. When this connection was confirmed by the drill, the three separate structures of Burgan, Magwa, and Ahmadi were accepted as being individual culminations on a larger, domal uplift, oriented NNE-SSW and measuring 35 by 20 km (23 by 13 mi) at the original oil-water contact. This very large producing area became known as the Greater Burgan field.

For 14 years, from the inception of oil production in mid-1946 to the commissioning of the Raudhatain field at the end of 1959, the Greater Burgan field was the sole source of crude oil in Kuwait. During 1947, the first full year of production, the field produced at an average rate of only 44,000 BOPD from a handful of wells on the Burgan dome, principally those wells drilled during the period 1938–1942 and then shut in for the balance of the war years.

By 1950, the Burgan dome contained 100 producing wells and the daily level of oil production had reached 345,000 barrels. By 1955, with oil from the Magwa and Ahmadi sectors making its contribution, the Greater Burgan field had passed the 1 million BOPD mark and, within the next 5 year period, had reached the level of 1.5 million BOPD, an incremental rate of increase of 100,000 BOPD per year. Cumulative production to 1960 amounted to just over 4 billion bbl. Estimates of recoverable oil reserves had risen in concert during this same period, doubling from the Kuwait Oil Company estimate of 11 billion bbl in 1950 to 22 billion bbl in 1954 (Levorsen, 1954), and more than doubling again to 47 billion bbl in 1960.

During the early development of the Burgan dome, wells were drilled approximately 2 km (1.25 mi) apart, at the intersections of a north-south/east-west alphabetic grid. Each primary location, identified by a well number followed by two coordinate letters, lay at the center of a hexagonal drainage unit covering 365 ha. (900 ac). Initial infill wells were drilled at the common corners of the hexagons, on drainage units covering 120 ha. (300 ac). In the crestal area, where extremely thick pay sections were present above bottom water, interior infill wells were eventually drilled on a 40 ha. (100 ac) pattern. The drilling was carried out using National 50-type rigs with conventional derricks, and a technique of "skidding" the drilling units from location to location was quickly developed to reduce the need for repetitive rig-building.

The drilling program involved the setting of 16 in. conductor pipe through the near-surface clastics; a string of 13⅜-in. surface casing into the Eocene Dammam Formation above 300 m (1000 ft); a string of 10¾-in. or 11¾-in. intermediate casing into the upper part of the "Maastrichtian limestone," now Bahrah Formation, about 210 m (700 ft) above the top of the oil zone; and a production string of 7-in. (occasionally 8⅝-in.) casing through the entire pay section.

The early wells were dual completions through casing and 3-in. tubing, with a packer set between the producing horizons. Most such dual completions involved the "First and Second sands," now Wara Formation, and the "Third sand," now the Third Sand Member of the Burgan Formation. In the crestal area of the Burgan dome, where a gross pay section of 310 m (1016 ft) was above water in the Burgan Formation alone (Table 1), the dual completions were made in the "Third and Fourth sands" (Third and Fourth Sand Members of the Burgan Formation), with a packer set opposite a shaly interval separating the two main sand developments.

During 1959, a new oil field was discovered at Minagish, 45 km (25 mi) west of Greater Burgan. The Minagish structure contained 20° API gravity crude oil in the Albian to Cenomanian sands, and 35° API gravity oil in a deeper reservoir of oolitic limestone developed within the early Early Cretaceous carbonate section known locally as the "Ratawi limestone." This was a new producing zone for Kuwait proper, although it had already tested oil at the Wafra field, located 25 km (15 mi) south of the Greater Burgan

Table 1. Structural relationships of producing zones in Burgan, Magwa, and Ahmadi sectors.

	Burgan ft (m)	Magwa ft (m)	Ahmadi ft (m)
Top Wara:	3245 (990)	3585 (1090)	3670 (1120)
Gas-oil contact:	not present	3780 (1150)	3850 (1170)
Top Third Sand:			
Upper:	3450 (1050)	3788 (1155)	3880 (1180)
Middle:	3510 (1070)	3902 (1190)	4028 (1230)
Lower:	3767 (1150)	4176 (1270)	4240 (1290)
Top Fourth Sand:	3926 (1200)	4260 (1300)	4356 (1330)
Oil-water contact:	4466 (1360)	4466 (1360)	4466 (1360)
Closure at oil-water contact:	1221 (370)	881 (270)	796 (240)
Gross gas section:	0 (0)	195 (60)	180 (55)
Gross oil section:	1221 (370)	686 (210)	616 (190)

field in the Kuwait-Sa'udi Arabian Neutral Zone. The producing section at Minagish would eventually be defined as the Minagish Member of the Raudhatain Formation, but on a local basis it was simply termed the "Minagish oolite" or "Minagish zone." The zone appeared to be of early Early Cretaceous, probably Berriasian-Valanginian, age.

The Minagish discovery caused some reconsideration of early Early Cretaceous oil possibilities at Greater Burgan, where equivalent section had already been penetrated in the Burgan deep test well drilled during 1950-1951. That well had penetrated 520 m (1718 ft) of detrital, pelletoid, and sub-oolitic limestones immediately overlying the Gotnia evaporites of presumed Late Jurassic age. This limestone section, of inferred early Early Cretaceous age, contained a limited column of heavy crude oil, trapped beneath the shales of the overlying Ratawi Formation. Since the deep test well had not been located at the apex of the Burgan dome, the possibility of establishing oil production in the early Early Cretaceous "Minagish zone" somewhere within the Greater Burgan structural complex could not be discounted.

Any search for such deeper production at Greater Burgan during the 1960s had to be fitted into an extensive exploration and field-development program already in progress. A number of projects competed for attention and investment: Greater Burgan itself; the Minagish discovery; the Raudhatain, Sabriyah, and Bahrah areas of northern Kuwait; and the recent discovery at Umm Gudair, close to the border of the Kuwait-Sa'udi Arabia Neutral Zone.

At Greater Burgan, intensive drilling had confirmed that the uppermost oil reservoir, the Wara Formation, was continuous across the structural saddles separating the individual culminations, so outlining a productive area covering hundreds of thousands of acres (Adasani, 1965). The crestal areas of the structural complex were extensively faulted; one prominent northwest-southeast fault trend, apparently forming a graben, clearly affected communication between the Burgan dome and the Magwa and Ahmadi structures at the level of the Burgan Formation (Figure 8). Closures on the individual culminations had been confirmed; the oil column in the Burgan dome at the structural high point of the field was 370 m (1221 ft) in thickness.

During the early 1960s, attempts were made to develop oil production from the carbonates of the Aruma and Hasa groups (Late Cretaceous to Paleogene) on the Burgan and Ahmadi structures. This section produced heavy oil at Wafra field, in the Neutral Zone, and carried oil shows in many wells at Greater Burgan; however, a six-well program at Burgan-Ahmadi yielded only noncommercial amounts of heavy oil and the effort was suspended.

Oil production from the Albian to Cenomanian reservoirs rose steadily, so that by the end of the decade of the 1960s Greater Burgan field was producing at a rate of 2.1 million BOPD and cumulative production had already exceeded 10 billion bbl (Table 2). At this point, the original concession, signed in 1934, still had 40 years to run. However, at the end of 1974 the government of Kuwait demanded and obtained a 60% level of participation in the oil operation. Within a further 2 years, Kuwait completed the process of nationalization, assuming 100% control over the assets and activities of the Kuwait Oil Company. A state oil entity, Kuwait Petroleum Corporation, was formed to manage the nationalized properties.

Oil production at Greater Burgan field rose steadily from 1946 to a peak of 2.4 million BOPD in 1972 and then began to decline (Figure 3). This did not represent declining field capacity, but rather production restraints imposed by the Kuwait government for political reasons. The declining oil production resulted in a reduced supply of associated natural gas, which had come into increasing demand for reservoir-pressure maintenance at Greater Burgan, for power generation and water desalination, and for the extraction of liquified petroleum gas (LPG) products for export. By 1972, the production of associated gas had reached 604 billion MCFG/year, of which almost half was utilized. By 1976, gas production had declined to 390 billion MCFG and by 1978 to 259 billion MCFG, while demand continued to rise. As the decline in associated gas production began to threaten the LPG export contracts, Kuwait

Table 2. Greater Burgan field production statistics.

Year	Average BOPD	Cumulative: Billions Bbl.
1946	16,438	
1947	43,836	
1948	128,767	
1949	246,575	
1950	356,164	
1951	547,945	0.489
1952	739,726	0.759
1953	794,520	1.049
1954	958,904	1.399
1955	1,095,890	1.799
1956	1,095,890	2.199
1957	1,139,726	2.615
1958	1,397,260	3.125
1959	1,389,041	3.632
1960	1,501,370	4.180
1961	1,517,808	4.734
1962	1,693,150	5.352
1963	1,783,562	6.003
1964	1,871,232	6.686
1965	1,868,493	7.368
1966	1,968,493	8.086
1967	1,986,301	8.811
1968	2,087,671	9.573
1969	2,134,247	10.352
1970	2,209,589	11.158
1971	2,345,205	12.014
1972	2,415,068	12.895
1973	2,209,589	13.701
1974	1,917,808	14.401
1975	1,534,246	14.961
1976	1,328,767	15.446
1977	1,378,082	15.949
1978	1,514,795	16.501
1979	1,730,263	17.132
1980	1,078,740	17.525
1981	731,117	17.792
1982	544,189	17.991
1983	688,140	18.241
1984	716,356	18.504
1985	680,322*	18.750
1986	979,600*	19.100

*Indicates estimated production.

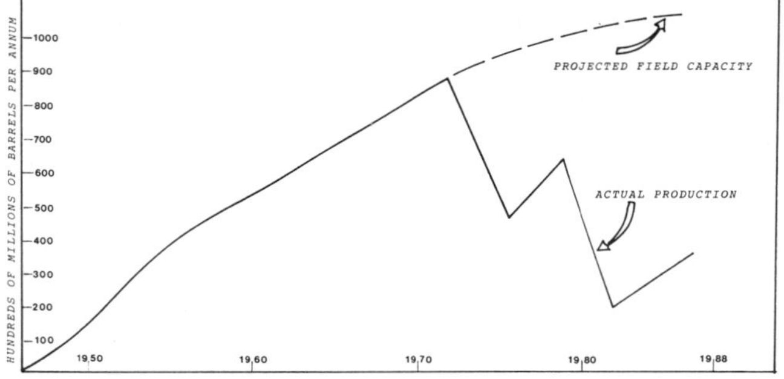
HUNDREDS OF MILLIONS OF BARRELS PER ANNUM
1000
900
800
700
600
500
400
300
200
100
19,50
19,60
19,70
19,80
19,88
PROJECTED FIELD CAPACITY
ACTUAL PRODUCTION

Figure 3. Oil production at Greater Burgan field, 1946-1986.

Petroleum Corporation began to explore for alternate sources of natural gas, preferably nonassociated and therefore not related to levels of oil production.

This search involved deep drilling to the Upper Permian (Kungurian to Tatarian) Khuff Formation, a carbonate section known to be a prolific source of wet gas in Bahrain, Qatar, and Sa'udi Arabia. The Khuff section could be predicted to be present in the subsurface of Kuwait, since the Late Permian transgression onto the Arabian platform had come from the east and northeast, the Permian deposits thinning west and southwest into the Arabian interior. It was understood that wells in the 6000 m (20,000 ft) depth range would be required to evaluate the Khuff section in the area of the Greater Burgan field.

Two attempts were made to drill to the Khuff Formation in the Magwa sector of Greater Burgan during 1977-1979. The first well blew out at a depth of 2670 m (8750 ft) while drilling at the level of the Cretaceous-Jurassic boundary, and drilling was terminated at that depth. The second well at Magwa was suspended at a depth of 3640 m (11,938 ft), in pre-Cretaceous section but well above the target Permian horizon. At the end of 1979, two additional deep wells were spudded in the Burgan sector of the field; both reached the target Khuff section, at depths of 5930 m (19,470 ft) and 6780 m (22,235 ft), respectively, but failed to establish commercial gas production. At this point, the Khuff drilling program was suspended in favor of a project to pipe in assured supplies of natural gas from the Rumaila field of southern Iraq.

But deeper drilling at Greater Burgan was successful in proving up additional reserves of light oil (31°-38° API gravity) in carbonate reservoirs of early Early Cretaceous (Neocomian) age (Minagish Oolite Member) and of late Early Jurassic (?Toarcian) age (Marrat Formation). It has been estimated that the pre-Albian reservoirs may add 150,000-200,000 BOPD to the Greater Burgan field producing capacity. Equally significant, the Jurassic discovery, made in 1984 in the Magwa sector of the field, may have added 10 billion bbl of oil to the field's recoverable reserve. This reserve has been estimated to amount to 75 billion bbl of oil or 87 billion bbl of oil equivalent, taking into account the field's content of associated gas (Carmalt and St. John, 1986). Even these figures may prove conservative, since they appear to refer only to the recoverable reserve contained in Cretaceous reservoirs. Of the great volume of recoverable oil contained in the Greater Burgan field, approximately 19 billion bbl had been produced through 1986 (Table 2).

DISCOVERY METHOD

The Burgan discovery of 1938 resulted from wildcat drilling in proximity to a major surface occurrence of bitumen. The bitumen formed a substantial lake beneath a few feet of eolian sand, while a shallow borehole had encountered plastic bitumen with traces of liquid oil saturating the near-surface clastics to a depth of at least 65 m (218 ft). Associated with the surface bitumen were deposits of the sour, gypseous earth known locally as *gach-i-turush,* a frequent indicator of hydrocarbon seepages in the Middle East, together with a small, active seepage of hydrocarbon gas.

The bitumen deposits of the desert plain of Burgan had been known to the local Arabs for centuries and to Europeans since at least 1913, when Pascoe (1913) made a geological report on the area. The question arises as to why the area was not drilled until 1938. The principal reason appears to have been the opinion prevailing among geologists working the Zagros foothills play of Iran that the Tertiary-Mesozoic section on the Arabian side of the Gulf would prove too thin and of too continental an aspect to offer serious petroleum potential. A secondary caveat, expressed by Cox (1932), concerned the nature of any oil that might be present—whether it might be too heavy to be of commercial interest.

Cox, whose 1932 report on the oil prospects of Kuwait was by far the most sophisticated up to that time, went on to say that should the currently drilling wildcat on Bahrain Island discover good-quality oil in Eocene or Cretaceous horizons, the prospects of Kuwait would be greatly enhanced (Owen, 1975). The Bahrain well did discover oil, and the Kuwait concession was then actively sought by Anglo-Iranian Oil Company (later British Petroleum Company) and Gulf Oil Corporation, the two finally combining as Kuwait Oil Company to obtain the concession in 1934.

Once the Kuwait concession was in the hands of an operating oil company, it was inevitable that the area of the Burgan bitumen deposits should be reexamined and programmed for drilling. The area was surveyed geophysically, a subsurface lead associated with the overlying bitumen deposits was outlined and developed into a drillable anomaly, and the structure was drilled and brought into production.

STRUCTURE

Geologic Setting

Precambrian igneous and metamorphic rocks are exposed along the eastern shore of the Red Sea from the Gulf of Aqaba, in the north, to the Gulf of Aden, in the south. These Precambrian outcrops extend eastward into the interior of the Arabian peninsula to form the Arabian shield (Figure 4). Arcing around the eastern flank of the Arabian shield lies the interior homocline (Aramco, 1975) or stable shelf (Beydoun and Dunnington, 1975), in which Paleozoic and Mesozoic strata dip basinward at rates reflecting the influence of the underlying basement rocks. East and southeast of the interior homocline or stable shelf lies the interior platform (Aramco, 1975) or unstable shelf (Beydoun and Dunnington, 1975), characterized by the presence of basically flat-lying Tertiary deposits that no longer obviously reflect basement configuration.

The area of the interior platform or unstable shelf includes the coastal zone of eastern Arabia, the subsurface of the Persian Gulf, and the greater part of the riverine lowlands of Mesopotamia in southeastern Iraq. Several distinct basinal areas are developed along the interior platform/unstable shelf, including the Rub al-Khali basin of the southeastern Gulf and adjacent land areas, and the North Arabian Gulf basin, lying at the head of the Gulf between the Arabian shore and the Zagros foreland zone of Iran and extending northwestwards into Iraq. The entire State of Kuwait, including the Greater Burgan field, lies within the interior platform or unstable shelf sector (Figure 4). Kuwait stands at the edge of the Arabian foreland, at the limit of the Mesopotamian sedimentary basin, and at the limit of the effects of the Zagros orogeny (Fox, 1959).

The Greater Burgan structural complex is one among several highs developed along a prominent anticlinal axis that trends north-northwest out of eastern Sa'udi Arabia through Kuwait and into southern Iraq. These individual highs, in which the principal oil accumulations of the area are localized, tend to be of large areal extent and substantial structural relief.

Adasani (1965) attributed structural growth in the area to subsidence of the North Arabian Gulf basin; to compressional stress generated by shortening of the bottom profile of the basin acting against the stable margins to produce faulting and supratenuous folding. Fox (1959) also accepted the possibility of supratenuous folding across a buried high or highs, but considered the Burgan and Magwa culminations to be uplifts probably caused by halokinesis at depth. Murris (1980) described Burgan as a combined halokinetic-basement horst-block structure. In each case, the proposed halokinesis assumed the mobilization of the deep-seated "Infracambrian" Hormuz Salt Series, known from regional evidence to be present in the subsurface of the North Gulf salt basin (AlSharhan and Kendall, 1986), which includes the area of Kuwait. There is evidence of salt-piercement to the present-day land surface at Jabal Sana'am in southern Iraq, 140 km (90 mi) north of Greater Burgan (Wyllie, 1926).

Within the field area, the large Burgan dome and the smaller Magwa dome exhibit ovoid outlines and quasi-radial fault patterns suggestive of vertical uplift (Figure 5). Well control indicates that the vertical growth has been long-sustained; all post-Jurassic formations show stratigraphic thinning toward the crests of the domes, as a result of reduced or non-deposition, penecontemporaneous erosion, or some combination of these processes (Adasani, 1965). The uplift and erosion varied in intensity with time and gave rise to a number of disconformities and to two major unconformities within the section (Figures 6 and 7).

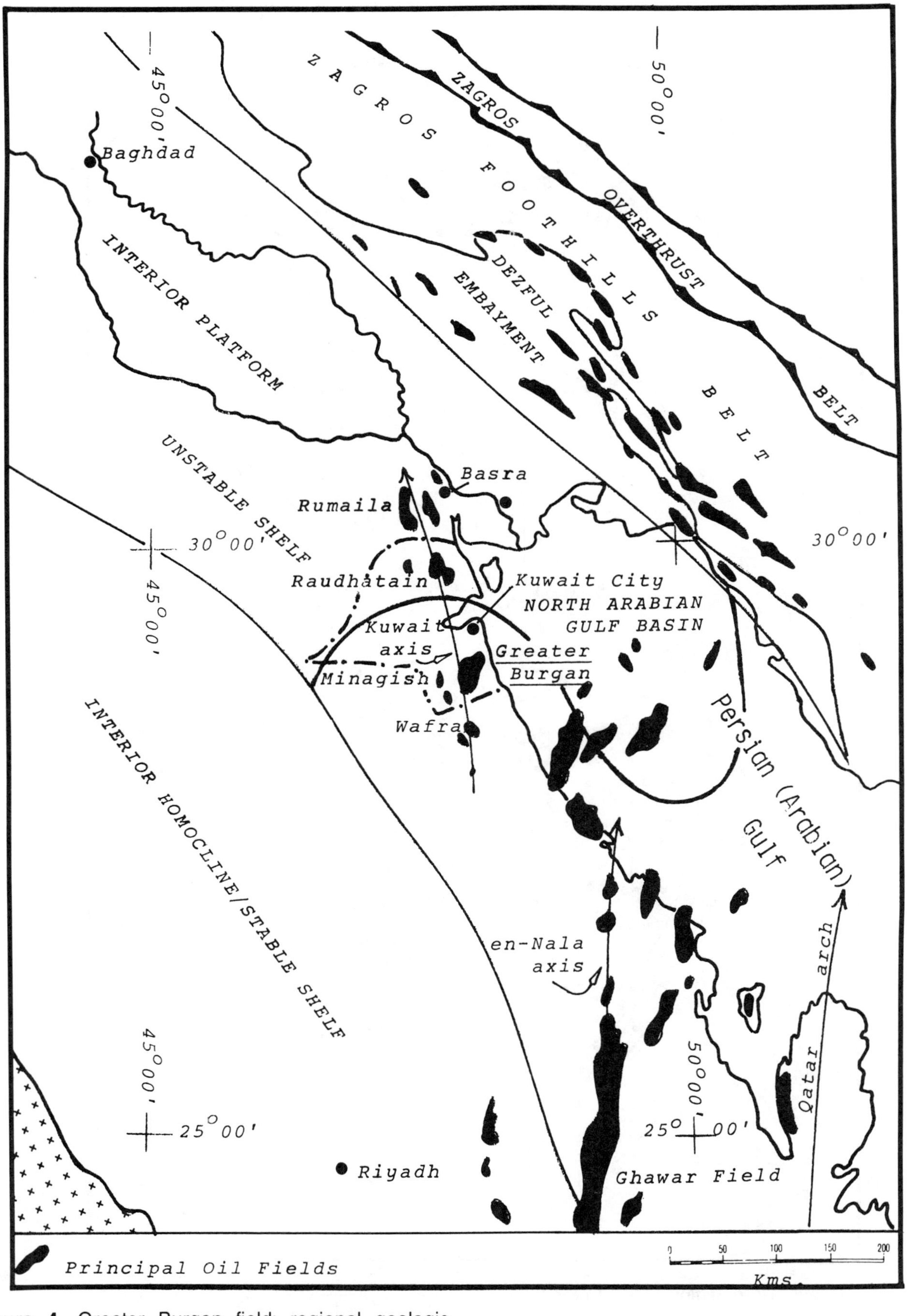
Baghdad
ZAGROS FOOTHILLS BELT
ZAGROS OVERTHRUST BELT
DEZFUL EMBAYMENT
INTERIOR PLATFORM
UNSTABLE SHELF
Basra
Rumaila
Raudhatain
Kuwait City
NORTH ARABIAN GULF BASIN
Kuwait axis
Greater Burgan
Minagish
Wafra
Persian (Arabian) Gulf
INTERIOR HOMOCLINE/STABLE SHELF
en-Nala axis
Qatar arch
Riyadh
Ghawar Field
45°00'
50°00'
30°00'
25°00'
Principal Oil Fields
0 50 100 150 200
Kms.

Figure 4. Greater Burgan field: regional geologic setting.

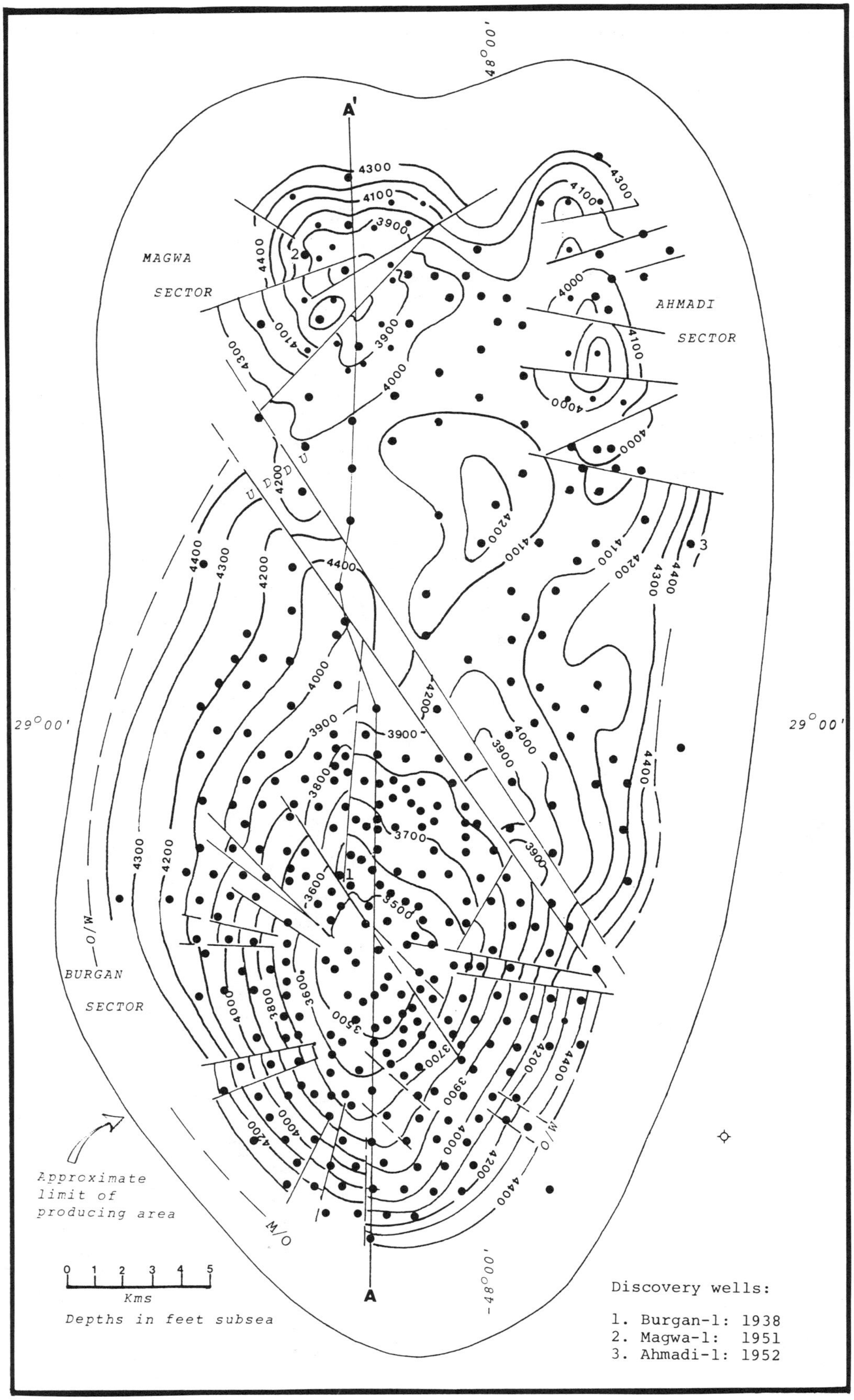

A'
A
MAGWA
SECTOR
AHMADI
SECTOR
BURGAN
SECTOR
48°00'
29°00'
29°00'
O/W
U
D
Approximate
limit of
producing area
0 1 2 3 4 5
Kms
Depths in feet subsea
Discovery wells:
1. Burgan-1: 1938
2. Magwa-1: 1951
3. Ahmadi-1: 1952

Figure 5. Greater Burgan field: structure on top of Mauddud Formation (modified from Adasani, 1965; Fox and Rollison, 1966).

PERIOD	EPOCH	GP	FORMATION	MEMBER	LITHOLOGY	EARLIER TERMINOLOGY
NEOGENE	MIOCENE-PLEIST	KUWAIT	KUWAIT SERIES		Continental sands, grits, gravels	KUWAIT SERIES
PALEOGENE	PALEOCENE-EOCENE	HASA	DAMMAM		Dolomitized nummulitic limestone; chert cap	EOCENE LIMESTONE
			RUS		Massive anhydrite, minor limestone and marl	EOCENE ANHYDRITE
			RADHUMA		Limestone, dolomitic; minor anhydrite	*Rotalia-Lockhartia* MARKER
CRETACEOUS	LATE	ARUMA	TAYARAT		Dolomitized reefal limestone, black shale	FIRST MAESTRICHTIAN SHALE MAESTRICHTIAN LIMESTONE
			BAHRAH		Limestone, cherty, part oolitic and detrital	
			GUDAIR		Limestone, detrital	SENONIAN LIMESTONE
		WASIA	MAGWA	MISHRIF	Limestone, dense, pyritic	
				RUMAILA	Limestone, dense, pyritic shaly towards base	
			AHMADI		Shale, green, brown and gray; minor limestone	CAP ROCK SHALES
			WARA		Sandstone, fine-grained; siltstone: shale	*Cythereis bahreini* LIMESTONE FIRST BURGAN SAND SECOND BURGAN SAND
			MAUDDUD		Limestone, organic, pseudo-oolitic, detrital	*Orbitolina concava* LIMESTONE
			BURGAN	THIRD SAND	Sand and sandstone, med-coarse-grained, shale	THIRD BURGAN SAND
				FOURTH SAND	Sand and sandstone as above, clean, massive	FOURTH BURGAN SAND
	EARLY	THAMAMA	SHUAIBA		Dolomitized rudist limestone, porous, cavernous	*Orbitolina discoidea* LIMESTONE
			ZUBAIR		Sandstone, fine-grained, interbedded with shale	ZUBAIR SANDS
			RATAWI		Shale, greenish-black, massive; minor limestone	RATAWI SHALE
				FUWARIS	Limestone, dolomitic, in part detrital	RATAWI LIMESTONE
			RAUDHATAIN	MINAGISH	Limestone, sub-oolitic to oolitic, porous	RATAWI OOLITE
				JAIDAN	Limestone, dense, finely-crystalline	
JURASSIC	LATE		GOTNIA		Anhydrite and halite; minor bituminous shale	ZEKRIT EVAPORITE SERIES
			NAJMAH		Limestone, oolitic	
	MIDDLE		SARGELU		Limestone, dense, oolitic; ostracod-rich gray shale	*Modiolus (Inoperna) plicatus* ZONE
	EARLY		ALAN	MARRAT	Limestone and anhydrite, nodular and interbedded	*Modiolus (Inoperna) aff. scalprum* ZONE
			MUS	MARRAT	Limestone, oolitic, part detrital	
			ADAIYAH	MARRAT	Limestone, dolomitic, interbedded with anhydrite	
			BUTMAH		Limestone, marl, shale, sandstone	

Regional unconformity

Minor unconformity or disconformity

Figure 6. Greater Burgan field: Mesozoic-Cenozoic stratigraphy.

The earlier of the two major unconformities reflects a period of uplift and erosion at the close of Turonian time. From the common occurrence of ooliths composed of hydrous ferric oxides, it has been proposed that the erosion took place under water but close to wave base (Owen and Nasr, 1958). This unconformity profoundly affects strata of the Albian to Turonian Wasia Group and separates the Wasia from the overlying Aruma Group of Coniacian to Maastrichtian age. The "post-Wasia unconformity" is an important horizon in the subsurface of all structural trends in the area, with much of the Cenomanian-Turonian section commonly eroded from structural crests. Across the crest of the Burgan dome, the thickness of the critical cap rock section (Cenomanian Ahmadi Formation), immediately overlying the first Cretaceous oil reservoir, has been reduced to a mere 25 m (80 ft) from its typical flank development of 75 m (250 ft), although some part of this thinning may have been depositional.

The second major erosional hiatus occurred during Paleogene time, when uplift on a regional scale terminated marine deposition along the eastern edge of the Arabian platform, initiating a cycle of erosion that would occupy late Eocene and all of Oligocene time. The resulting unconformity, cut into carbonates of middle Eocene age, is of regional significance along much of the present-day western shore of the Gulf. Regional tilting toward the north and northeast occurred at this time; post-unconformity Neogene deposits thicken northward from a mere veneer in southeastern Kuwait (6 m, 20 ft, across the crest of the Ahmadi structure) to a maximum of 850 m (2800 ft) in southern Iraq. The Eocene unconformity has been locally rejuvenated by post-Eocene growth of structures along the Kuwait axis, and the silicified

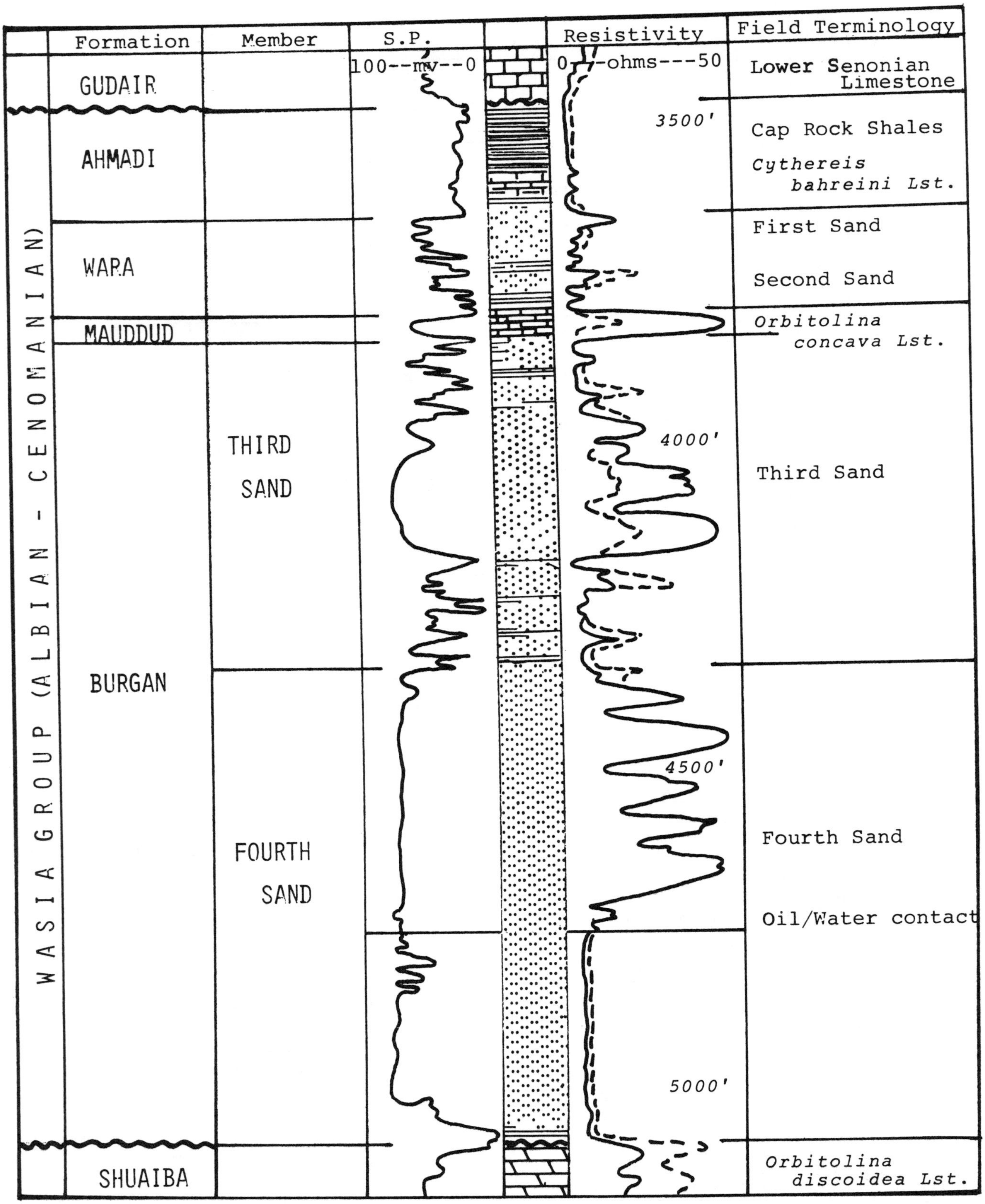
Formation
Member
S.P.
Resistivity
Field Terminology
100--mv--0
0--ohms---50
GUDAIR
AHMADI
WARA
MAUDDUD
BURGAN
SHUAIBA
THIRD SAND
FOURTH SAND
WASIA GROUP (ALBIAN - CENOMANIAN)
3500'
4000'
4500'
5000'
Lower Senonian Limestone
Cap Rock Shales
Cythereis bahreini Lst.
First Sand
Second Sand
Orbitolina concava Lst.
Third Sand
Fourth Sand
Oil/Water contact
Orbitolina discoidea Lst.

Figure 7. Greater Burgan field: Type log, Wasia Group.

erosional surface reflects deeper structure in modified form. The Greater Burgan structure continued to grow independently throughout Neogene time; the post-unconformity Kuwait Group (Miocene to Pleistocene continental clastics) shows thinning toward the structural crests in common with older stratigraphic units. Indeed, both the Burgan and Magwa domes are subtly expressed at the present-day land surface, while the Ahmadi structure is more boldly expressed as a topographic ridge.

The Paleogene uplift that terminated marine deposition in Kuwait was a precursor of the much more intense Zagros tectonic episode, which occupied much of Miocene and all of Pliocene time and which created the northwest–southeast-trending fold belt forming the tectonized northeastern margin of the Middle East basin (Ala, 1982). The effects of the Zagros orogeny along the Arabian margin were limited by distance. Within the Greater Burgan field area, renewed vertical uplift of the Burgan and Magwa domes may have been stimulated, while in the eastern part of the structural complex, post-Eocene folding apparently played a significant role in the shaping of the Ahmadi structure. The shape of the Ahmadi fold, narrow and linear in contrast to the rounded outlines of Burgan and Magwa (Figure 4), seems at least suggestive of compressional force. The effects of the post-Wasia unconformity are less severe at Ahmadi than at Magwa, suggesting less growth on the Ahmadi high prior to the end of Turonian time. On the other hand, the Ahmadi structure is overlain by the thinnest recorded development of Kuwait Group (near-surface) clastics, indicating substantial uplift during post-unconformity, Neogene, time.

During much of Late Cretaceous time, the Kuwait area underwent general subsidence; although growth persisted at Burgan and Magwa something approaching a balance between structural uplift and sedimentary deposition resulted in a thinner sedimentary section while creating relatively few hiatuses in the sequence. This pattern persisted into Paleogene time, until the uplift and subsequent erosion that created the widespread middle Eocene unconformity along the margin of the Arabian platform. In the basinal area further toward the northeast, subsidence and sedimentation persisted until interrupted by the Mio-Pliocene Zagros orogeny. This Neogene tectonism contributed to the structural growth of the Ahmadi sector of the Greater Burgan structural complex. The age of the folding imposes a constraint upon the time of migration of crude oil into the Greater Burgan structural trap.

STRATIGRAPHY

At the time of the Burgan oil discovery, no regional stratigraphic framework was yet in place for the eastern part of the Arabian peninsula. Only a handful of control wells had been drilled within the region; the mapping of the outcrop belts in Sa'udi Arabia was not yet completed; and the relationship of the surface exposures to the subsurface formations penetrated in wells drilled further to the east was still being investigated.

The entire State of Kuwait was blanketed by flat-lying Neogene deposits, so that no relevant outcrops were available to guide the early drilling. As a consequence, the initial terminology for the subsurface formations encountered at Burgan was informal, including such terms as "Cap Rock shales" and "Burgan sands" (Figures 6 and 7). By 1955, a stratigraphic nomenclature for Kuwait and southern Iraq had been formalized; this was published by Owen and Nasr (1958) and refined by Adasani (1965).

Published estimates of the total thickness of sedimentary section underlying the Greater Burgan area range from 7280 m (23,900 ft) (Adasani, 1965) to 8530 m (28,000 ft) (Kamen-Kaye, 1970). Recent deep drilling to the Permian Khuff Formation, at depths of 5930 to 6780 m (19,470 to 22,235 ft), suggests that the total sedimentary section present may exceed 9150 m (30,000 ft). The oldest sediments present at depth are probably the salt-bearing Hormuz Series, variously described as of late Proterozoic, Infracambrian, or Infracambrian to Cambrian age (Ala, 1974; Murris, 1980; AlSharhan and Kendall, 1986). Salt intrusions of Hormuz origin appear to underlie producing structures in eastern Sa'udi Arabia, Qatar, Bahrain, and southern Iraq (Murris, 1980; AlSharhan and Kendall, 1986; Wellings, in Owen, 1975). Available data suggest that Hormuz-age salt may have played a significant role in the structural evolution of the Greater Burgan field (Fox, 1959; Murris, 1980).

A clastic section ranging in age from Cambrian to Permian–Carboniferous may be present above the Hormuz Series. Along the edge of the Arabian platform, this Paleozoic clastic assemblage is transgressively overlain by shallow-water carbonates and evaporites of the Khuff Formation of Upper Permian (late Kungurian to mid-Tatarian) age (Aramco, 1975). This Permian section has been penetrated in the deepest wells drilled to date in the Greater Burgan field.

Carbonates and evaporites, alternating with nonmarine clastics, were laid down during Triassic time. Coarse clastics persisted into the Early Jurassic, to be succeeded by 910 m (3000 ft) of limestone with subordinate anhydrite and shale, including the Marrat Formation of probable Toarcian age (Aramco, 1975) and the Sargelu Formation of Bajocian–Bathonian age. The Middle–Late Jurassic contact at Greater Burgan is unconformable (Adasani, 1965). The Late Jurassic section is developed in the "Gotnia facies" characteristic of the North Arabian Gulf basin and comprises a basal unit of detrital to oolitic limestone, overlain by a thick 410 m (1366 ft) section of interbedded halite and anhydrite with occasional thin bituminous limestones. A unit of anhydrite 75 m (250 ft) thick occurs at the top of the evaporite section, and this may correlate with the Hith anhydrite of Sa'udi Arabian terminology. The evaporites of the Gotnia Formation were assigned a probable Callovian to Tithonian age by Jaber (1975) and a post-lower Kimmeridgian age by AlSharhan and Kendall (1986). The contact between the Late Jurassic Gotnia evaporites and the overlying limestones of the lower part of the Early Cretaceous Thamama Group appears to be essentially conformable in the Greater Burgan area.

The Cretaceous section at Greater Burgan has a drilled thickness of 1890 m (6200 ft) and falls naturally into the threefold division common to eastern Arabia. In ascending order, the Cretaceous formations are organized into the Thamama, Wasia, and Aruma groups, separated by one minor and one major unconformity and corresponding to deposits of Neocomian to Aptian, Albian to Turonian, and Coniacian through Maastrichtian age (Figure 6).

The Thamama Group includes, in ascending order, the Raudhatain, Ratawi, Zubair, and Shuaiba formations. The Raudhatain Formation, originally known as the "Ratawi limestone," consists of 520 m (1700 ft) of neritic limestones of probable Berriasian-Valanginian age, marking a return to more normal marine conditions at the close of Jurassic time. Adasani (1965) has divided the Raudhatain into three members, of which the middle member—the Minagish Oolite Member—is an important exploration target in the Kuwait area. The Raudhatain Formation is overlain by the shales of the Ratawi Formation, which are in turn overlain by the thick sands of the Zubair Formation, of lower Aptian age and an important oil-producing horizon in northern Kuwait and southern Iraq.

The Zubair sands are overlain with apparent conformity by a limestone unit, the Shuaiba Formation, indicating an abrupt, if brief, return to marine conditions at the close of Aptian time. The Shuaiba is 60 m (200 ft) thick at Greater Burgan and is highly porous, even cavernous, in its upper part, suggesting the weathering of an exposed surface. The formation does not produce at Greater Burgan but is immediately overlain by the Burgan Formation, oldest formation of the Albian to Turonian Wasia Group and lowermost unit of the Albian to Cenomanian oil-producing interval in the Greater Burgan field. The Shuaiba–Burgan contact has been considered conformable (Owen and Nasr, 1958), but the indicated subaerial weathering of the Shuaiba surface suggests at least a disconformity. The contact is clearly unconformable further to the south, in Sa'udi Arabia (Aramco, 1975).

The Wasia Group reaches a total thickness of 760 m (2500 ft) at Greater Burgan and comprises, in ascending order, the Burgan, Mauddud, Wara, Ahmadi, and Magwa formations. The Burgan and Wara formations comprise the late Early to early Late Cretaceous oil reservoirs of the Greater Burgan field and are treated at greater length in the *Reservoir* section. The Mauddud Formation, lying between the Burgan and Wara Formations and representing a brief return to shallow-marine conditions at the end of Albian time, was assigned a Cenomanian age by Owen and Nasr (1958) but is regarded as late Albian by Loutfi and Jaber (1970) and AlShamlan (1975).

The Ahmadi Formation lies conformably above the oil-producing Wara Formation. Originally termed the "Cap Rock shales," the Ahmadi is composed predominantly of shales, gray in the lower part, green and brown in the upper. Thin stringers of marly limestone occur toward the base of the formation, typically rich in the ostracod *Cythereis bahreini* and locally known as the "CB limestone." In its full flank development at Greater Burgan, the Ahmadi Formation is 76 m (250 ft) thick, but this thickness has been reduced by post-Wasia erosion (and possibly by depositional thinning) to only 25 m (80 ft) across the crest of the Burgan dome. For obvious reasons, the top of the Ahmadi Formation was a critical subsurface marker during the early phase of drilling in the crestal area.

Across the crest of the Burgan dome the Ahmadi Formation is overlain with marked unconformity by carbonates of the Coniacian–Campanian Gudair Formation, oldest unit of the Late Cretaceous Aruma Group. Off-structure, the Ahmadi thickens to its full development and is overlain by limestone with minor shale of the Magwa Formation, of Cenomanian to Turonian age and the youngest formation of the Wasia Group. The Magwa Formation has been eroded from the crest of the Burgan dome, is present in attenuated form across the crest of the Magwa dome, and is considerably thicker across the crest of the Ahmadi structure, providing a useful measure of structural growth in the three main sectors of the Greater Burgan field during early Late Cretaceous time. The Magwa Formation at Greater Burgan can be subdivided into two members, and these have been correlated with the Rumaila and Mishrif formations of southern Iraq (El-Naggar and Al-Rifaiy, 1972, 1973).

The Late Cretaceous Aruma Group, overlying the post-Wasia unconformity, varies in thickness from 260 m (850 ft) at crestal to almost 460 m (1500 ft) at flank locations across the Greater Burgan structure. The Aruma Group consists predominantly of shallow-water limestones and is divided into three formations of which the lowest, the Senonian–Campanian Gudair, is detrital and varies widely in thickness from positive to negative areas of the structural complex, reflecting a continuation of the structural growth so apparent during late Early Cretaceous to early Late Cretaceous time. The Gudair is overlain by the Maastrichtian Bahrah Formation, of which the detrital to oolitic lower member also varies erratically in thickness, while the upper member is of much more uniform thickness, suggesting reduced structural growth. The Bahrah is conformably overlain by shallow-marine, possibly reefoid carbonates of the late Maastrichtian Tayarat Formation, youngest formation of the Aruma Group (Owen and Nasr, 1958).

The late Maastrichtian Tayarat Formation is conformably overlain by shallow-water carbonates of the Radhuma Formation, lowermost unit of the Paleocene to Eocene Hasa Group; the Radhuma is in turn conformably overlain by the massive anhydrites of the Rus Formation, of inferred middle Eocene age. Both formations show stratigraphic thinning across the Greater Burgan structure, from a total thickness of 680 m (2250 ft) at flank to 480 m (1600 ft) at crestal locations, indicating continued structural growth during early Paleogene time. The Rus evaporites are disconformably overlain by nummulitic limestones of the Lutetian Dammam Formation, 180-240 m (600-800 ft) thick across the Greater Burgan structure and uppermost unit of the Hasa Group. Marine deposition in the area was terminated by regional uplift at the close of middle Eocene time (Aramco, 1975), initiating a period of erosion that would persist through late Eocene and all of Oligocene time. The resulting unconformity is of regional significance along the western side of the Persian (Arabian) Gulf.

The eroded and silicified surface of Dammam Formation limestone is directly overlain by a section of continental sands and gravels that extends upward to the present-day land surface of the Greater Burgan area. These post-unconformity continental clastics are of inferred Miocene-Pleistocene age and are assigned to the Kuwait Group of Owen and Nasr (1958) and the Kuwait Series of Gregory (1929) and Cox (1932). The Kuwait Group thins across the crestal areas of the Greater Burgan structure to a minimum of 6 m (20 ft) at Ahmadi, where the underlying middle Eocene erosional surface was exposed in a quarry during 1965 (Fuchs et al., 1968).

TRAP

The hydrocarbon trap at Greater Burgan field is structural, a domal uplift of approximately oval shape with the long axis oriented almost north-south, measuring 35 by 20 km (23 by 13 mi). Superimposed on this large uplift are three culminations: the Burgan dome, which forms the main body of the field; the Magwa dome to the north; and the more linear Ahmadi structure to the east of Burgan-Magwa (Figure 5).

Regional dip in the area is toward the north, so that critical closure at Greater Burgan is at the southern, regionally updip, end. In spite of the great structural relief present, rates of dip on all flanks of the Greater Burgan structure are uniformly low, in the order of 2° to 3° (Adasani, 1965; Milton, 1967). During the early stage of field development, these low flank dips caused speculation that structural closure toward the south might prove insufficient to contain the thick oil column already proven in the crestal area of Burgan dome; that the critical southern closure might prove partially stratigraphic. Such was not the case; in spite of the low rates of dip, the structure was so large that all flanks plunged away beyond their intersection with the oil-water contact, an essentially plane surface cutting across the Cretaceous structure at a subsea elevation of 1360 m (4466 ft).

The Burgan dome has experienced the greatest degree of structural growth; the Albian-Cenomanian producing section has a closure of 370 m (1221 ft) at the level of the original oil-water contact at Burgan, 270 m (881 ft) at the Magwa dome, and 240 m (796 ft) at the Ahmadi culmination (Adasani, 1965). The vertical and lateral seal for the Albian-Cenomanian hydrocarbon accumulation at Greater Burgan is the Ahmadi Formation (originally "Cap Rock shales"), a section of shales with minor limestones up to 75 m (250 ft) thick at flank locations but reduced by erosion and/or depositional thinning across the Burgan structural crest to 25 m (80 ft), "little more than a membrane in relation to the size of the structure and its petroleum reserves" (Kent and Warman, 1972).

RESERVOIR

Fox (1959) defined the Albian-Cenomanian oil accumulation at Greater Burgan as comprising a common reservoir of very large area but relatively small depth, upon which are superimposed three subsidiary reservoir structures isolated from one another. Adasani (1965) confirmed the existence of the three separate subsidiary structures—Burgan, Magwa, Ahmadi—and defined the field as possessing three major pay zones: Wara (originally the First and Second sands), Burgan Third, and Burgan Fourth sands.

Although the three subsidiary reservoir structures share a common oil column and a common oil-water contact, some variations occur within the field-wide hydrocarbon system. The crude oil at the Burgan dome, structurally highest of the three culminations, was originally undersaturated, while at the structurally lower Magwa and Ahmadi crests, small primary gas caps were present in the uppermost (Wara) reservoir. The gas caps at Magwa and Ahmadi were of comparable thickness, but the gas-oil contact was lower by 20 m (70 ft) at Ahmadi, due to the lesser structural amplitude of the Ahmadi fold. The crude oil immediately below the gas cap at Ahmadi was of slightly higher API gravity than that at Magwa, while the crude oil below the gas cap at Magwa was of slightly higher API gravity than the undersaturated oil at the Burgan crest. This variation in API gravity was confined to the highest levels of the three reservoir culminations. Fox (1959) suggested that the variations in the distribution of oil and gas between the three structures might have been a function of differential entrapment during migration.

The Wara Formation, containing the highest producing sands, is continuous across the entire area of the Greater Burgan structure but is also the weakest and most isolated of the reservoirs. The Third and Fourth sand reservoirs, both members of the Burgan Formation, comprise a single mammoth reservoir (Adasani, 1965), acting together in reservoir performance and containing the bulk of the oil reserve. The Third Sand Member reservoir actually comprises three sections, readily distinguishable on borehole logs (Figure 7). The middle unit, composed of almost pure quartz sand, is the major producing section of the Third Sand Member. The Fourth Sand Member reservoir is a thick body of pure quartz sand, comparable to the middle unit of the Third Sand reservoir.

Because of its great areal extent and structural relief, the Burgan dome has a thick and extensive development of Third Sand–Fourth Sand above water (Figure 8). The top of the Third Sand at the Burgan dome crest is 310 m (1016 ft) and the top of the Fourth sand 165 m (540 ft) above the field oil-water contact. Because of the successively lesser structural amplitudes of the Magwa and Ahmadi crests, lesser sections of these Burgan Formation reservoir units are above the oil-water contact. At Magwa, the top of the Third Sand is 200 m (678 ft) and the top of the Fourth Sand only 60 m (200 ft) above water. At Ahmadi, the top of the Third Sand is 180 m (586 ft) and the top of the Fourth Sand only 30 m (100 ft) above water. Because of these structural conditions, and because the quality of the reservoir crude oil deteriorates as the oil-water contact is approached, the Fourth Sand Member reservoir has yielded relatively little production in the Magwa and Ahmadi sectors of Greater Burgan field.

The structural relationships of these producing zones in the Burgan, Magwa, and Ahmadi sectors of the field are given in Table 1. The figures refer to locations near the structural crests of the Greater Burgan field, and all elevations are subsea. Even at the highest structural point of the field, the base of the Fourth Sand Member lies well below the field oil-water contact.

Reservoir Stratigraphy and Characteristics

The Albian-Cenomanian producing interval at Greater Burgan field forms part of a clastic section 410 m (1350 ft) in thickness, containing one minor interval of carbonate. Of this total section, 91%—370 m (1221 ft)—is above bottom water at the crest of the Burgan dome, the largest and structurally highest of the three culminations of the Greater Burgan structure (Figure 7).

The entire reservoir section is of shallow-water origin, littoral to deltaic in the lower part, littoral to lagoonal in the upper part. The clastic series appears to have been deposited in a gently subsiding shelf environment between an active basinal area to the northeast and the stable massif of the Arabian shield with its rim of Paleozoic sediments to the southwest. The sands in the lower part of the reservoir section become progressively coarser in grain toward the west and southwest, pointing to a source in that direction. Toward the north and northeast, the sands grade basinward into marly and shaly sections (Owen and Nasr, 1958; Fox, 1959; Milton and Davies, 1965; James and Wynd, 1965; Aramco, 1975).

The sands show good to excellent sorting, particularly in the lower part of the section. Combined with an overall lack of heavy minerals, this suggests that they may be second generation sands. The presence of glauconite, amber and other resins, lignite, and occasional plant remains indicates a shallow, near-shore environment of deposition. The associated shales also carry resins, lignite, and plant remains, together with a sparse micro- and macrofauna.

The upper part of the clastic section, the Wara Formation, is of early Late Cretaceous (Cenomanian) age. It is underlain by a thin unit of limestone (Mauddud Formation, originally "Orbitolina concava limestone"), which represents a minor marine incursion at the close of late Early Cretaceous (Albian) time. The entire clastic series overlies a carbonate unit of Early Cretaceous (Aptian) age, the Shuaiba Formation. The lower part of the clastic assemblage, between the top of the Shuaiba and the base of the Mauddud, has been assigned an inferred late Early Cretaceous (Albian) age (Owen and Nasr, 1958; Loutfi and Jaber, 1970). This lower part of the clastic section comprises the Burgan Formation (Figure 7).

The Burgan Formation is 360 m (1179 ft) thick in its type section at Greater Burgan field, with 80% of this thickness being sand and sandstone of excellent reservoir quality. Some 310 m (1015 ft) of the formation lie above water at the crest of the Burgan dome. The Burgan Formation is divided into two members, of which the lower, the Fourth Sand Member, immediately overlies the leached surface of Shuaiba carbonate (Figure 7).

The Fourth Sand Member consists of a thin (6 m, 20 ft) basal shaly unit, succeeded by 210 m (700 ft) of pure quartz sandstone, medium- to coarse-grained, well-sorted, very clean and soft, containing little in the way of secondary cementation. This sand section is so unconsolidated that recovery of core material for core analysis presented serious problems prior to the advent of diamond coring. The Fourth Sand is an excellent petroleum reservoir; its porosity averages 23%, permeabilities commonly exceed 4000

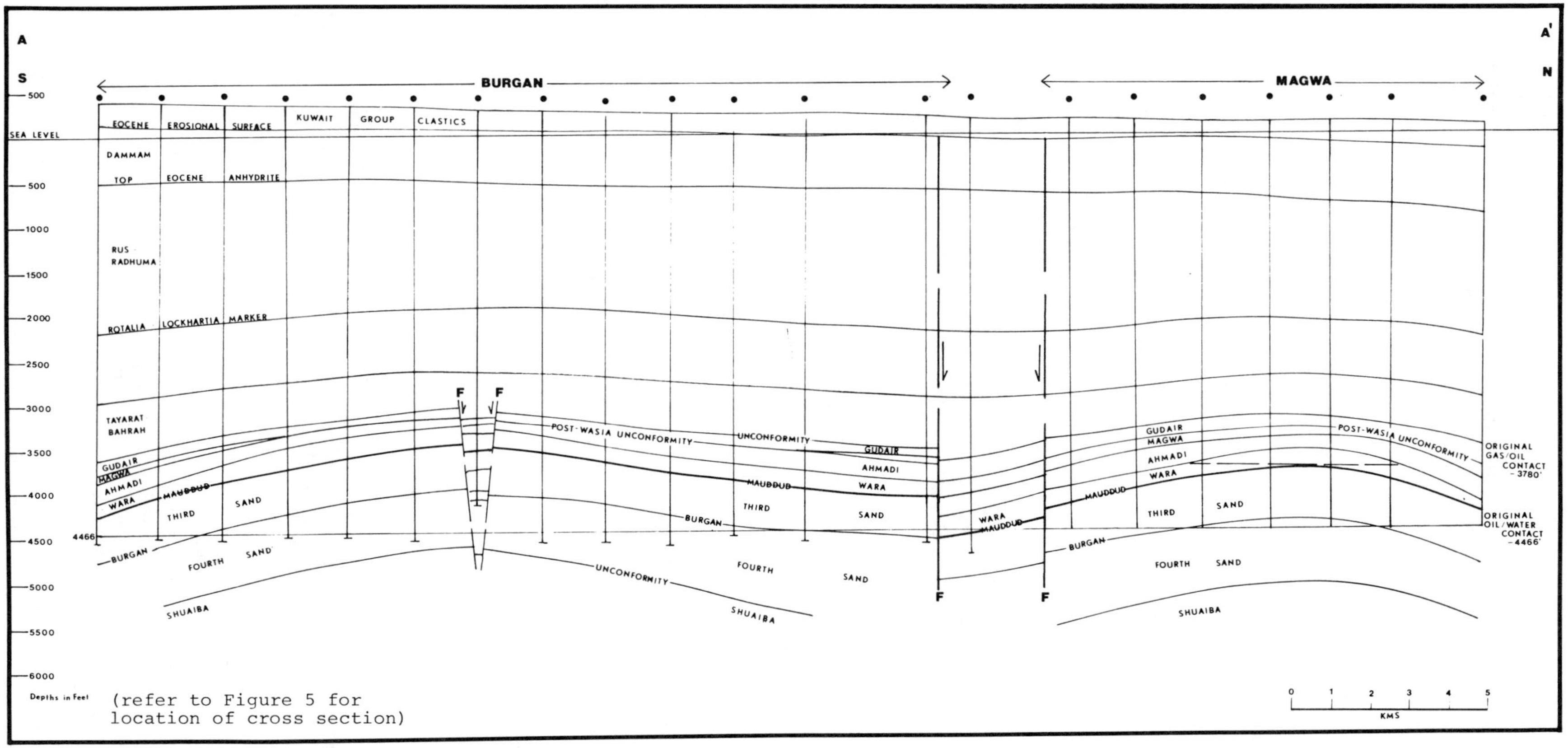
A
A'
S
N
BURGAN
MAGWA
SEA LEVEL
500
500
1000
1500
2000
2500
3000
3500
4000
4500
5000
5500
6000
Depths in Feet
4466
EOCENE EROSIONAL SURFACE
KUWAIT GROUP CLASTICS
DAMMAM
TOP EOCENE ANHYDRITE
RUS RADHUMA
ROTALIA LOCKHARTIA MARKER
TAYARAT BAHRAH
GUDAIR
MAGWA
AHMADI
WARA
MAUDDUD
THIRD SAND
BURGAN
FOURTH SAND
SHUAIBA
UNCONFORMITY
POST-WASIA UNCONFORMITY
F
ORIGINAL GAS/OIL CONTACT -3780'
ORIGINAL OIL/WATER CONTACT -4466'
0 1 2 3 4 5
KMS
(refer to Figure 5 for location of cross section)

Figure 8. Greater Burgan field: north-south structural cross-section.

md, and interstitial water (irreducible water saturation, S_{wi}) is as low as 4%. All of the Fourth Sand Member section above water is comprised of net pay.

The overlying Third Sand Member of the Burgan Formation contains three subdivisions (Figure 7), of which the lower and upper consist of interbedded glauconitic sands and dark gray shales and the middle unit of pure quartz sand similar to that of the Fourth Sand Member. The entire Third Sand Member is 145 m (475 ft) thick, with the middle, pure-sand section averaging 75 m (250 ft) in thickness. Porosity, permeability, and interstitial water values in the middle unit of the Third Sand Member are similar to those already quoted for the Fourth Sand Member. The interbedded glauconitic sands of the lower unit of the Third Sand Member show porosity in excess of 20%, permeability in the range of 600 md, and interstitial water saturation of about 25%. Porosities remain above 20% in the glauconitic sands of the upper unit of the Third Sand Member, with permeability averaging 380 md and interstitial water, 31%. Net pay within the Third Sand Member may reach 110 m (350 ft).

Throughout the Greater Burgan field area, the Burgan Formation is conformably overlain by the thin limestone of the Mauddud Formation. The Mauddud is a compact, even dense, calcarenitic, locally detrital and pseudo-oolitic limestone 5 to 10 m (20 to 30 ft) thick, containing abundant microfossils. Although it is commonly oil stained, the Mauddud produces only locally at Greater Burgan and is not considered a significant reservoir, because of generally low permeability (average 10–15 md).

The Mauddud carbonate separates the extremely powerful reservoir of the combined Third and Fourth sands (Burgan Formation) from the much weaker reservoir of the overlying Wara Formation. The oil column appears to be continuous across all the reservoirs, and the oil-water contact is certainly common, but the Wara reservoir is partially isolated by its own basal shales and the underlying Mauddud limestone. The Wara reservoir exhibits producing characteristics distinct from those of the deeper Burgan Formation reservoir.

The sands of the Wara Formation are quartzose but are finer grained and less well sorted than those of the Burgan Formation and are associated with greater amounts of finer-grained clastics, siltstones and shales. The lower part of the Wara consists of gray, glauconitic and lignitic shales that enclose a single, prominent, lenticular body of fine-grained glauconitic sand, the so-called Second sand of the original oil field terminology (Figure 7). The upper part of the Wara (originally "First sand") consists of interbedded fine-grained sandstones, siltstones, and gray shales, with glauconite, amber, and lignite. The Wara Formation ranges in thickness from 40 m (140 ft) to 50 m (180 ft) across the Greater Burgan structure. Up to 60% of the total thickness comprises reservoir sand. Porosities in the cleaner sand sections of the Wara reservoir reach 30% and average about 24%. Permeabilities vary in magnitude from a few to thousands of millidarcies, reflecting marked variations in rock type within the formation. Interstitial water saturations vary widely, with an average in the range of 18–20%.

The Wara Formation was assigned a Cenomanian age by Owen and Nasr (1958) and a lower Cenomanian age by Loutfi and Jaber (1970). It is conformably overlain by the gray shales with minor limestone of the Ahmadi Formation, also of Cenomanian age, which forms the seal for the Greater Burgan field hydrocarbon accumulation.

Oil and Field Characteristics

Crude oil produced from the Wara Formation–Burgan Formation reservoirs at Greater Burgan field is of asphaltic-paraffinic mixed base; average API gravity approximately 31.8°; sulfur content 2.5 wt%; Saybolt universal viscosity 52 sec at 100°F; pour point 0°F. There is a geothermal gradient of 2°F/100 ft across the oil column, with a temperature of 145°F at the oil-water contact (Adasani, 1965). The original reservoir pressure was 2075 psig @ 1220 m (4000 ft).

The average API gravity of the field production is misleading. The slight variation in API gravity between the crestal areas of the three reservoir culminations has been noted. In addition, the crude oil within each reservoir shows a decrease in API gravity and solution gas-oil ratio, and an increase in sulfur content and viscosity, with increasing depth. At the crest of the Wara reservoir, API gravity is 36°; at a subsea elevation of 1340 m (4400 ft) in the same reservoir, the API gravity has fallen to 29°. In the Third Sand Member reservoir (Burgan Formation), the variation is 34° to 28° API; in the Fourth Sand Member reservoir, from 32° to 29° API. Below 1340 m (4400 ft) subsea, the crude oil becomes heavy, viscous, and undersaturated. A tar mat is present at the oil-water contact, and pockets of tarry crude occur below this, within the bottom-water zone. The bottom water of the field is extremely saline, averaging 150,000 ppm/total dissolved solids.

The Wara reservoir at the Burgan dome went on production during 1946 at individual well rates of 800–1300 BOPD, producing GOR 528 MCFG/bbl. A decade later, the reservoir was yielding 100,000 BOPD from 86 wells but was also experiencing serious pressure decline, indicating no active water drive. Pressure maintenance by gas injection was initiated during 1961, eventually reaching 100 mmcf/day and resulting in a marked improvement in cumulative production per pound of pressure drop (Adasani, 1965). The Wara reservoir at Magwa/Ahmadi went on stream during 1953, at comparable well rates but higher GORs, reflecting the presence of free gas at the crests of those subsidiary reservoir culminations. During its first 20 years of productive life, the Wara reservoir yielded 538 million bbl.

It became apparent early that a large part of the field's reserve of oil was contained within the Third Sand Member reservoir. By the early 1960s, this reservoir was yielding more than 1 million BOPD from 250 wells, average per-well rate 4877 BOPD, producing GOR 461 MCFG/bbl. Cumulative production from this reservoir during its first 20 years was 5.25 billion bbl. Because of reduced structural amplitude at Magwa and Ahmadi, the productive thickness and limits proved more restricted in those sectors of the field. This applied also to the Fourth Sand Member reservoir, which had a substantial section above water in the Burgan dome, far less in the Magwa and Ahmadi culminations (Adasani, 1965). By the mid-1960s, the Fourth Sand reservoir had produced just under 1 billion bbl of oil, almost all of this production coming from the Burgan dome.

The presence of a strong and efficient water drive in the Third Sand/Fourth Sand reservoir at the Burgan dome became apparent as the initial pressure decline leveled off and stabilized (Lack, 1959). Oil production from the Third Sand Member was 65 million bbl per psi drop in reservoir pressure, and for the Fourth Sand Member, 8 million bbl per psi drop (Adasani, 1965). In spite of indicated fault separation at deeper reservoir levels (Figures 5 and 8), the active water drive in the Third Sand/Fourth Sand reservoir was confirmed by pressure behavior at Magwa and Ahmadi. This suggested that the producing Albian (Burgan Formation) sands at Greater Burgan formed part of a very widespread subsurface aquifer.

Origin and Migration of the Oil

The recoverable oil reserve at Greater Burgan appears to be at least 75 billion bbl, the greater part contained within sand reservoirs of Albian to Cenomanian age. Recovery factors of one-third to one-half have been assumed; the original oil in place may therefore have amounted to between 150 and 225 billion bbl.

The Greater Burgan structure is the largest of several individual culminations developed along a structural ridge that plunges north through Kuwait into southern Iraq (Figure 4). Other oil fields along this 240 km (150 mi) trend include, from south to north, Fuwaris, Wafra, Umm Gudair, Minagish, Bahrah, Sabriyah, Raudhatain, Rumaila, and Zubair. None of these fields approaches Greater Burgan in size, but all carry oil in Cretaceous reservoirs, oil of similar fundamental composition possibly suggestive of a common origin (Strong, 1959; Kent and Warman, 1972). Total recoverable Cretaceous reserves along the Kuwait axis will certainly exceed 120 billion bbl, suggesting an original oil in place of 240-360 billion bbl. The generation and accumulation of so large a volume of liquid hydrocarbons must be of interest to all petroleum geologists.

Ayres et al. (1982) concluded that the crude oil present in Late Jurassic (Tithonian) Arab Formation reservoirs in Sa'udi Arabia had its origin in the deeper Dhruma-Tuwaiq Mountain formations of Bajocian to Oxfordian age. The Dhruma-Tuwaiq Mountain section is equivalent in age to the Sargelu Formation of the northern Gulf area. At its type section, the Sargelu is described as a bituminous, phosphatic marl; Murris (1980) regarded the euxinic facies of the Sargelu (laminated bituminous lime muds and marls, laid down during a major southward expansion of the Lurestan basin during late Oxfordian time) as a major Jurassic source in the northern Gulf. These deposits could certainly have provided a source for oil in the Early Jurassic (Marrat Formation) reservoir at Greater Burgan. Brennan (1983) discounted this Jurassic (Sargelu) source for oil in the Cretaceous reservoirs, because of the intervening thick development of Late Jurassic Gotnia Formation halite and anhydrite.

Ayres et al. (1982) also suggested that the oil in late Early to early Late Cretaceous (Albian to Cenomanian) reservoirs in northern Sa'udi Arabia had its origin in source rocks of Early Cretaceous Berriasian-Valanginian age. In southeastern Kuwait, it is reasonable that these sediments might have provided a source for the oil in the "Minagish oolite zone," which is of similar Early Cretaceous age. The Berriasian-Valanginian sediments are less satisfactory as a possible source for the oil in the main Albian to Cenomanian reservoirs at Greater Burgan, owing to the presence of the thick and apparently competent shales of the Ratawi Formation, lying between the Minagish reservoir and the Albian to Cenomanian clastics (Burgan and Wara formations).

The Cretaceous depocenter lay northeast of Greater Burgan, and the clastics of the Burgan and Wara formations give way in that direction to deeper-water deposits, now exposed in surface section along the Zagros mountain front (Fox, 1959; Kent and Warman, 1972). These deeper-water deposits are the dark, bituminous shales and limestone of the Kazhdumi Formation, about 210 m (700 ft) thick at outcrop; the Kazhdumi is Albian to Cenomanian in age and is the shale equivalent of the Burgan-Nahr Umr section of Kuwait and southern Iraq (James and Wynd, 1965) (Figure 9). Ala (1982) identified the Kazhdumi as containing mature organics and as providing the source for the hydrocarbons in the Bangestan Group (Cretaceous) and Asmari Formation (Oligo-Miocene) reservoirs in southwestern Iran. Ayres et al. (1982) agreed that lateral migration from a Kazhdumi source could have charged the Albian clastic reservoirs in such northern Gulf oil fields as Safaniyah-Khafji, which occupy a geological setting comparable to that of Greater Burgan. The cross-section used by Ayres et al. (1982) to illustrate this source-trap relationship would apply equally well to the Greater Burgan field (Figure 10).

Migration of hydrocarbons from a northeasterly source might also explain, through a process of differential entrapment, the presence of free gas in the crests of the more easterly and northeasterly

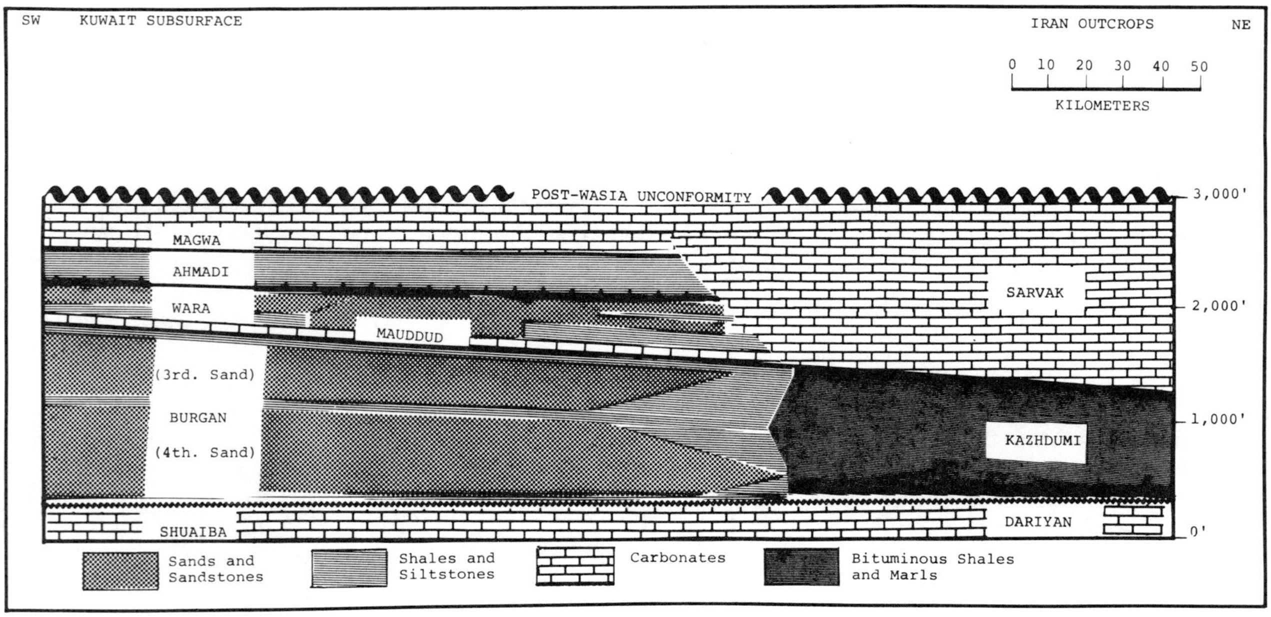
SW
KUWAIT SUBSURFACE
IRAN OUTCROPS
NE
0 10 20 30 40 50
KILOMETERS
POST-WASIA UNCONFORMITY
3,000'
2,000'
1,000'
0'
MAGWA
AHMADI
WARA
MAUDDUD
SARVAK
(3rd. Sand)
BURGAN
(4th. Sand)
KAZHDUMI
SHUAIBA
DARIYAN
Sands and Sandstones
Shales and Siltstones
Carbonates
Bituminous Shales and Marls

Figure 9. Stratigraphic relationship, Wara-Burgan formations (sand) and Kazhdumi Formation (shale), Kuwait-Iran area (Kuwait subsurface stratigraphy after Owen and Nasr, 1958; Iran outcrop data after James and Wynd, 1965).

culminations of the Greater Burgan structure, and the variation in oil gravities from crest to crest (Fox, 1959). In support of this, Strong (1959) pointed out that the well-known stratification of the crude oil in the common part of the oil column at Greater Burgan was tilted slightly toward the northeast, heavier crudes occurring further down-flank in that part of the structural complex.

Both Jurassic and Cretaceous oil sources were apparently present in the deeper parts of the North Arabian Gulf basin. The richness of the sources remains to be explained. Windley (1977) related this to the presence, throughout Early Cretaceous time, of a narrow Tethyan seaway between Laurasia and Gondwanaland, cutting across what is now Iran and including the area of the present Persian Gulf. There were extensive marine transgressions during this period, accompanied by a dramatic increase in the numbers and variety of marine organisms, particularly plankton. This favored the deposition of rocks rich in organic components.

Windley (1977) suggested that 60% of all known oil reserves may have been generated during the Albian through Turonian interval of the Cretaceous Period, between 110 and 80 Ma. In the case of the North Arabian Gulf basin, it should be noted that, in addition to the large reserves contained in Cretaceous reservoirs, the oil in Tertiary reservoirs along the Iran foothills belt is widely held to be largely of Cretaceous origin (Kent and Warman, 1972). However, deposition of even the richest of source rocks does not result in the immediate generation of hydrocarbons; maturation of the organic content is required, and this is normally a function of increasing depth of burial of the source rocks. In the main depositional area of the Kazhdumi Formation, the minimum depth of burial required to initiate generation of hydrocarbons (about 1400 m, 4600 ft) was not attained until Eocene time (Ala, 1982). Kent and Warman (1972) dated the oil migration into the Greater Burgan structure as post-Cretaceous, Eocene, or even later, reasoning that the thin Ahmadi Formation shale section overlying the Burgan crest could not have contained the pressure of the underlying oil accumulation unless (or until) this was balanced by the deposition of additional post-Ahmadi section.

Some breaching of the Ahmadi Formation cap rock did occur across the crestal area of the Burgan dome. The presence of inspissated oil in the overlying Late Cretaceous and post-Cretaceous section and the deposits of bitumen at the surface above the subsurface structure confirms this breaching effect. This loss of reservoir oil was probably through minor tensional faults that breached the attenuated shale seal as the structure continued to grow through Cretaceous and Tertiary time. No such upward loss of hydrocarbons occurred at the crest of the Ahmadi culmination, where the effects of post-Turonian erosion—the Wasia-Aruma unconformity—were far less severe than at the Burgan dome.

The indicated age of the Ahmadi fold is itself evidence for later rather than earlier migration of the oil. Formational thicknesses suggest that the Ahmadi structure is largely post-Cretaceous in age, a product in part of the Zagros orogeny that reached a climax in Miocene to Pliocene time. Since the Ahmadi structure shared an original common oil-

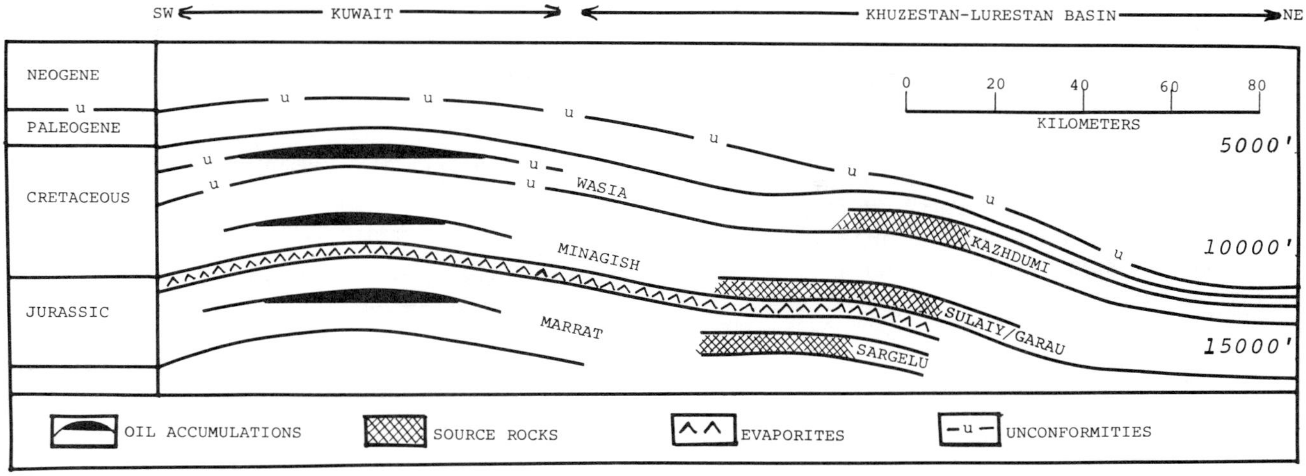

SW ← KUWAIT →
← KHUZESTAN-LURESTAN BASIN → NE
NEOGENE
PALEOGENE
CRETACEOUS
JURASSIC
0 20 40 60 80
KILOMETERS
5000'
10000'
15000'
WASIA
MINAGISH
MARRAT
KAZHDUMI
SULAIY/GARAU
SARGELU
OIL ACCUMULATIONS
SOURCE ROCKS
EVAPORITES
UNCONFORMITIES

Figure 10. Diagrammatic section illustrating relationship between Kazhdumi and older source rocks and existing oil accumulations in southeastern Kuwait (modified from Ayres et al., 1982).

water contact with the Burgan and Magwa domes, it is apparent that the entire Greater Burgan structure must have received its charge of hydrocarbons during the same period. In the case of the Ahmadi sector, this could not have occurred until late Tertiary time. The believed Kazhdumi Formation source beds did not reach maturation until Eocene time and probably did not generate significant quantities of hydrocarbons before the close of Paleogene time (Ala, 1982). The available evidence suggests that migration of hydrocarbons into the Greater Burgan structure was a Neogene event.

EXPLORATION AND DEVELOPMENT CONCEPTS

The hydrocarbon accumulation at Greater Burgan is contained in a composite domal fold of large areal extent, great structural relief, and moderate subsurface depth. Structural growth has been long-sustained, and even the near-surface Miocene–Pliocene clastics show stratigraphic thinning across the structural crests. Yet surface expression of the Greater Burgan structure is minimal, largely because of the unconsolidated nature of the surficial deposits, which include eolian sands, playa silts, and fluvial sands and gravels.

Geologists were first attracted to the area by reports of the presence of surface impregnations of bitumen on the plain of Burgan. The existence of these surface shows was balanced by a prevailing belief that the sedimentary section along the Arabian side of the Gulf was likely to be thin and nonmarine. The 1932 discovery of oil in marine Cretaceous strata at Bahrain Island dispelled this notion, and the bitumen deposits at Burgan in Kuwait became immediately interesting.

Once the Burgan area was accepted as prospective, the subsurface prospect was confirmed by geophysical surveys, gravity, and magnetics, followed by limited reflection seismic. The techniques of the time (1937–1938) were adequate to outline the area of interest, except across the Ahmadi ridge sector, where the subsurface structure was eventually confirmed by shallow drilling. Modern geophysical techniques could certainly have overcome the near-surface weathering problems experienced at Ahmadi. Satellite imaging, had it been available at the time, could have helped in the positioning of the geophysical surveys, since tonal variations at the surface reflect the outline of the Greater Burgan field (Halbouty, 1980).

There was therefore nothing uniquely difficult in the outlining of the Greater Burgan field, once the presence of oil at depth was suspected. Here the bitumen deposits were critical; in its origin, this was a play based on the presence of surface shows of oil and gas.

The Burgan discovery well was completed after penetrating only 6 m (20 ft) into a pay section of unprecedented thickness and quality. Subsequent drilling showed the uppermost of the field reservoirs to be 40 m (145 ft) thick, to be increasingly shaly toward its base, and to be underlain by a section of oil-stained but unproductive limestone. Drilling could feasibly have been suspended at that level, in the belief that the clastic pay section had been completely penetrated; this could have delayed the discovery of the deeper and more prolific Burgan Formation reservoirs. Fortunately, this did not occur; succeeding wells drilled through the Mauddud Formation limestone section and the Early Cretaceous (Albian) Third and Fourth sand reservoirs were delineated. Following intensive development of these zones, a deep test well was drilled in the indicated crestal area of the Burgan dome, to evaluate early Early Cretaceous and Jurassic section. Production was eventually established in these older intervals, at Greater Burgan and elsewhere in southeastern Kuwait.

The first nine wells on the Burgan dome were drilled from 1938-1942 and then shut in for the duration of World War II. During the first postwar year, 1945-1946, eight of these nine wells were worked over and brought into production; pipelines were laid and a tank farm was constructed; and exports of oil were commenced through a temporary marine terminal at the Persian Gulf shore 24 km (15 mi) east of the Burgan dome. Thereafter, the Burgan dome was actively developed, and the adjoining Magwa and Ahmadi structural culminations were drilled as wildcat prospects and later incorporated into the Greater Burgan field. The development of the Greater Burgan field presented no problems that could not be solved using the technology of the day.

REFERENCES CITED

Adasani, M., 1965, The Greater Burgan field: 5th. Arab Petroleum Congress, p. 7-27.

Ala, M. A., 1982, Chronology of trap formation and migration of hydrocarbons in Zagros sector of southwest Iran: American Association of Petroleum Geologists Bulletin, v. 66, n. 10, p. 1535-1541.

AlShamlan, A A., 1975, Petrographic and microfacies analyses of the Mauddud Formation in Kuwait: 9th. Arab Petroleum Congress, Dubai (U.A.E.), Paper 126 (B-3), p. 1-14.

AlSharhan, A. S., and G. C. St. C. Kendall, 1986, Precambrian to Jurassic rocks of Arabian Gulf and adjacent areas: their facies, depositional setting, and hydrocarbon habitat: American Association of Petroleum Geologists Bulletin, v. 70, n. 8, p. 977-1002.

Arabian American Oil Company, 1975, Eastern Arabia and adjacent areas, *in* Schlumberger, Well Evaluation Conference, Arabia, p. 9-25.

Ayres, M. G., M. Bilal, R. W. Jones, L. W. Slentz, M. Tartir, and A. O. Wilson, 1982, Hydrocarbon habitat in main producing areas, Saudi Arabia: American Association of Petroleum Geologists Bulletin, v. 66, p. 1-9.

Beydoun, Z. R., and H. V. Dunnington, 1975, The petroleum geology and resources of the Middle East: Beaconsfield, Scientific Press Ltd., 99 p.

Brennan, P., 1983, Hydrocarbon habitat in main producing areas, Saudi Arabia: discussion: American Association of Petroleum Geologists Bulletin, v. 67, p. 2147-2151.

Carmalt, S. W., and B. St. John, 1986, Giant oil and gas fields, *in* M.T. Halbouty, ed., Future petroleum provinces of the world: American Association of Petroleum Geologists Memoir 40, p. 11-54.

Cox, P. T., 1932, A report on the oil prospects of Kuwait Territory, *in* A.H.T. Chisholm, The first Kuwait oil concession: London, Frank Cass & Co., Note 49, p. 143-155.

El-Naggar, Z. R., and I. A. Al-Rifaiy, 1972, Stratigraphy and microfacies of type Magwa Formation of Kuwait, Arabia, Part 1: Rumaila Limestone Member: American Association of Petroleum Geologists Bulletin, v. 56, p. 1464-1493.

El-Naggar, Z. R., and I. A. Al-Rifaiy, 1973, Stratigraphy and microfacies of type Magwa Formation of Kuwait, Arabia, Part 2: Mishrif Limestone Member: American Association of Petroleum Geologists Bulletin, v. 57, p. 2263-2279.

Fox, A. F., 1959, Some problems of petroleum geology in Kuwait: Institute of Petroleum Journal, London, v. 45, p. 95-110.

Fuchs, W., T. E. Gattinger, and H. F. Holzer, 1968, Explanatory text to the synoptic geologic map of Kuwait: Geological Survey of Austria, Vienna, 87 p.

Gregory, J. W., 1929, The structure of Asia: London, Methuen & Co.

Halbouty, M. T., 1980, Geologic significance of Landsat data for 15 giant oil and gas fields, *in* M. T. Halbouty, ed., Giant oil and gas fields of the decade 1968-1978: American Association of Petroleum Geologists Memoir 30, p. 7-38.

Halbouty, M. T., A. A. Meyerhoff, R. E. King, R. H. Dott, H. D. Klemme, and T. Shabad, 1970, World's giant oil and gas fields, geologic factors affecting their formation, and basin classification, *in* M. T. Halbouty, ed, Geology of giant petroleum fields: American Association of Petroleum Geologists Memoir 40, p. 502-528.

Heim, A., 1924, the question of petroleum in eastern Arabia (Koweit, Hasa, Bahrein), *in* A. H. T. Chisholm, The first Kuwait oil concession: London, Frank Cass & Co., Note 25, p. 103-109.

Jaber, A. S., 1975, Stratigraphy of the Jurassic succession in the off-shore Saudi Arabia-Kuwait Partitioned Zone: 9th Arab Petroleum Congress, Dubai (U.A.E.), Paper 124 (B-3), p. 1-13.

James, G. A., and J. G. Wynd, 1965, Stratigraphic nomenclature of Iranian Oil Consortium Agreement area: American Association of Petroleum Geologists Bulletin, v. 49, p. 2182-2245.

Kamen-Kaye, M., 1970, Geology and productivity of Persian Gulf synclinorium: American Association of Petroleum Geologists Bulletin, v. 54, p. 2371-2394.

Kent, P. E., and H. R. Warman, 1972, An environmental review of the world's richest oil-bearing region—the Middle East: 24th. International Geological Congress, sec. 5, p. 142-152.

Lack, H. C., 1959, Some problems of petroleum geology in Kuwait: discussion: Institute of Petroleum Journal, London, v. 45, p. 95-110.

Levorsen, A. I., 1954, Geology of petroleum: San Francisco, W. H. Freeman and Co., p. 32-33.

Loutfi, G., and A. S. Jaber, 1970, Geology of the Upper Albian-Campanian succession in the Kuwait-Saudi Arabia Neutral Zone offshore area: 7th. Arab Petroleum Congress, Kuwait, Paper 62 (B-3), p. 1-14.

Milton, D. I., 1967, Geology of the Arabian Peninsula: Kuwait: U.S. Geological Survey Professional Paper 560F, p. F1-F7.

Milton, D. I., and C. C. S. Davies, 1965, Exploration and development of the Raudhatain field: Institute of Petroleum Journal, London, v. 51, n. 493, p. 17-28.

Murris, R. J., 1980, Middle East: stratigraphic evolution and oil habitat: American Association of Petroleum Geologists Bulletin, v. 64, p. 597-618.

Owen, E. W., 1975, Trek of the oil finders: a history of exploration for petroleum: American Association of Petroleum Geologists Memoir 6, p. 1335-1340.

Owen, R. M. S., and S. N. Nasr, 1958, Stratigraphy of the Kuwait-Basra Area, *in* L.G. Weeks, ed., Habitat of oil: American Association of Petroleum Geologists, p. 1252-1278.

Pascoe, E. M., 1913, Prospects of obtaining oil near Kuwait, Persian Gulf, *in* A. H. T. Chisholm, The first Kuwait oil concession: London, Frank Cass & Co., Note 5, p. 88-89.

Strong, T. M. W., 1959, Some problems of petroleum geology in Kuwait: discussion: Institute of Petroleum Journal, London, v. 45 p. 95-110.

Van Bellen, R. C., H. V. Dunnington, R. Wetzel, and D. M. Morton, 1959, Lexique Stratigraphique International, vol. III Asie, fasc. 10A, Iraq.

Wellings, F. E., 1965, Iraq and Asiatic Mediterranean countries, *in* E.W. Owen, Trek of the oil finders: American Association of Petroleum Geologists Memoir 6, p. 1305-1309.

Windley, B. F., 1977, The evolving continents: Chichester, John Wiley & Sons, 399 p.

Wyllie, B. K. N., 1926, Anglo-Persian Oil Company report on Kuwait, *in* A. H. T. Chisholm, The first Kuwait oil concession: London, Frank Cass & Co., Note 31, p. 111-118.

Appendix 1. Field Description

Field name .... *Greater Burgan field*

Ultimate recoverable reserves .... *75 billion bbl*

Field location:

- **Country** .... *Kuwait*
- **State** .... *Kuwait*
- **Basin/Province** .... *Arabian*

Field discovery:

- **Year field discovered** .... *1938 (Wara Formation only)*
- **Year second pay discovered** .... *1939–1942 (Burgan Formation, 3rd & 4th Sands)*
- **Third pay** .... *1951 (Minagish—first oil tested): 4th. 1984 (Jurassic)*

Discovery well name and general location
(i.e., Jones No. 1, Sec. 2T12NR5E; or Smith No. 1, 5 mi west of Sheridan, Wyoming):

- **First pay** .... *Burgan-1: 28°55′N, 47°56′E*
- **Second pay** .... *Burgan-2 through Burgan-9*
- **Third pay** .... *Burgan 113-HNC (first oil from Lower Cretaceous)*

Discovery well operator .... *Kuwait Oil Company (Gulf Oil/British Petroleum)*
(if more than one pay in field, list operators of discovery well in other pays)

- **Second pay** .... *Kuwait Oil Company (Gulf Oil/British Petroleum)*
- **Third pay** .... *Kuwait Oil Company (Gulf Oil/British Petroleum)*

IP in barrels per day and/or cubic feet or cubic meters per day:

- **First pay** .... *4386 BOPD on 1-in. choke*
- **Second pay** .... *Not specified; average rate (3rd Sand) 4877 BOPD*
- **Third pay** .... *NA*

All other zones with shows of oil and gas in the field:

Age	Formation	Type of Show
Eocene-Paleocene	*Hasa Group*	*Heavy oil & bitumen*
Late Cretaceous	*Aruma Group*	*Heavy oil & bitumen*
Late Jurassic	*Gotnia Evaporites*	*Heavy oil & H_2S*

Geologic concept leading to discovery and method or methods used to delineate prospect, e.g., surface geology, subsurface geology, seeps, magnetic data, gravity data, seismic data, seismic refraction, nontechnical:

Discovery of extensive surface impregnations of bitumen associated with minor gas seepage, confirmed by pitting and shallow drilling. Subsurface prospect confirmed by limited geophysical coverage prior to drilling.

Structure:

Province/basin type (see St. John, Bally, and Klemme, 1984)

Bally 221; Klemme IICa

Tectonic history

Located in gently subsiding shelf environment from Paleozoic well into Mesozoic time. Vertical uplift on structure from Jurassic through Tertiary, possibly involving deep-seated halokinesis. Part of structural complex apparently modified by late Tertiary orogenesis.

Regional structure

On outer edge of Arabian platform between Arabian shield/stable shelf to southwest and Mesopotamian (Zagros) geosyncline to northeast.

Local structure

One of several major culminations developed along regional anticlinal ridge plunging gently north through Kuwait into Iraq. Long-sustained vertical uplift indicated by crestal thinning of stratigraphic units.

Trap

Trap type(s)

Structural trap: domal uplift with three individual culminations. Oil column is continuous across intervening structural saddles but some fault separation occurs in lower part of reservoir section, i.e., within Burgan Formation.

Basin stratigraphy (major stratigraphic intervals from surface to deepest penetration in field):

Age	Formation	Depth to Top
Pleistocene-Miocene	*Kuwait Group*	*Surface*
Eocene-Paleocene	*Hasa Group*	*200 ft (61 m)*
Late Cretaceous	*Aruma Group*	*2600 ft (792 m)*
	*Wasia Group: Magwa**	*4000 ft (1219 m)**
	Ahmadi	*3400 ft (1036 m)*
	Wara	*3500 ft (1067 m)*
Early Cretaceous	*Mauddud*	*3670 ft (1119 m)*
	Burgan	*3700 ft (1127 m)*
	Thamama Group	*5000 ft (1524 m)*
Late Jurassic	*Gotnia/Najmah*	*8700 ft (2652 m)*
Middle Jurassic	*Sargelu*	*10,000 ft (3048 m)*
Early Jurassic	*Marrat*	*?11,500 ft (3505 m)*

** Magwa Formation eroded from main crestal area of structure.*

Location of well in field .... *Discovery well located on crestal area of main dome.*

Reservoir characteristics: *(Discovery well proved Wara pay only; deeper pays proved by subsequent wells.)*

Number of reservoirs *5*

Formations .... *Wara, Mauddud**, Burgan, Minagish, Marrat*
*** Mauddud reservoir produces only in Ahmadi sector of field.*

Ages .... *Early Late Cretaceous to Early Jurassic*

Depths to tops of reservoirs

Wara .... *-3245 ft (-990 m)*
Mauddud .... *-3425 ft (-1044 m)*
Burgan .... *-3450 ft (-1052 m)*
Minagish .... *-6500 ft (-1981 m) (est.)*

Gross thickness (top to bottom of producing interval) .... *1221 ft (372 m)*
Wara-Burgan pay section only

Net thickness—total thickness of producing zones

Average .... *Wara-Burgan 900 ft (274 m)*
Maximum .... *NA*
Average .... *Minagish 450 ft (137 m) (est. only)*
Maximum .... *NA*

Lithology

Uppermost reservoir (Wara) is alternation of fine-grained sandstones, siltstones, and shales. Lower reservoir (Burgan) consists of medium- to coarse-grained, well-sorted, clean quartz sand with minor shales. Minagish reservoir consists of buff, soft, porous oolitic limestone.

Porosity type .... *Intergranular in Wara-Burgan; intergranular to ooliclastic in Minagish*
Average porosity .... *Wara reservoir 24%; lower reservoir (Burgan) 23%*
Average permeability .... *Wara 100-500 md; Burgan-3rd. 380-4000 md; Burgan-4th. 4000 md; Minagish reservoir 30%/1000 md (est.)*

Seals:

Upper

Formation, fault, or other feature .... *Ahmadi Formation*
Lithology .... *Shale, gray-green, calcareous*

Lateral

Formation, fault, or other feature .... *Ahmadi Formation*

Lithology .... *Shale, gray-green, calcareous*

Source:

Formation and age .... *Kazhdumi Formation (Albian)*

Lithology .... *Dark, bituminous shale with subordinate argillaceous limestone*

Average total organic carbon (TOC) .... *NA*

Maximum TOC .... *NA*

Kerogen type (I, II, or III) .... *II*

Vitrinite reflectance (maturation) .... *NA*

Time of hydrocarbon expulsion .... *Late Eocene, possibly to Miocene*

Present depth to top of source .... *Presently exposed at surface*

Thickness .... *690 ft (210 m)*

Potential yield .... *NA*

Appendix 2. Production Data

Field name .... *Greater Burgan field*

Field size: .... *(Area within Wara Formation o/w contact exceeds 300 mi^2)*

Proved acres .... *131,000 ac (53,000 ha.)*

Number of wells all years .... *600+*

Current number of wells (as of year) .... *300 producing (approx.)*

Well spacing .... *200 to 66 ⅔ ac (81 to 27 ha.)*

Ultimate recoverable .... *75,000 million bbl (10.3 billion metric tons)*

Cumulative production .... *19,000 million bbl (2.6 billion metric tons)*

Annual production .... *350 million bbl (48 million metric tons)*

Present decline rate .... *NA*

Initial decline rate .... *NA*

Overall decline rate .... *NA*

Annual water production .... *NA*

In place, total reserves .... *200,000 million bbl (24.5 billion metric tons)*

In place, per acre-foot .... *1675 bbl (230 metric tons)*

Primary recovery .... *NA*

Secondary recovery .... *NA*

Enhanced recovery .... *NA*

Cumulative water production .... *NA*

Drilling and casing practices:

Amount of surface casing set .... *300–500 ft (90–150 m)*

Casing program .... *Wara-Burgan wells: 13⅜ in. into Dammam Formation;*
9⅝ in. into Ahmadi Formation
7 in. to below oil/water contact

Drilling mud .... *Based on locally produced brackish water*

Bit program .... *NA*

High pressure zones .... *No high pressure zones in Tertiary-Cretaceous section*

Completion practices:

Interval(s) perforated .... *Substantial intervals; completion dual or single*

Well treatment .... *Not required in Wara-Burgan reservoirs*

Formation evaluation:

Logging suites .... *Early wells, SP-ES (16 in./64 in. normal) + ML; later wells utilized GR-N, MLL, induction log; CCL widely used*

Testing practices .... *Early wells "washed in" and burned off; later wells completed and production tested*

Mud logging techniques .... *NA*

Oil characteristics:

Type (Tissot and Welte Classification in "Petroleum Formation and Occurrence," 1984, Springer-Verlag, p. 419) .... *Paraffinic-asphaltic mixed base*

API gravity .... *28°-34°; field average 31.8°*

Base .... *Mixed asphaltic-paraffinic*

Initial GOR .... *Wara 528 MCFG/bbl; Burgan 426-461 MCFG/bbl*

Sulfur, wt% .... *2.5 (increases vertically through oil column)*

Viscosity, SUS .... *52 sec at 100°F (38°C)*

Pour point .... *0°F (-18°C)*

Gas-oil distillate .... *NA*

Field characteristics:

Average elevation .... *250-385 ft (76-117 m)*

Initial pressure .... *2075 psi @ -4000 ft (281 kg/cm^2 @ 1219 m)*

Present pressure .... *NA*

Pressure gradient .... *NA*

Temperature .... *136°F @ 4000 ft (58°C @ 1320 m)*

Geothermal gradient .... *0.02°F/ft (0.03°C/m)*

Drive .... *Water drive supplemented by gas injection*

Oil column thickness .... *1221 ft (370 m)*

Oil-water contact .... *-4466 ft (-1360 m)*

Connate water .... *Wara 18% average; Burgan-3rd 4-31%; Burgan-4th 4%*

Water salinity, TDS .... *154,000 ppm/TDS (Cl^- 95,000; Na^+/K^+ 46,000)*

Resistivity of water .... *0.05 ohm-m @ 100°F (38°C)*

Bulk volume water (%) .... *NA*

Transportation method and market for oil and gas:

Short pipeline to export terminal at Persian (Arabian) Gulf littoral, thence by tanker to market.

Bodalla South Field

J. A. SALOMON, S. L. KEENIHAN, and A. P. CALCRAFT
LASMO Oil Company Australia Limited, Brisbane, Queensland

FIELD CLASSIFICATION

BASIN: Eromanga
BASIN TYPE: Cratonic Sag
RESERVOIR ROCK TYPE: Sandstone
RESERVOIR ENVIRONMENT OF DEPOSITION: Fluvio-Deltaic
RESERVOIR AGE: Jurassic
PETROLEUM TYPE: Oil
TRAP TYPE: Anticline

LOCATION

The Bodalla South oil field is located in southwestern Queensland, Australia, approximately 1000 km west of Brisbane.

The field is situated within the southern part of the Jurassic–Cretaceous Eromanga basin generally at the eastern end of a productive trend that follows the southern subcrop edge of the underlying Permian Cooper basin (Figure 1).

The field is 160 km from the Jackson oil field, which is the largest accumulation (100 MMSTOIP) discovered in the area to date. The closest pipeline facilities are located at the Jackson oil field, and most of the production from the Bodalla South oil field is road hauled to that terminal.

The topography of the area surrounding the oil field is relatively flat with an average elevation of 200 m above sea level; maximum elevation is 300 m.

The area is semi-arid, and land use is dominantly pastoral with the raising of sheep for wool being the main activity.

HISTORY OF EXPLORATION

The surface geology of the area poorly expresses the underlying structure. Tertiary duricrust outcrops sporadically throughout the area, and it is possible only occasionally to map weak surface anticlines based on very low dips. As a consequence, the exploration of the area has relied heavily on geophysical methods to provide information on the subsurface geology.

Geophysical exploration of the area commenced in 1942 with a reconnaissance gravity survey conducted by Shell Development Pty. Ltd. This survey and subsequent gravity work up until 1963 provided valuable information concerning the subsurface geology. Some of the more important features to emerge were:

1. Warrabin gravity low.
2. Pinkilla–Canaway anticlinal ridge, which corresponds to surface topographic features and forms the eastern boundary of the Warrabin low.
3. The Tallyabra gravity high, a somewhat less distinct feature, trending NNW-SSE and forming the western boundary of the Warrabin low.

The location of these features is shown on Figure 2.

The first seismic survey in the area was recorded by the Bureau of Mineral Resources in 1958. This work and the surveys which followed were restricted to confirming that some of the more obvious anticlinal features observed by the gravity surveys persisted at depth.

Of relevance to the Bodalla South area were surveys recorded by Alliance Oil Development Australia N.L. (Alliance) (Mulready, 1971) and British Petroleum Development Australia Pty. Ltd. (B.P.) (Farley, 1967) in 1966. These led to the identification of the Bodalla Prospect, which was subsequently drilled in 1967. At this time, exploration of the area was chiefly aimed at locating prospects in the Permian–Triassic sediments of the Cooper basin (see Figure 3 for stratigraphic chart). These sediments had been proven to be hydrocarbon-bearing in the South Australian portion of the basin (350 km west). In addition, Chandos No. 1 some 25 km to the north of the Bodalla Prospect had recovered minor amounts of oil from the upper part of the Triassic sequence. The results of drilling in the region suggested that the Permian–Triassic sediments thinned significantly onto the anticlinal trends as shown in the cross section, Figure 4 (see Figure 2 for location of section). The Bodalla Prospect was chosen for drilling because it offered a thicker Permian–Triassic section that was hoped would increase the likelihood of encountering thick, porous sands. It was on the downthrown side of the Tallyabra fault and relied on fault seal for closure. Figures 5A and B show the seismic interpretation of the "C" (intra Lower Cretaceous) and the "P" (top Permian) seismic events prior to the drilling of the well.

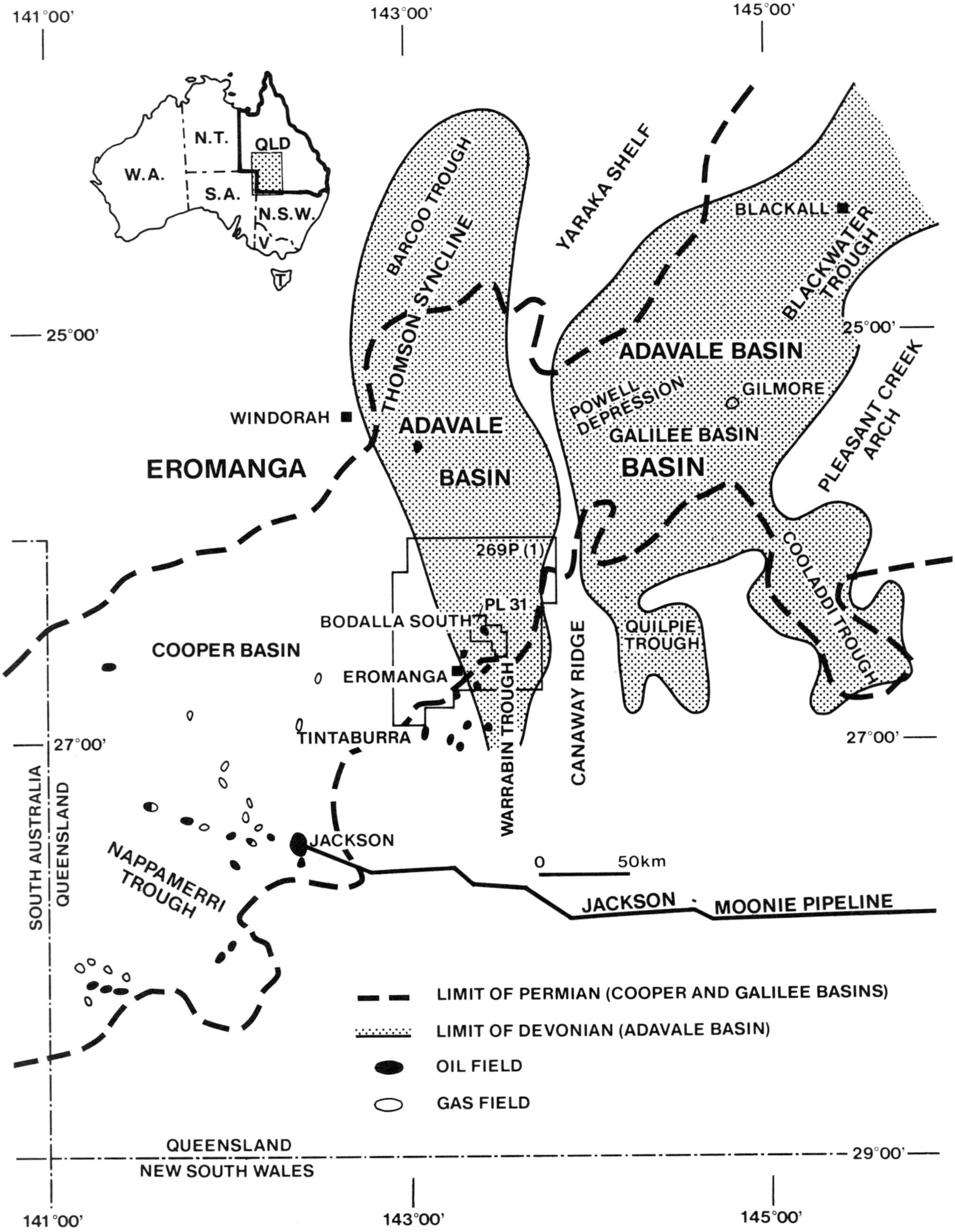
141°00'
143°00'
145°00'
N.T.
QLD
W.A.
S.A.
N.S.W.
V
T
BARCOO TROUGH
THOMSON SYNCLINE
YARAKA SHELF
BLACKALL
BLACKWATER TROUGH
25°00'
ADAVALE BASIN
POWELL DEPRESSION
GILMORE
GALILEE BASIN
BASIN
PLEASANT CREEK ARCH
WINDORAH
ADAVALE
BASIN
EROMANGA
269P (1)
PL 31
BODALLA SOUTH
COOPER BASIN
EROMANGA
QUILPIE TROUGH
COOLADDI TROUGH
TINTABURRA
27°00'
WARRABIN TROUGH
CANAWAY RIDGE
SOUTH AUSTRALIA
QUEENSLAND
NAPPAMERRI TROUGH
JACKSON
0
50km
JACKSON - MOONIE PIPELINE
LIMIT OF PERMIAN (COOPER AND GALILEE BASINS)
LIMIT OF DEVONIAN (ADAVALE BASIN)
OIL FIELD
GAS FIELD
QUEENSLAND
NEW SOUTH WALES
29°00'

Figure 1. Regional map showing major structural features and basin limits. Authority to Prospect 269P (1) was awarded to the Joint Venture by the Queensland Government for petroleum exploration. Petroleum Lease (PL) 31 was excised from the ATP after discovery of the Bodalla South field to allow permanent production.

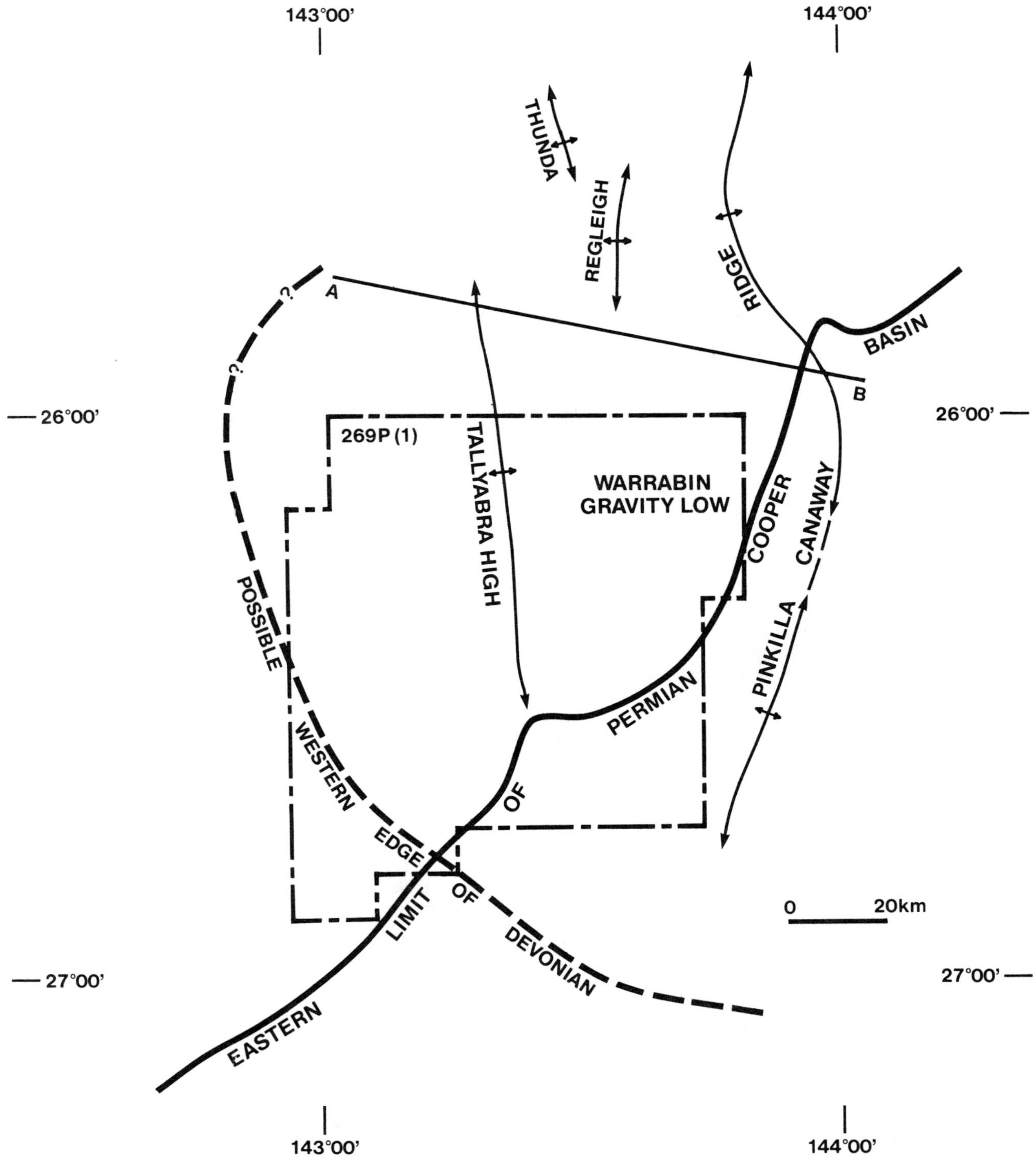
143°00'
144°00'
THUNDA
REGLEIGH
RIDGE
BASIN
A
B
?
26°00'
269P (1)
TALLYABRA HIGH
WARRABIN
GRAVITY LOW
COOPER
CANAWAY
PINKILLA
PERMIAN
POSSIBLE
WESTERN
EDGE
OF
DEVONIAN
OF
LIMIT
EASTERN
0
20km
27°00'

Figure 2. Early mapping of subsurface geology and structural elements as determined by gravity surveys conducted by Shell Development Pty. Ltd. between 1942 and 1963.

The Bodalla No. 1 well was drilled by B.P. to a total depth of 2682 m RT between September and November 1967. The well was plugged and abandoned as a dry hole on 6 November 1967.

Lack of success and the cessation of government incentives effectively stopped active exploration in the area from 1971 to 1980.

During 1980, the area was awarded by the Queensland Government Department of Mines to Beach Petroleum and Hudbay Oil (now LASMO Oil) as Authority to Prospect (ATP) 269P and a concerted exploration effort commenced again. The objectives of the initial farm-in work were Devonian reefal bodies that were thought to exist in the eastern part

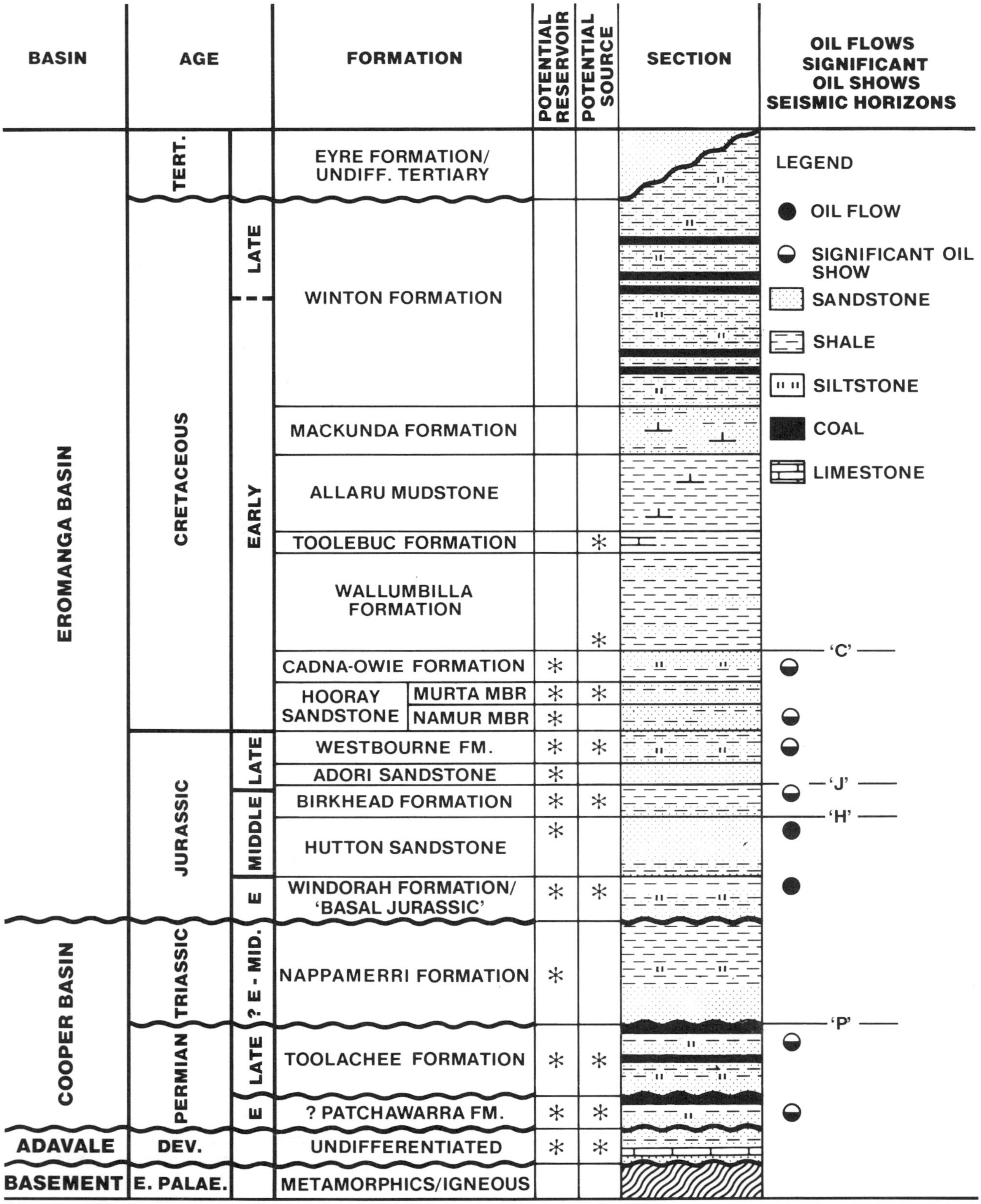
BASIN
AGE
FORMATION
POTENTIAL RESERVOIR
POTENTIAL SOURCE
SECTION
OIL FLOWS
SIGNIFICANT OIL SHOWS
SEISMIC HORIZONS
EROMANGA BASIN
TERT.
CRETACEOUS
LATE
EARLY
EYRE FORMATION/ UNDIFF. TERTIARY
WINTON FORMATION
MACKUNDA FORMATION
ALLARU MUDSTONE
TOOLEBUC FORMATION
WALLUMBILLA FORMATION
CADNA-OWIE FORMATION
HOORAY SANDSTONE
MURTA MBR
NAMUR MBR
JURASSIC
LATE
MIDDLE
E
WESTBOURNE FM.
ADORI SANDSTONE
BIRKHEAD FORMATION
HUTTON SANDSTONE
WINDORAH FORMATION/ 'BASAL JURASSIC'
COOPER BASIN
TRIASSIC
? E - MID.
NAPPAMERRI FORMATION
PERMIAN
LATE
E
TOOLACHEE FORMATION
? PATCHAWARRA FM.
ADAVALE
DEV.
UNDIFFERENTIATED
BASEMENT
E. PALAE.
METAMORPHICS/IGNEOUS
LEGEND
OIL FLOW
SIGNIFICANT OIL SHOW
SANDSTONE
SHALE
SILTSTONE
COAL
LIMESTONE
'C'
'J'
'H'
'P'

Figure 3. Generalized stratigraphic column for the central Eromanga basin. The Bodalla South field produces from the Hutton Sandstone and the Windorah Formation.

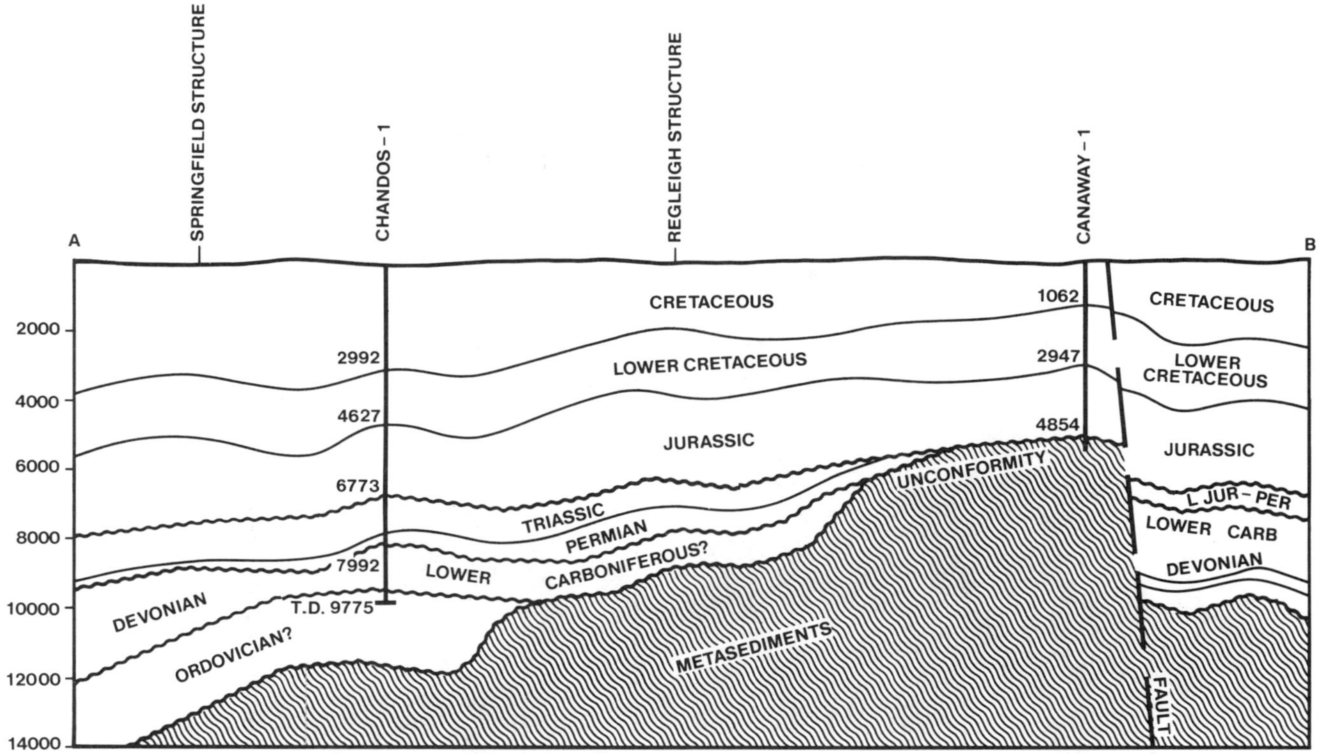
SPRINGFIELD STRUCTURE
CHANDOS-1
REGLEIGH STRUCTURE
CANAWAY-1
A
B
2000
4000
6000
8000
10000
12000
14000
CRETACEOUS
LOWER CRETACEOUS
JURASSIC
TRIASSIC
PERMIAN
LOWER CARBONIFEROUS?
DEVONIAN
ORDOVICIAN?
METASEDIMENTS
UNCONFORMITY
L JUR-PER
LOWER CARB
FAULT
2992
4627
6773
7992
T.D. 9775
1062
2947
4854

Figure 4. Early structural cross section of ATP 269P(1) area. Note the Permian-Triassic thinning onto the basement ridge.

of the ATP, which overlies a remnant of the Devonian Adavale basin. Following seismic acquisition and the drilling of a well during 1981-1982 (Eromanga No. 1) to Devonian objectives, this play concept was abandoned in preference for exploration for younger Eromanga basin targets.

Active exploration for Eromanga basin objectives in the area dates from the discovery of the Jackson oil field in 1981 (Halyburton and Robertson, 1984). This was followed in 1983 by the Tintaburra oil field (Newton, 1986), which is located close to the southwestern corner of ATP 269P(1). These discoveries reinforced the potential of the Eromanga basin sequence and, since both were located along the southern edge of the Permian Cooper basin, the possible importance of this basin edge in the search for Eromanga basin oil accumulations. ATP 269P(1) appeared to be ideally located with respect to this Permian basin edge (see Figure 5) and as a result exploration effort in the area since 1982 has been concentrated on finding prospects within the Eromanga basin sequence.

Discovery

During 1982 and 1983 a large seismic effort (1542 km) was undertaken and Bodalla South No. 1, drilled during 1984, was the first well to test a closure identified on these seismic data (see Figure 6). Bodalla South No. 1 was completed in July 1984 as a dual zone producer from Middle Jurassic Hutton Sandstone and Lower Jurassic Windorah Formation reservoirs (see Figure 3). These zones produced 888 and 2057 BOPD on open-hole drill-stem tests, respectively. The original oil in place is estimated to be 10.3 MMSTB, i.e., 4.9 MMSTB in the Hutton Sandstone and 5.4 MMSTB in the Windorah Formation.

Post-Discovery

Following the discovery well, six wells were drilled between 1984 and 1986. The first oil was marketed at the end of 1984, initially transported by truck over 850 km to the Moonie oil terminus in the Surat basin in eastern Queensland (see Figure 1).

All locations of wells on the Bodalla South oil field were based on seismic mapping. As a consequence a fixed well spacing has not eventuated, but wells generally are 600 m apart corresponding roughly to an 80 ac (32 ha.) spacing. This spacing is regarded as adequate to drain the field.

The oil field has produced 3.41 million bbl of oil up to the end of 1987 (0.85 mmbbl from Hutton Sandstone and 2.56 mmbbl from Windorah Formation) in association with 2.14 million bbl of water. The ultimate recovery varies for the two reservoir zones. The Windorah Formation is expected to yield

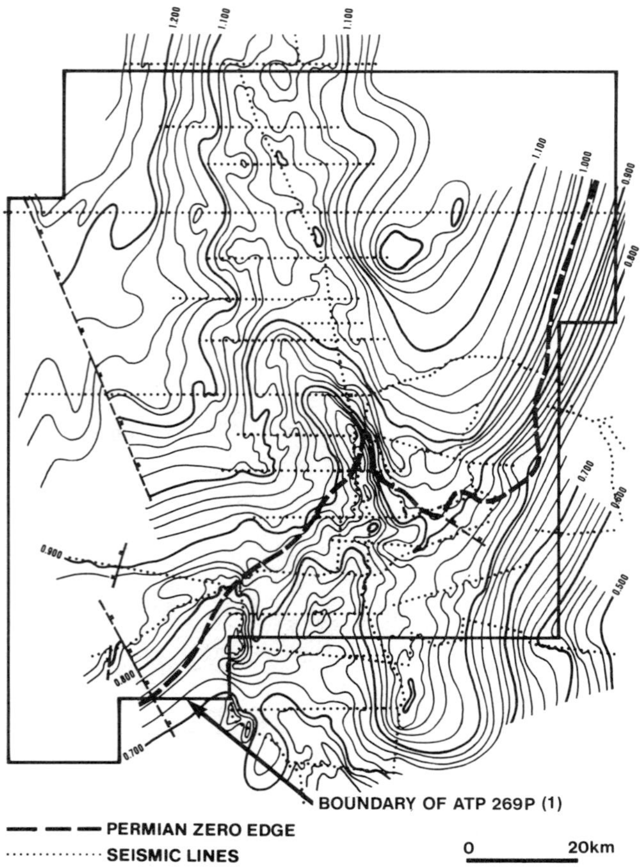

BOUNDARY OF ATP 269P (1)
PERMIAN ZERO EDGE
SEISMIC LINES
0
20km

Figure 5A. 1967 Two-way time structure map of "C" (intra Lower Cretaceous) horizon prior to drilling of the Bodalla No. 1 well.

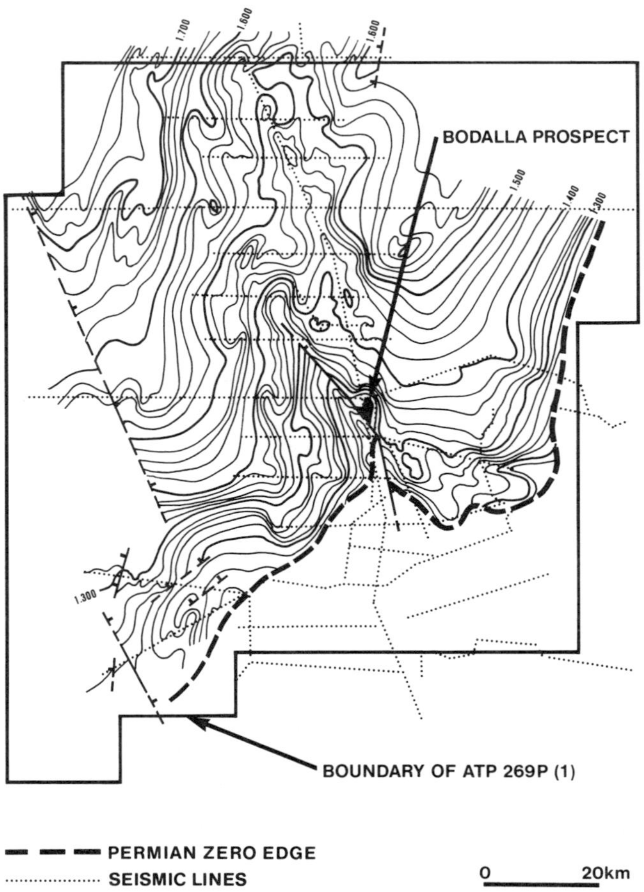

BODALLA PROSPECT
BOUNDARY OF ATP 269P (1)
PERMIAN ZERO EDGE
SEISMIC LINES
0
20km

Figure 5B. 1967 Two-way time structure map of "P" (top Permian) horizon prior to drilling of the Bodalla No. 1 well.

60% of the original oil in place while the Hutton Sandstone is likely to produce about 40% of the original oil in place. This oil field is currently producing 1500 BOPD with 3900 BWPD. Disposal of the large volumes of produced water is facilitated by the hot, dry climate through especially constructed evaporation ponds. The oil field has recently gone onto artificial lift (jet pumping) in an effort to increase the recovery rate and to maintain oil productivity as the water cut increases.

DISCOVERY METHOD

The oil field was discovered through the mapping of an anticlinal closure on seismic data.

Exploration, however, was originally attracted to the area by the presence of large elongate Tertiary anticlines. These features are poorly manifested at the surface but were known prior to the acquisition of seismic data, being observed on magnetic and gravity data and supported by a limited amount of surface data. The location of the first seismic lines in the area was aimed at delineating these features.

Regional dip over the whole of ATP 269P(1) is toward the northeast as is shown by a map of the Base Mesozoic event (Figure 6). Effective entrapment of hydrocarbons therefore relies on locating areas where this dip is disrupted either by rollover of the beds or by stratigraphic changes. As an adjunct to this requirement, an empirical relationship has been noted suggesting that oil accumulates either on or updip from the subcrop of the Triassic-Permian sediments of the Cooper basin. This factor has tended to concentrate the exploration effort of ATP 269P(1) in the southern half of the area.

Following the discovery of oil at Bodalla South, ATP 269P(1) has undergone relatively active exploration effort (by Australian standards) with the recording of over 5000 km of seismic data and the drilling of 18 exploration wells. This has resulted in the discovery of three additional fields. To date, exploration effort has been primarily through seismic acquisition followed by drilling. During this process, a number of important problems affecting exploration have been identified. These can be summarized as follows:

1. Regional velocity variations—velocities of the Mesozoic section decrease toward the southeast.
2. Local velocity variations—apparently random in

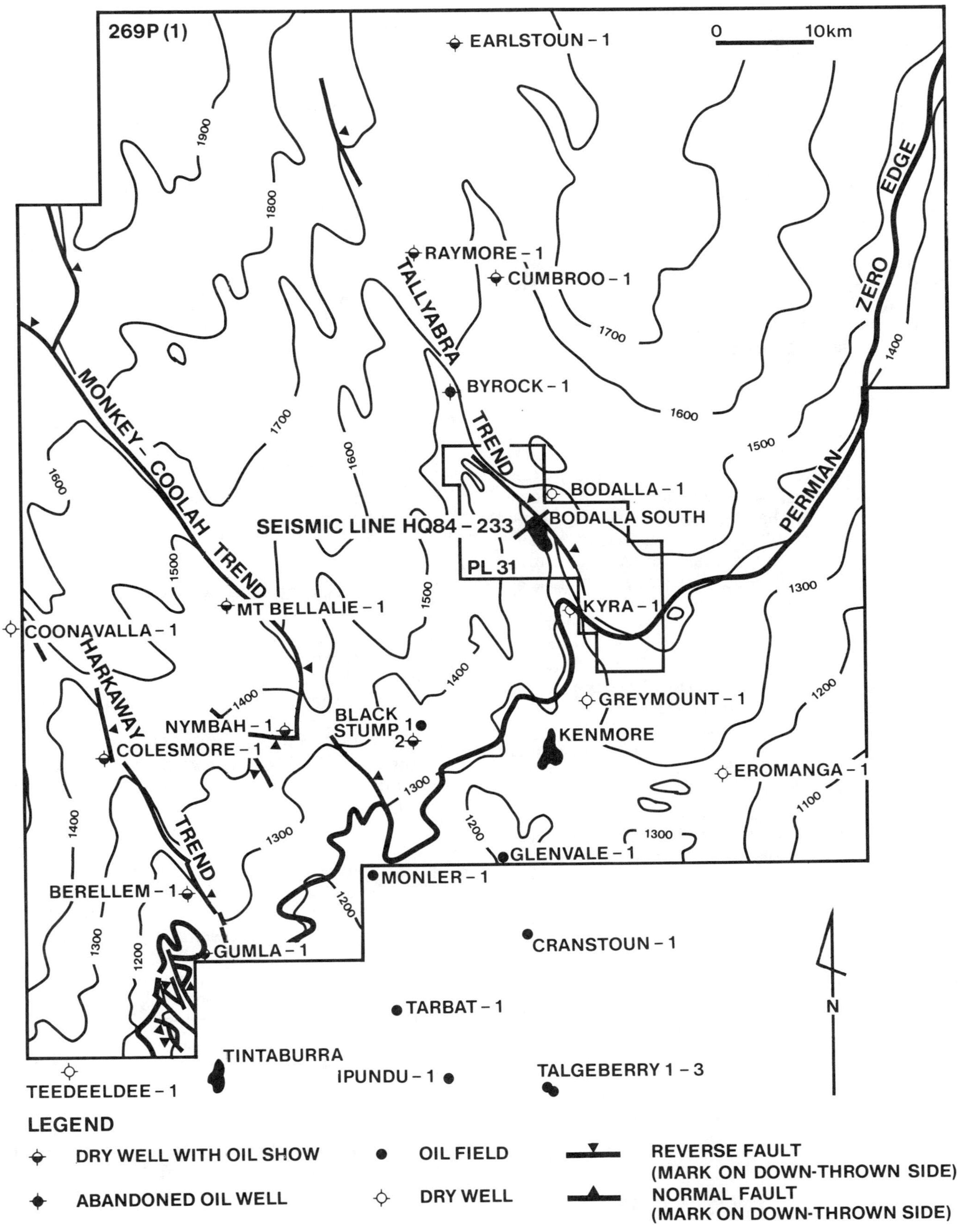
269P (1)
0
10km
EARLSTOUN - 1
RAYMORE - 1
CUMBROO - 1
TALLYABRA
TREND
BYROCK - 1
MONKEY - COOLAH
TREND
ZERO
EDGE
PERMIAN
BODALLA - 1
BODALLA SOUTH
SEISMIC LINE HQ84 - 233
PL 31
KYRA - 1
MT BELLALIE - 1
COONAVALLA - 1
HARKAWAY
TREND
NYMBAH - 1
BLACK STUMP 1
2
COLESMORE - 1
GREYMOUNT - 1
KENMORE
EROMANGA - 1
GLENVALE - 1
MONLER - 1
BERELLEM - 1
GUMLA - 1
CRANSTOUN - 1
TARBAT - 1
TINTABURRA
IPUNDU - 1
TALGEBERRY 1 - 3
TEEDEELDEE - 1
N
1900
1800
1700
1600
1500
1400
1300
1200
1100
LEGEND
DRY WELL WITH OIL SHOW
ABANDONED OIL WELL
OIL FIELD
DRY WELL
REVERSE FAULT
(MARK ON DOWN-THROWN SIDE)
NORMAL FAULT
(MARK ON DOWN-THROWN SIDE)

Figure 6. Current structure map (two-way time) on top Permian-base Mesozoic horizon showing major trends, location of fields and wells. The boundaries of ATP 269P(1) and PL 31 are shown.

nature but can cause errors in depth prediction of ± 20 m.
3. Stratigraphic changes—rapid facies variations at the top of the Hutton Sandstone can eradicate an entire pay section.
4. Near-surface seismic static problems in the order of 10 msec.

STRUCTURE

The Bodalla South oil field is contained within an anticlinal structure on the western, upthrown flank of the northwest–southeast-trending Tallyabra trend. This trend is a simple asymmetric anticline involving deep-seated high-angle reverse faulting downthrown to the east and gradually dipping away to the west (see Figures 7 and 8).

According to Finlayson et al. (1988), the Eromanga basin region underwent at least four phases of tectonic activity: (1) in the Carboniferous, (2) in the Late Permian, (3) during the Triassic, and (4) during the middle Tertiary. All episodes are described as crustal shortening events, occurring as reactivations over the older northwest–southeast-trending basement features but having different compressional directions. As a result, these translated into strike-slip movements in some places. Major unconformities follow the Carboniferous, Permian, and Triassic episodes, and extensive regional erosion followed the Carboniferous, Triassic, and Tertiary episodes.

The development of structural closure in the Bodalla South oil field is due primarily to compression during the middle Tertiary. However, pre-existing closure is mapped on intra-Devonian reflectors underlying the Bodalla South oil field. It is possible that any topographic expression of this closure persisting after the extensive erosion and planation of the Late Triassic caused drape closure through the Jurassic. One suggestion that has resulted after studying the distribution of dry and producing structures is that an early component of structuring is necessary for the accumulation of hydrocarbons. Certainly if present, the amplitude of any early structuring is below the resolution limits of seismic and the current well control.

Tertiary tectonism produced extensional faulting over anticlines in the less competent sediments within higher parts of the stratigraphic section (Winton and Toolebuc Formation) and compressional faulting lower in the section (below the "C" horizon).

Figures 9A and B show the structural outline of the field at the Top Hutton Sandstone and the Top Windorah Formation, the two main pay zones of the oil field.

STRATIGRAPHY

The Jurassic to Lower Cretaceous section of the Eromanga basin is typified by repeated depositional cycles from high-energy braided stream deposits to lower energy fluvial, lacustrine, and marsh deposits. These have produced alternating sand-shale packages that can be readily correlated lithologically for great distances across the Eromanga basin. Each lithostratigrahic unit comprises a separate formation as shown on Figure 3. Seismic data show a number of readily mappable reflectors that adequately show the gross structural configuration of the area. Isochore maps between these horizons show little departure from a gentle basinward thickening trend.

A geohistory plot for the Bodalla South oil field (Figure 10) shows relatively slow deposition of the sequence from Windorah Formation to Cadna-Owie Formation and rapid burial during the latter part of the Early Cretaceous (Winton, Mackunda, Toolebuc, and Wallumbilla Formations and Allaru Mudstone). Tertiary deposition was not substantial with the main occurrence being the development of a number of deep weathering profiles (Senior and Mabbutt, 1979). Maximum depths of burial have not substantially exceeded present depths.

Hydrocarbons in the Eromanga basin have been produced from almost every horizon from the Windorah Formation to the Cadna-Owie Formation (see Figure 3). Almost all of the significant oil accumulations occur in the vicinity of the underlying Cooper basin Permian and Triassic zero edges. This empirical relationship has resulted in a focus of exploration along this northeast–southwest-trending feature. As a general rule, hydrocarbons occur higher in the stratigraphic section toward the basin's margins. To date only minor oil recoveries have been recorded from the underlying Cooper basin sequence in this locality.

The Hutton Sandstone produces the bulk of the Eromanga basin's oil and is regionally consistent in terms of distribution and quality. Hutton Sandstone reservoirs in the area occur at depths of between 1000 and 1500 m. The fields have been discovered by drilling structural closures; however, significant modification of the structural configuration at the top of the reservoir is often caused by stratigraphic variations that are difficult to predict on a field scale and are often below the resolution of the seismic exploration methods currently available.

The Windorah Formation consists of interbedded sandstone and shale, each bed being generally less than 10 m thick. The environment of deposition of the sandstones ranges from high-energy to lower-energy fluvial, to deltaic. Reservoir quality varies vertically and laterally depending on the facies.

Good-quality reservoir sands are widespread in the Adori Sandstone and the Namur Member of the Hooray Sandstone. Lesser quality sands occur sporadically within the Birkhead Formation, the Westbourne Formation, the Murta Member of the Hooray Sandstone, and the Cadna-Owie Formation. Good shows have been recorded in almost all of these sections, with the only exception being the Adori Sandstone.

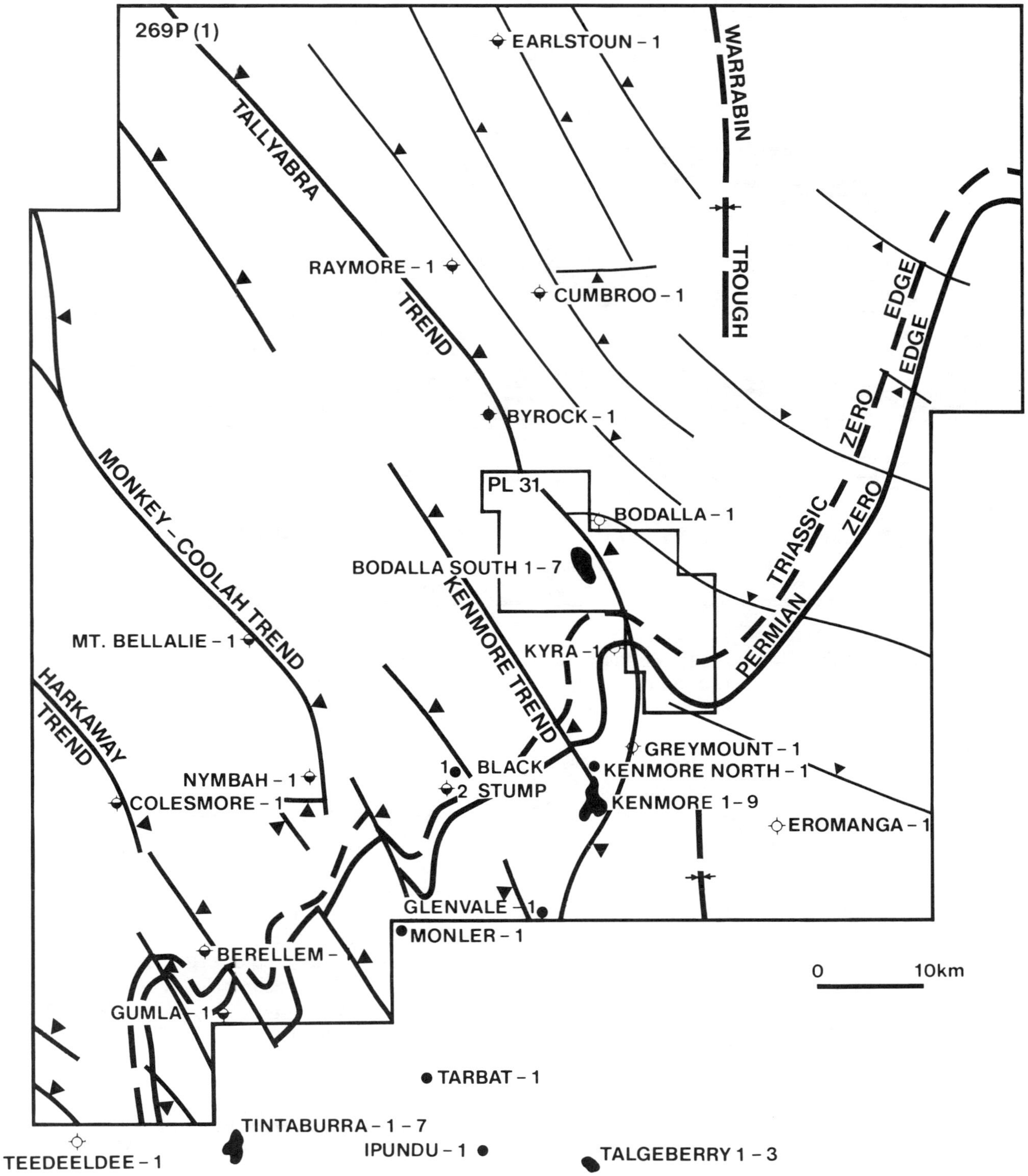
269P (1)
EARLSTOUN - 1
WARRABIN
TROUGH
TALLYABRA
TREND
RAYMORE - 1
CUMBROO - 1
BYROCK - 1
PL 31
BODALLA - 1
BODALLA SOUTH 1 - 7
EDGE
EDGE
ZERO
ZERO
TRIASSIC
PERMIAN
MONKEY - COOLAH TREND
KENMORE TREND
KYRA - 1
MT. BELLALIE - 1
HARKAWAY
TREND
GREYMOUNT - 1
KENMORE NORTH - 1
NYMBAH - 1
COLESMORE - 1
BLACK
STUMP
KENMORE 1 - 9
EROMANGA - 1
GLENVALE - 1
MONLER - 1
BERELLEM - 1
GUMLA - 1
0
10km
TARBAT - 1
TINTABURRA - 1 - 7
TEEDEELDEE - 1
IPUNDU - 1
TALGEBERRY 1 - 3

Figure 7. Deep-seated structural elements within ATP 269P(1). The trends are mapped from a seismic reflector (Cooladdi dolomite) within the Adavale basin sequence. Comparison with Figure 6 shows the influence of these trends on the overlying sedimentary sequences.

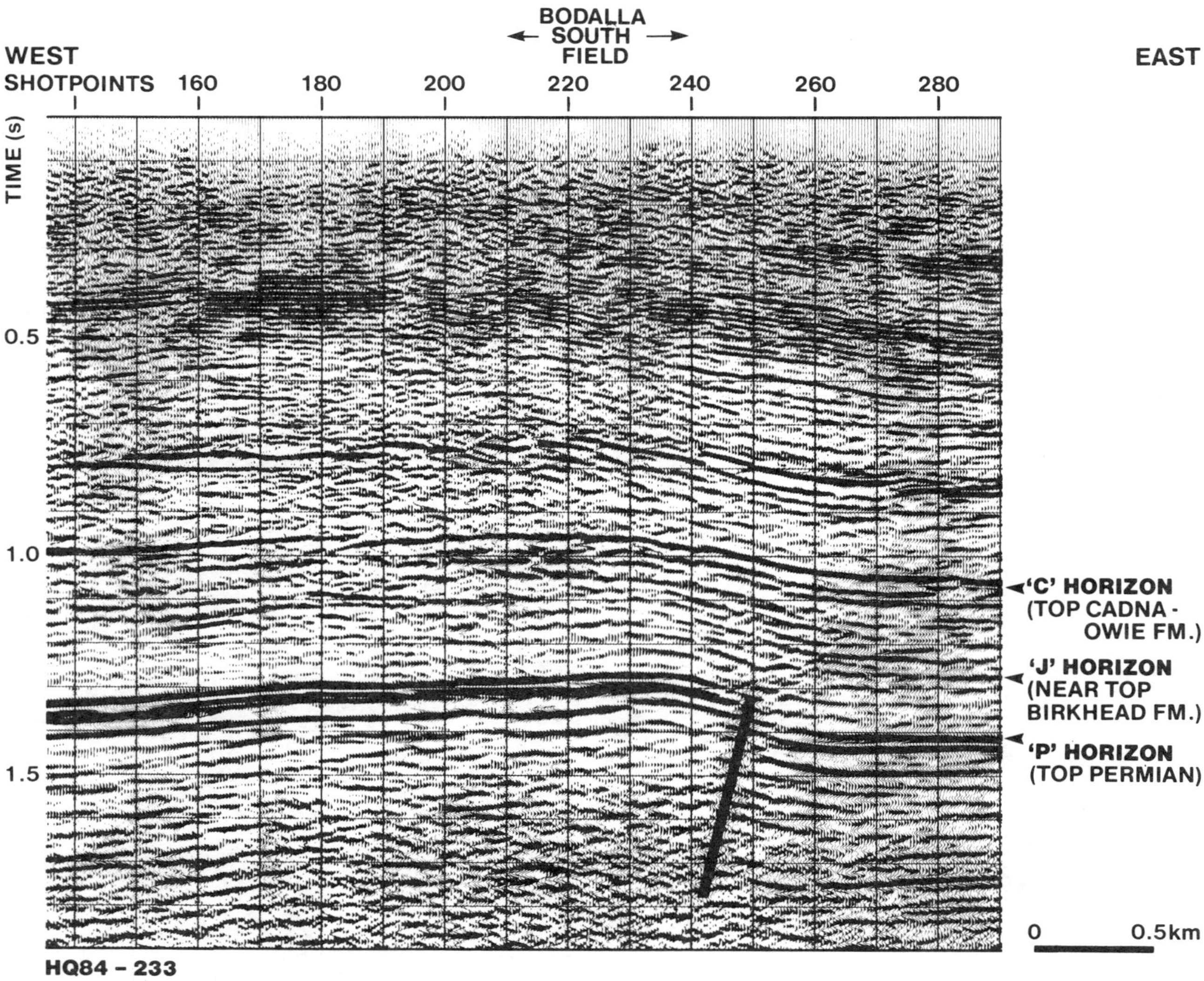
BODALLA SOUTH FIELD
WEST
EAST
SHOTPOINTS 160 180 200 220 240 260 280
TIME (s)
0.5
1.0
1.5
'C' HORIZON (TOP CADNA-OWIE FM.)
'J' HORIZON (NEAR TOP BIRKHEAD FM.)
'P' HORIZON (TOP PERMIAN)
0 0.5km
HQ84 - 233

Figure 8. Typical Seismic line over the Bodalla South field (see Figure 6 for location). The top of the Hutton Sandstone lies approximately 40 msec below the "J" horizon while the top Windorah Formation lies approximately 110 msec below the "J" horizon.

Field sizes in the region are generally small, i.e., in the 1-10 million bbl (recoverable) range.

SOURCE FOR OIL

Debate over the source of the hydrocarbons found in the Eromanga basin sequence continues between proponents of Permian and Jurassic sources. The Permian advocates cite the consistent relationship between oil fields and the Triassic zero edge (where the Triassic shales no longer act as a barrier to vertical migration of Permian sourced hydrocarbons) and the general lack of any mud gas other than methane recorded in drilling the Jurassic section. Proponents of a Jurassic source point to the richness of the Jurassic source rocks and a number of apparently isolated reservoir sands containing oil or oil shows, and question the apparent lack of significant oil accumulations in Triassic sandstones given that they lie between the proposed Permian source rocks and the Jurassic reservoirs.

Geochemical analyses have so far failed to produce universally accepted evidence for either being the sole source, although biomarkers appear to point to a significant Jurassic component.

Jurassic shales have total organic carbon (TOC) contents that usually exceed 1% and often exceed 2%. Kerogens are of type II-III and contain considerable amounts of oil-prone exinites at generally marginal maturities, i.e., vitrinite reflectance values exceed 0.7% only in the deepest parts of the basin. Permian source rocks are more mature with vitrinite reflectance generally exceeding 0.8%. The Permian rocks contain a number of coal seams that have a similar kerogen type to that of the Jurassic, with in general only slightly lower exinite contents being recorded.

A present-day measured geothermal gradient of

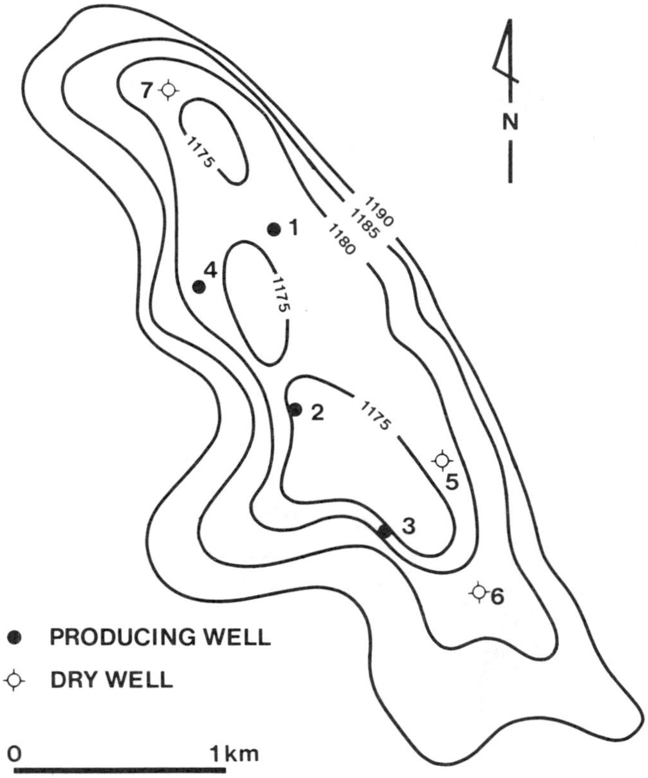
N
7
1
4
2
3
5
6
1175
1190
1185
1180
PRODUCING WELL
DRY WELL
0
1km

Figure 9A. Top Hutton Sandstone two-way time structure map. The No. 1 well was the discovery well; other wells were drilled sequentially.

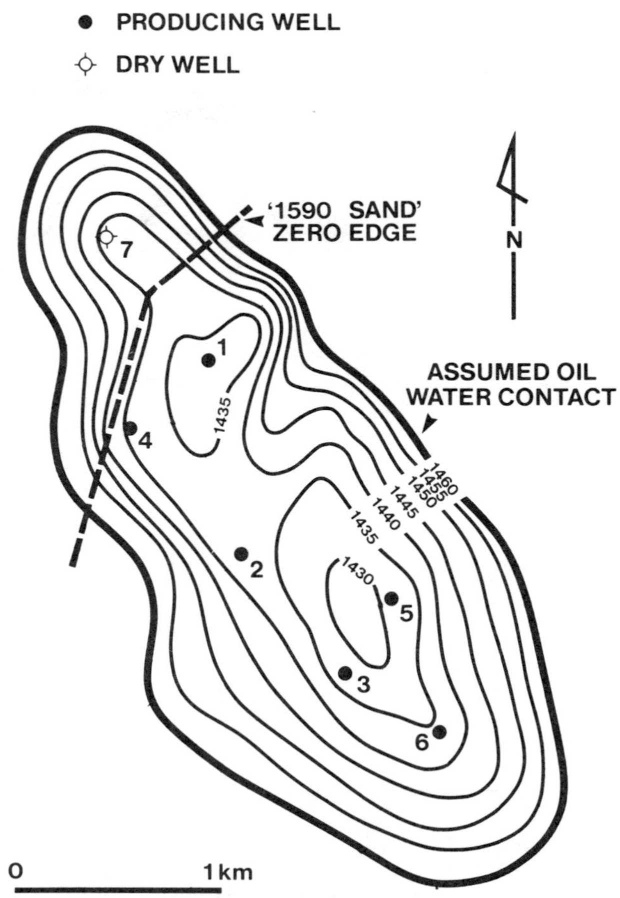
PRODUCING WELL
DRY WELL
'1590 SAND' ZERO EDGE
N
ASSUMED OIL WATER CONTACT
7
1
4
2
5
3
6
1435
1460
1455
1450
1445
1440
1430
0
1km

Figure 9B. Top Windorah Formation structure map (meters ss) showing the assumed oil-water contact and the approximate location of a "1590 sand" zero edge line in the northern part of the field.

4.7°C/100 m is much higher than is required to produce the observed maturity levels, and in order to match measured and modeled maturities lower gradients for the past must be used. Fission track analysis indicates that the present temperature gradient has existed only for the last 1-5 Ma (Duddy, 1987). The presence of major regional aquifers—i.e., the Namur, Adori, and Hutton Sandstones—most probably causes "dog-legs" in heat flow through the section although this has not been fully evaluated.

TRAP

The Bodalla South oil field was discovered by drilling a structural closure on a regional anticline (the Tallyabra trend) as delineated by the seismic method. While the structure plays an important part in the entrapment, stratigraphic variations cause significant modifications to the reservoir configuration.

In the case of the Hutton Sandstone, facies variations related to the fluvial environment of deposition alter the structural levels of the top of the reservoir from well to well, while in the case of the Windorah Formation reworking of a distributory mouth-bar sand produced a highly porous sand body of limited areal extent over part of the structural closure. These features are discussed further in the following section.

RESERVOIRS

In general terms, both oil pools are reservoired in quartzose sandstone in fluvial or fluvio-lacustrine environments. As would be expected, facies variations within this framework result in a number of depositional styles on a local scale. The Hutton Sandstone reservoir occurs at depths of between 1452 and 1474 m. Net hydrocarbon columns at the producing wells vary between 3 and 8.8 m within a gross sequence approximately 10 m thick. The Windorah Formation oil sand was intersected in wells at depths between 1588 and 1595 m. The main reservoir occurs as a uniform sand with an average net pay of 3.6 m.

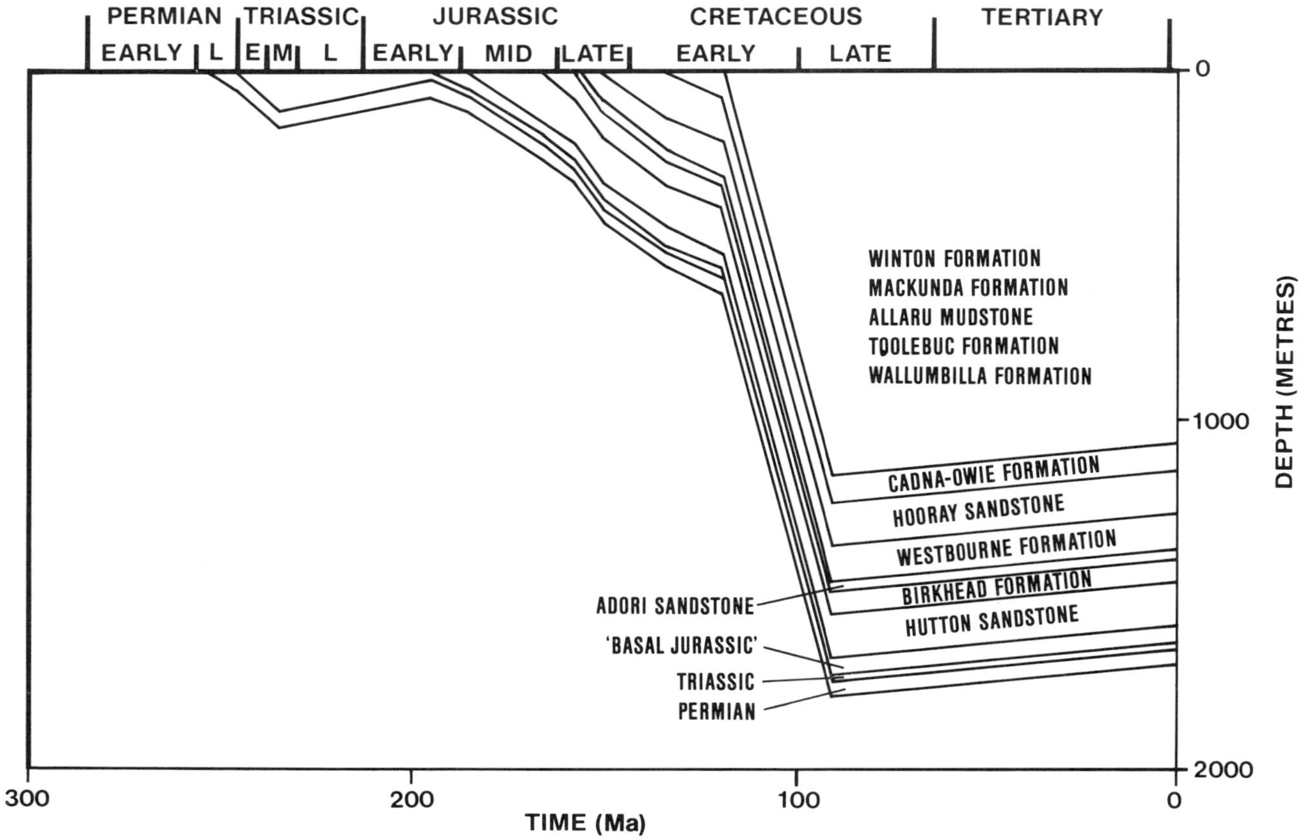
PERMIAN
TRIASSIC
JURASSIC
CRETACEOUS
TERTIARY
EARLY
L
E
M
L
EARLY
MID
LATE
EARLY
LATE
0
1000
2000
DEPTH (METRES)
WINTON FORMATION
MACKUNDA FORMATION
ALLARU MUDSTONE
TOOLEBUC FORMATION
WALLUMBILLA FORMATION
CADNA-OWIE FORMATION
HOORAY SANDSTONE
WESTBOURNE FORMATION
BIRKHEAD FORMATION
HUTTON SANDSTONE
ADORI SANDSTONE
'BASAL JURASSIC'
TRIASSIC
PERMIAN
300
200
100
0
TIME (Ma)

Figure 10. Geohistory plot for the Bodalla South field. Rapid subsidence in the early Cretaceous occurred; deposition ceased in the Late Cretaceous. Time scale after Harland et al. (1982).

Reservoir Stratigraphy

Hutton Sandstone

The Hutton Sandstone occurs in the top 10 m of a thick (130 m) sequence of stacked fluvial sandstones that for the most part were deposited by high-energy braided streams. The uppermost sandstones (which host the oil accumulation) show a waning in energy conditions with a general decrease in grain size and intervals showing variable, but dominantly upward-fining characters (Figures 11A and B). These intervals occur at different structural levels from well to well with resultant variations in porosity, permeability, and water saturations across the reservoir zone. Oil-water contacts occur at different structural levels within the wells (Figure 11B). Lithostratigraphic and chronostratigraphic correlation is difficult at the top of the Hutton Sandstone, and in order to simplify matters the top of the Hutton Sandstone is picked at the top of continuous porosity (with the exception of the Number 5 well discussed below) and is referred to as the "Porous Hutton" (Figures 11A and B).

Porosities in the oil column range from 18% to 20%. Permeabilities commonly show much larger variations from a minimum of around 100 md to in excess of 1 darcy. In general, permeabilities decrease toward the upper parts of the reservoir.

Loss of porosity at the top of the reservoir coincides with a gradational mineralogical change from quartzose sandstones of the Hutton Sandstone to lithic sandstones of the Birkhead Formation; the latter provide the seal to the reservoir. The quartzose sandstones are thought to have been derived from stable cratonic areas to the south and west while it is postulated that the lithic sandstones were derived from volcanic arc complexes some distance to the east (Watts, 1987).

Deposition through most of the medium- to coarse-grained Hutton Sandstone occurred in a braided-stream environment. Decreased energy levels resulted in deposition of finer-grained sediment at the top of the unit and the stabilization of channels and development of levees. Watts (1987) suggested that the overlying Birkhead sandstones were deposited in a shallow lake with the development of distributary channels and mouth bars at the transition from Hutton Sandstone to Birkhead Formation (i.e., the reservoir interval).

The variations observed in the oil-water contact can then be attributed to differential pore pressures precluding entry of hydrocarbons into less permeable

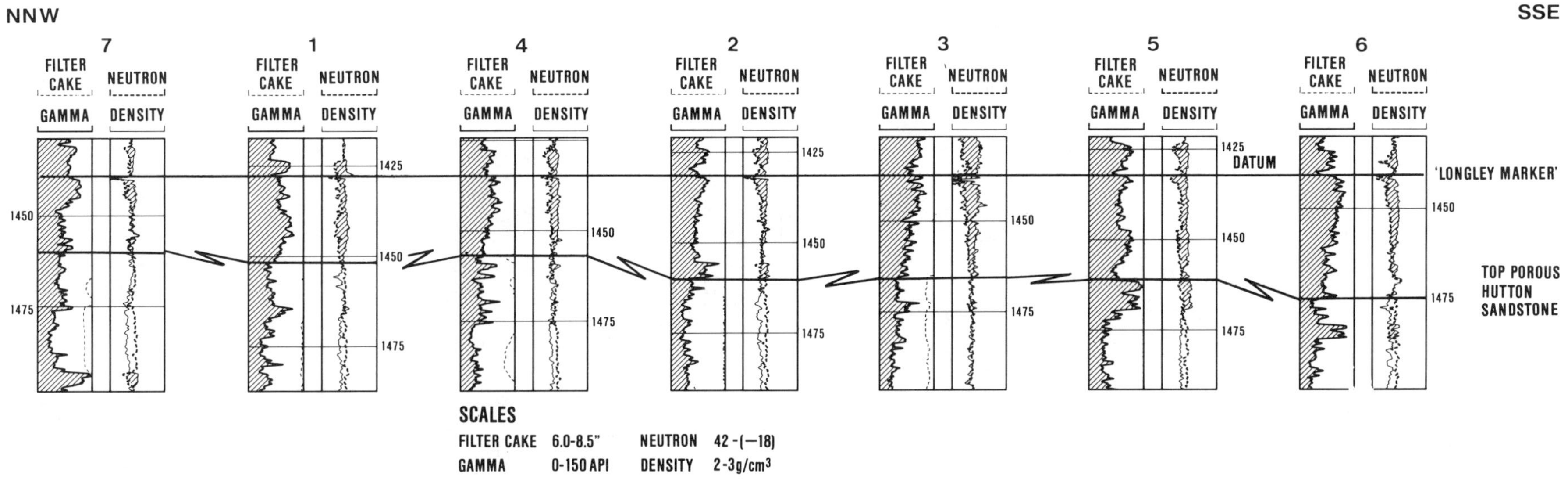
NNW
SSE
7
1
4
2
3
5
6
FILTER CAKE
NEUTRON
GAMMA
DENSITY
1425
1450
1475
DATUM
'LONGLEY MARKER'
TOP POROUS HUTTON SANDSTONE
SCALES
FILTER CAKE 6.0-8.5"
GAMMA 0-150 API
NEUTRON 42-(−18)
DENSITY 2-3g/cm³

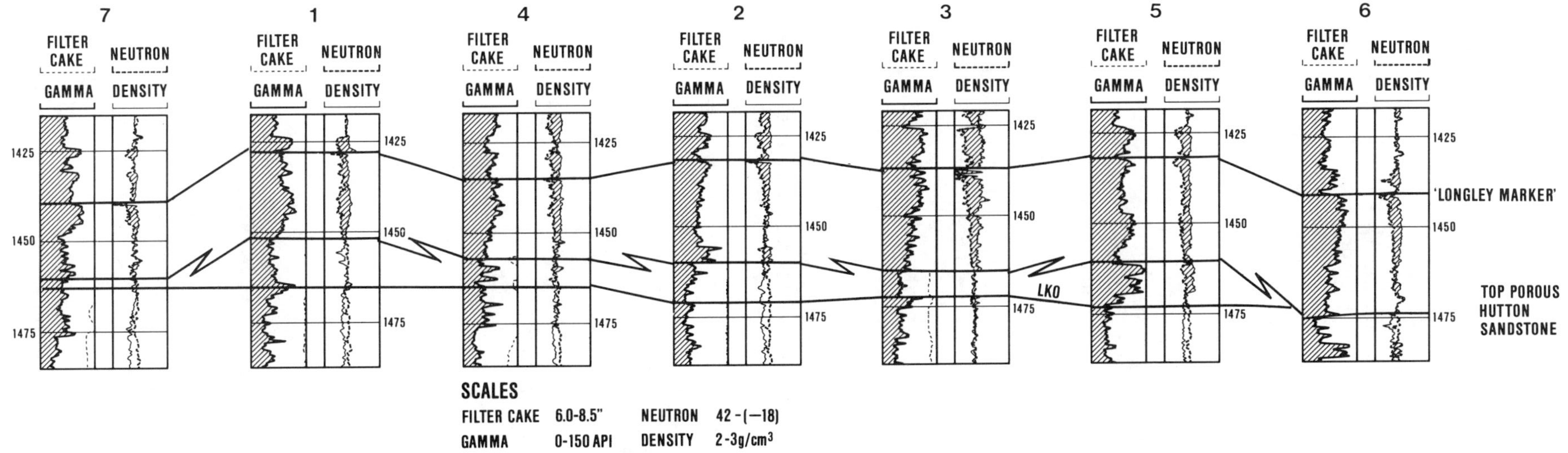
7
1
4
2
3
5
6
FILTER CAKE
NEUTRON
GAMMA
DENSITY
1425
1450
1475
LKO
'LONGLEY MARKER'
TOP POROUS HUTTON SANDSTONE
SCALES
FILTER CAKE 6.0-8.5"
GAMMA 0-150 API
NEUTRON 42-(−18)
DENSITY 2-3g/cm³

Figure 11A. Stratigraphic well correlation section hung on mid-Birkhead coal ("Longley Marker"). Note porosity climbs stratigraphically higher to the north. The well locations can be seen on Figures 9A and 9B.

Figure 11B. Structural well correlation section. Note variability in lowest known oil (LKO) levels. See Figures 9A and 9B for well locations.

zones. The trapping mechanism in this model is predominantly structural closure. Alternatively, the upward-fining beds can be interpreted as a meandering stream phase with the oil reservoired in a number of isolated point bar sands. Such sand bodies could have flat but separate oil-water contacts. It is possible to place the observed depths to lowest known oil at the Bodalla South oil field into this context. Thus wells 7, 1, and 4 to the north may have a common oil-water contact in one sand body, and wells 2, 3, and 5 to the south may be in a second point bar with the oil water contact 4 m lower (Figure 11B).

This model involves an important component of stratigraphic trapping and suggests that similar additional reservoirs may be located in the vicinity.

Support for the meander belt hypothesis is given by the interpretation of the shale plug at the Number 5 well at 1461 m (see Figure 11) as an oxbow lake deposit. A 0.5 m thick quartzose sandstone overlies the shale plug and places the top porous Hutton at the top of this sandstone.

Top porous Hutton climbs stratigraphically to the north (Figure 11A) suggesting that the proposed meander system migrated in this direction.

Diagenetic effects have resulted in reduced reservoir quality. These include silica overgrowths, some carbonate cement, and authigenic clays, which most probably resulted from breakdown of feldspars. These clays coat the sand grains, line the pore walls, and can occlude pore spaces. Kaolinite is dominant with illite and mixed layer illite/smectite also present.

Core studies show that the sandstone is sensitive to fresh water, and formation damage is caused by mobile kaolin particles. A number of the wells have had successful workovers using clay stabilization polymers and underbalance perforating (Dolan et al., 1988).

A strong water drive from the underlying regional Hutton Sandstone aquifer is present. This is one of the main aquifers within the Great Artesian basin, which includes the Eromanga basin and its adjacent basins. High water cuts accompany the oil production.

Windorah Formation

The Windorah Formation is informally divided into a number of subunits based on environmental interpretation of cores and logs (Figure 12). The unit unconformably overlies the Triassic Nappamerri Formation; the division between Triassic and Jurassic rocks is not necessarily associated with an obvious lithological change but is picked from palynological data where available. The Windorah Formation consists of interbedded sands and shales and while oil shows have been observed in all Windorah Formation sands, the major production at the Bodalla South oil field is established in the uppermost "1590 sand." Very minor contributors to oil production are the "lower main sand" at wells 1 and 5 and the uppermost "intra sand" at the Number 5 well.

Palynological dating divides the unit into a lower *C. classoides* zone and an upper *C. turbatus* zone. This division coincides with a climatic change from semi-arid to humid conditions and at Bodalla South is a dividing line between a lower fluvial-dominated regime and an upper lacustrine-dominated regime. The basal sands within the fluvial sequence are not productive at Bodalla South. In the lower part of the lacustrine sequence "intra sands" show small upward-coarsening gamma ray characters in a number of the wells, which are suggestive of either shoreline sands or distal progradational delta facies.

These "intra" sands are laterally extensive and can be correlated in wells away from the oil field; however, their laminated nature results in variable porosity and permeability from well to well. Bodalla South No. 5 is the only well to show porosity development (neutron-density crossover) in the "intra" sands and has 0.5 m of net pay derived from log analysis.

The main reservoir ("1590 sand") lies at the top of a distinctive 10 m thick upward-coarsening deltaic cycle and is sealed by 1.5–3.5 m of interlaminated lacustrine shale and silt. Prodelta shales at the base of the *C. turbatus* zone grades up to distal sands, then to delta front sands and distributary channels as the result of a regressive phase within an overall lacustrine transgression (see Figure 12). The "1590 sand" appears to be the result of a secondary phase of winnowing. Saxena (1979) describes the development of such sands as being due to delta lobe abandonment by cut-off of sediment supply and ensuing modification of the upper distributary bar sands by wave or current action as the transgression advances. Elongate sand bodies oriented parallel to the depositional strike of the abandoned delta lobe result. Such reworked sands are clean and well sorted, hence produce very good reservoirs. A shale band separating the reworked sand from the delta sands can develop in a landward lagoonal setting, represented in this case by the "tight band" (see Figure 13).

The reworked sand ("1590 sand") shows a distinct blocky log character and is present in the southern half of the field but is seen to thin or be influenced by other depositional processes in the Number 4 well and to be absent in Number 7. An oil-water contact has not been intersected in any of the wells; rather, oil is present down to the tight band, or in the case of the Number 4 well, down to a thin shale bed that coalesces with the "tight band" and forms a bottom seal in the vicinity of the well. This is the lowest known oil in the field and in the initial field development it was believed to be the oil-water contact. However, the volume of oil produced very quickly exceeded the total reserves calculated based on that assumption. A reservoir simulation modeled

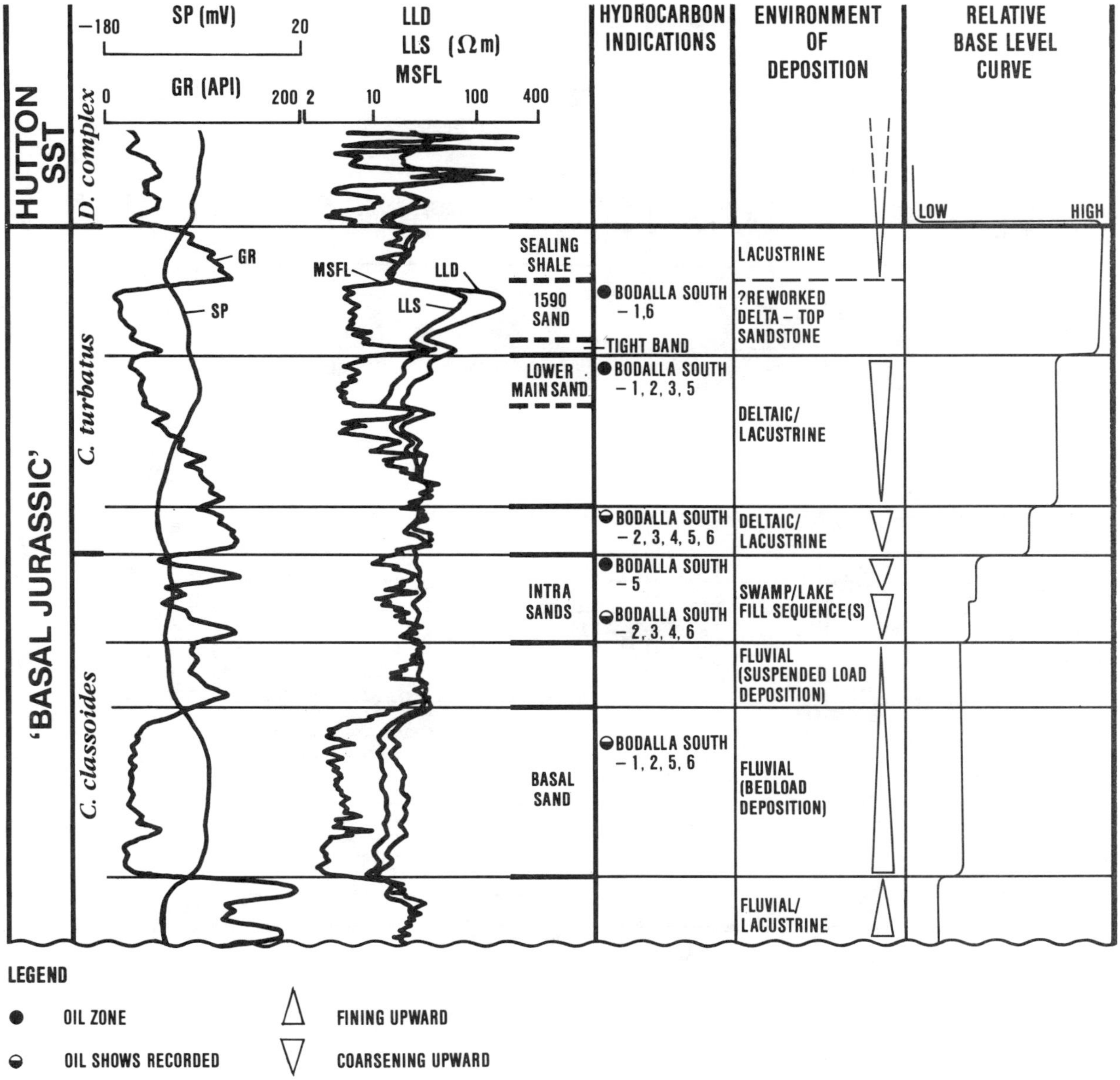
HUTTON SST
'BASAL JURASSIC'
D. complex
C. turbatus
C. classoides
SP (mV)
−180
20
GR (API)
0
200
LLD
LLS (Ωm)
MSFL
2
10
100
400
GR
SP
MSFL
LLD
LLS
SEALING SHALE
1590 SAND
TIGHT BAND
LOWER MAIN SAND
INTRA SANDS
BASAL SAND
HYDROCARBON INDICATIONS
BODALLA SOUTH – 1, 6
BODALLA SOUTH – 1, 2, 3, 5
BODALLA SOUTH – 2, 3, 4, 5, 6
BODALLA SOUTH – 5
BODALLA SOUTH – 2, 3, 4, 6
BODALLA SOUTH – 1, 2, 5, 6
ENVIRONMENT OF DEPOSITION
LACUSTRINE
?REWORKED DELTA – TOP SANDSTONE
DELTAIC/ LACUSTRINE
DELTAIC/ LACUSTRINE
SWAMP/LAKE FILL SEQUENCE(S)
FLUVIAL (SUSPENDED LOAD DEPOSITION)
FLUVIAL (BEDLOAD DEPOSITION)
FLUVIAL/ LACUSTRINE
RELATIVE BASE LEVEL CURVE
LOW
HIGH
LEGEND
OIL ZONE
OIL SHOWS RECORDED
FINING UPWARD
COARSENING UPWARD

Figure 12. Typical wireline log response over the Windorah Formation showing palynological divisions, environmental interpretations, and hydrocarbon shows. The well is Bodalla South No. 3.

on a simple anticlinal reservoir structure with an oil-water contact set 20 m lower, and a pinchout edge along the northwestern flank (between wells Number 4 and 7; see Figure 9B) and strong edge-water drive resulted in a very good match between actual and predicted dates of water breakthrough.

The "lower main sand" was perforated in wells 1 and 5; however, the quality of this zone varies considerably across the oil field, with reasonable porosity and permeability being found only in the Number 3 well. The reservoir quality declines rapidly below this hydrocarbon-bearing interval.

Although the wells are all located high up on the structure, very high sweep efficiencies related to consistent permeability distribution and an excellent natural water drive result in high recovery factors in the order of 60%. Hydraulic communication with the Hutton Sandstone aquifer is inferred by the strong water drive. This connection is either due to a pinchout of the sealing shale that thins to the east or to the Hutton Sandstone being thrown down against the "1590 sand" along the Tallyabra fault.

During field development, low and deteriorating well productivity occurred as a result of extensive and increasing formation damage. Very high skin factors were calculated for a number of wells, and actual productivities showed significant deviation from the ideal. Studies including X-ray diffraction and SEM on the cores indicated that migration of mobile clays and fines was the source of formation damage. Workovers using tubing conveyed perforated systems with an underbalance exceeding 1000

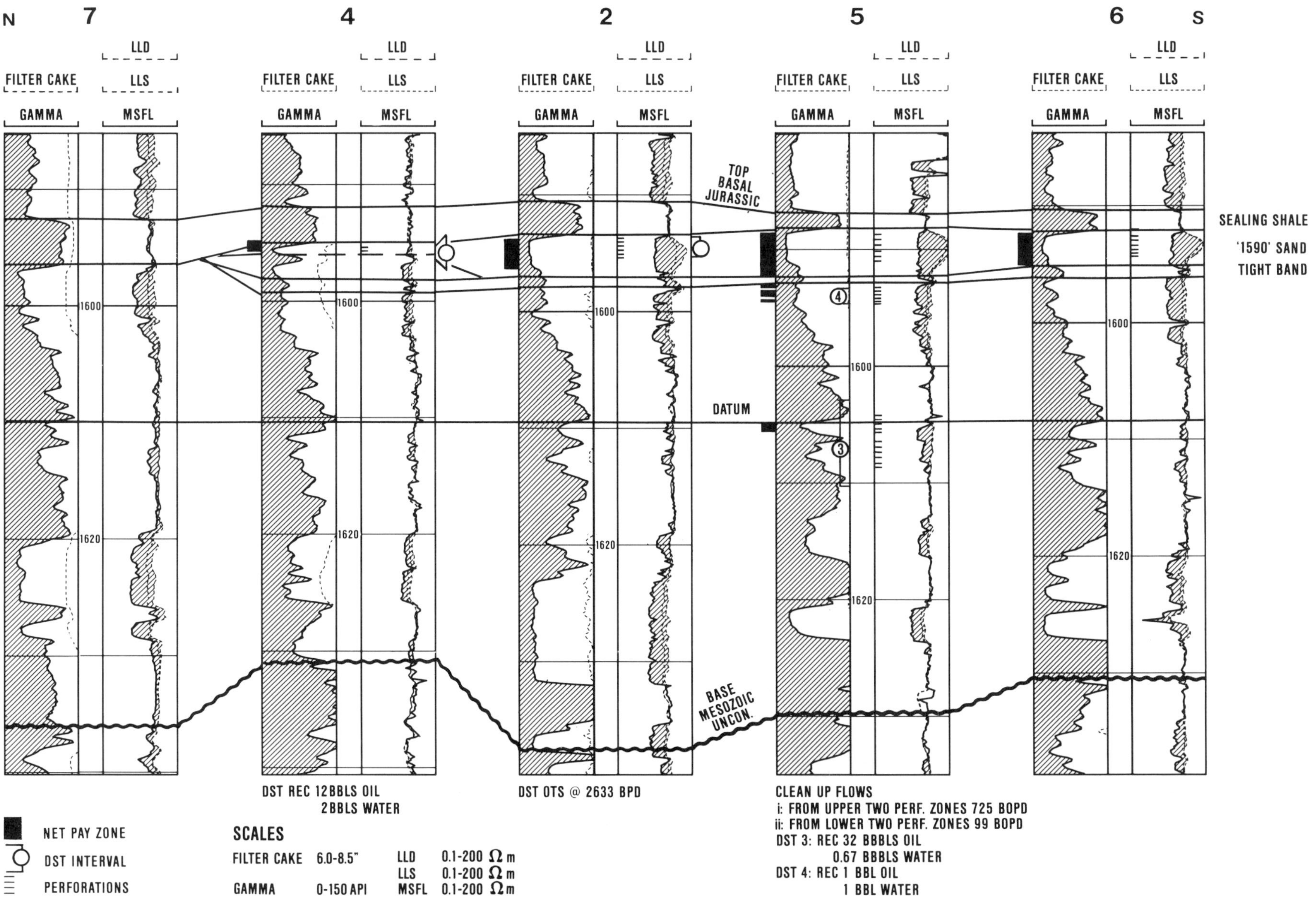

N
7
4
2
5
6
S
LLD
LLS
MSFL
FILTER CAKE
GAMMA
1600
1620
TOP BASAL JURASSIC
DATUM
BASE MESOZOIC UNCON.
SEALING SHALE
'1590' SAND
TIGHT BAND
DST REC 12BBLS OIL
2BBLS WATER
DST OTS @ 2633 BPD
CLEAN UP FLOWS
i: FROM UPPER TWO PERF. ZONES 725 BOPD
ii: FROM LOWER TWO PERF. ZONES 99 BOPD
DST 3: REC 32 BBBLS OIL
0.67 BBBLS WATER
DST 4: REC 1 BBL OIL
1 BBL WATER
NET PAY ZONE
DST INTERVAL
PERFORATIONS
SCALES
FILTER CAKE 6.0-8.5"
GAMMA 0-150 API
LLD 0.1-200 Ω m
LLS 0.1-200 Ω m
MSFL 0.1-200 Ω m

Figure 13. Stratigraphic well correlation of the Windorah Formation showing net pay zones, drill-stem tests, and perforated intervals.

psi resulted in substantially reduced skin factors and significantly improved production rates (30 BOPD to 970 BOPD and 165 to 1520 BOPD, respectively, for wells 1 and 5) (Dolan et al., 1988).

Diagenetic effects are similar to those observed in the Hutton Sandstone; however, the larger pore spaces reduce the blocking effect of migrating fines. Minor secondary porosity development due to dissolution of feldspars and corrosion of quartz grains has been noted.

FORMATION EVALUATION

Wireline log analyses were carried out using a shaly sand model. Water resistivities were selected on the basis of R_{wa} calculations or measured water resistivities as appropriate. This approach has yielded satisfactory results for the Windorah Formation reservoir but not for the apparently much fresher and shalier Hutton Sandstone reservoir.

Hutton Sandstone

The resistivity contrast between the oil reservoir and the underlying artesian aquifer is low (Figures 14 and 15). Consequently, conventional log analysis techniques compute water saturations in the 70% to 90% range in zones that flowed clean oil at almost 900 barrels per day on drill-stem test. Test data and capillary pressure measurements on core suggest that water saturations in the 30% to 40% range were more realistic.

Subsequent wells were fully cored over the Hutton Sandstone oil reservoir and a short program of XRD and CEC analyses showed that the sandstone had a clay content of less than 5%. The dominant clay type is kaolinite, but minor amounts of sericite are also present. The CEC of the bulk rock was rather low at 0.01 meq/100 g.

Special core analysis was carried out on cores from wells 1, 2, and 3, and capillary pressure measurements were normalized using a *J* function as illustrated by Figure 16. This plot indicates that there is an empirical relationship between *J* and water saturation and that the inherent assumptions of the *J* function approach are met.

The value of *J* can be converted from laboratory to field conditions, and if this is done then

$$J = \frac{h}{72}\sqrt{\frac{k}{\phi}}$$

where h = height above a free water level
k = permeability in millidarcies
ϕ = porosity (fractional)

Porosity and permeability were measured by routine core analysis. The free water level, at 1314 m subsea in this case, was chosen by trial and error until the best fit between residual oil saturations measured from core, the resistivity character of the logs, and the water saturations obtained from the *J* function was obtained for all wells. The results are illustrated by Figure 17.

In cases where core data were unavailable, net pay was determined from microlog separation and porosity from the nuclear logs. Permeabilities were estimated from flow rates, and water saturations were then estimated from the *J* function.

Several inferences can be made from this log evaluation of the Hutton Reservoir. First, if the low resistivity of the Hutton oil reservoir is due to neither clay effects nor boundary effects but to a change in water resistivity between the oil zone and the aquifer beneath as assumed here, then the oil migrated into the reservoir prior to the establishment of the aquifer. However, if the water resistivities are the same, then traditional saturation equations do not work properly in the very fresh and relatively hot (95°C) environments encountered in Bodalla South. Recent work carried out by Gravestock (in press) suppports this view. His work on the Waxman-Smits equation and Hutton Sandstone oil reservoirs in South Australia shows that the effective electrical activity of clay is much greater than expected if the formation waters are fresh and hot. This is attributed to the temperature dependence of the equivalent conductivity of the clay ("B") part of BQv in the Waxman-Smits equation (refer to Waxman and Smits, 1968) being much greater than previously recognized for low salinity brines.

The Windorah Formation

This reservoir is a clean, upward-coarsening sandstone with good resistivity response (Figure 12). Minimum water saturations of 20–30% were computed by a simple shaly sand model, and this was consistent with inferences derived from capillary pressure measurements.

A decrease in resistivity at the top of the reservoir is believed to be due to boundary effects influencing the resistivity measurements. This is supported by core analyses which indicate that permeability and porosity increase at the top of the sandstone and by the microresistivity traces which possess much greater vertical resolution and show smaller boundary effects.

OIL AND FIELD CHARACTERISTICS

Oil produced from both the Hutton and Windorah Formation reservoirs are very similar in type and quality. The oils are paraffinic crudes, generated from moderately mature terrestrial source rocks. Mild water washing has removed some of the lighter normal alkanes.

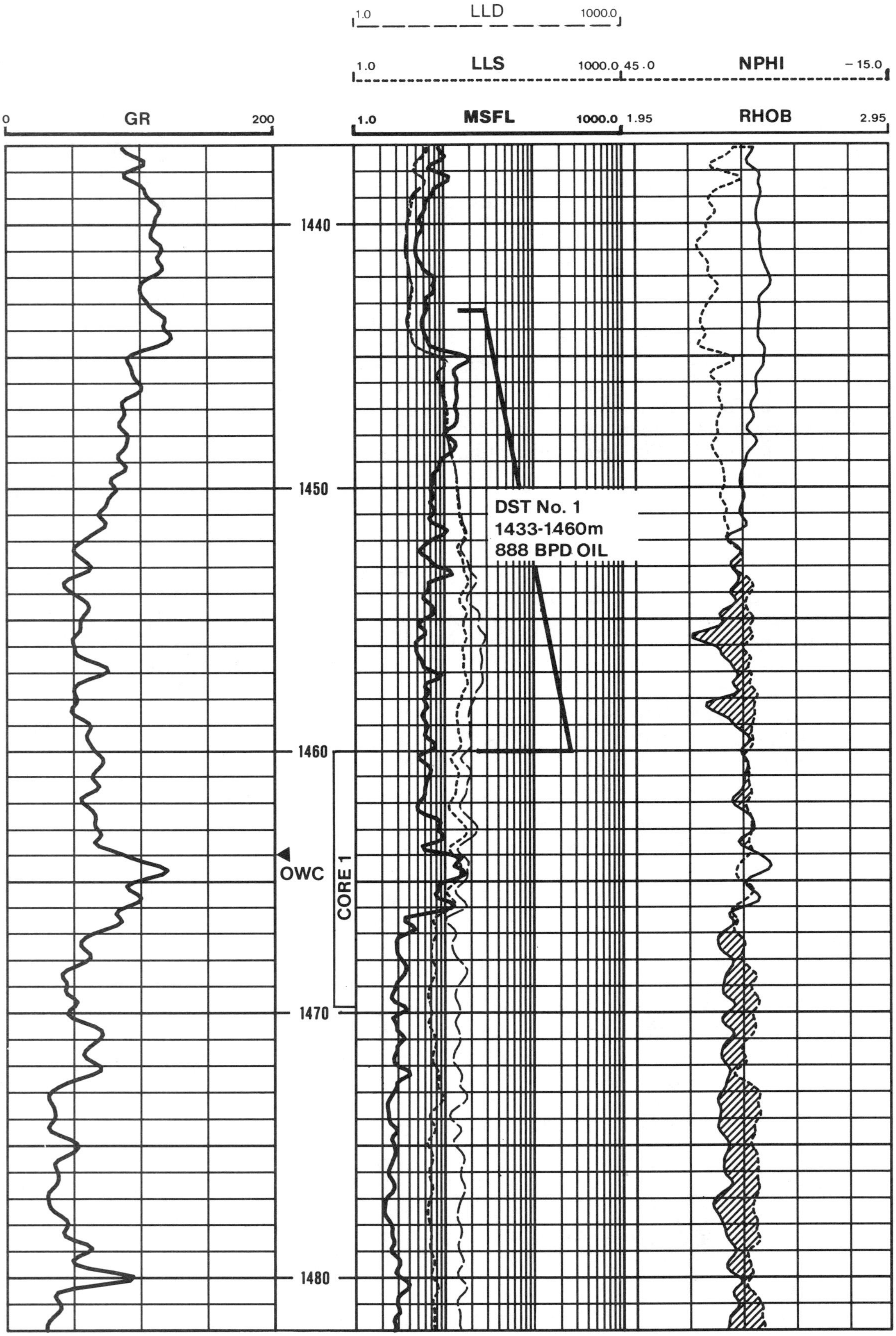
1.0
LLD
1000.0
1.0
LLS
1000.0
45.0
NPHI
-15.0
0
GR
200
1.0
MSFL
1000.0
1.95
RHOB
2.95
1440
1450
1460
1470
1480
DST No. 1
1433-1460m
888 BPD OIL
OWC
CORE 1

Figure 14. Hutton Sandstone log display Bodalla South No. 1. Drill-stem test and core zones are shown. Note the very low response in the LLS and LLD logs over the oil zone (cf. Figure 15).

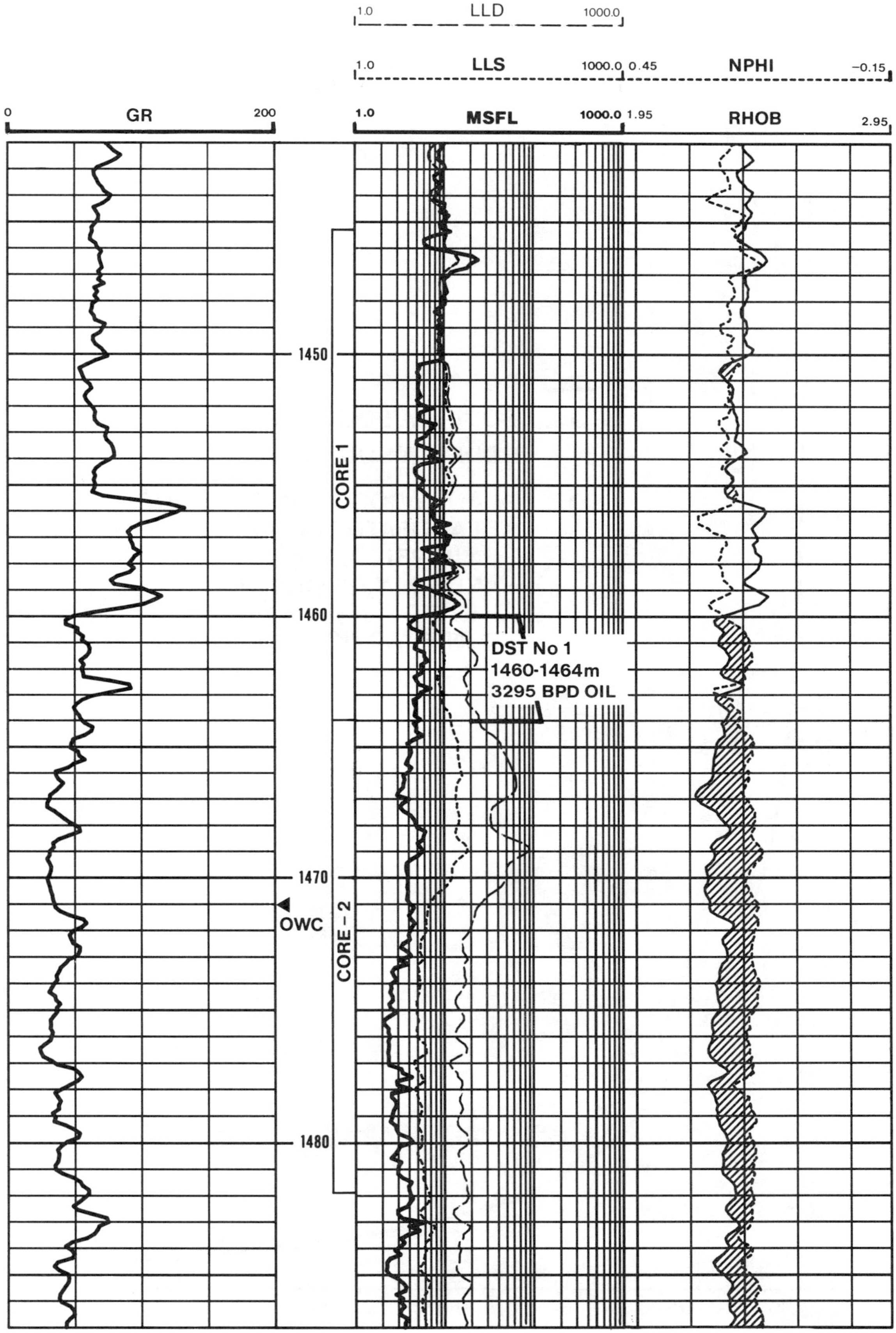
1.0 LLD 1000.0
1.0 LLS 1000.0 0.45 NPHI -0.15
0 GR 200
1.0 MSFL 1000.0 1.95 RHOB 2.95
1450
1460
1470
1480
CORE 1
CORE - 2
OWC
DST No 1
1460-1464m
3295 BPD OIL

Figure 15. Hutton Sandstone log display Bodalla South No. 2. Note the typical LLS, LLD response over the oil zone.

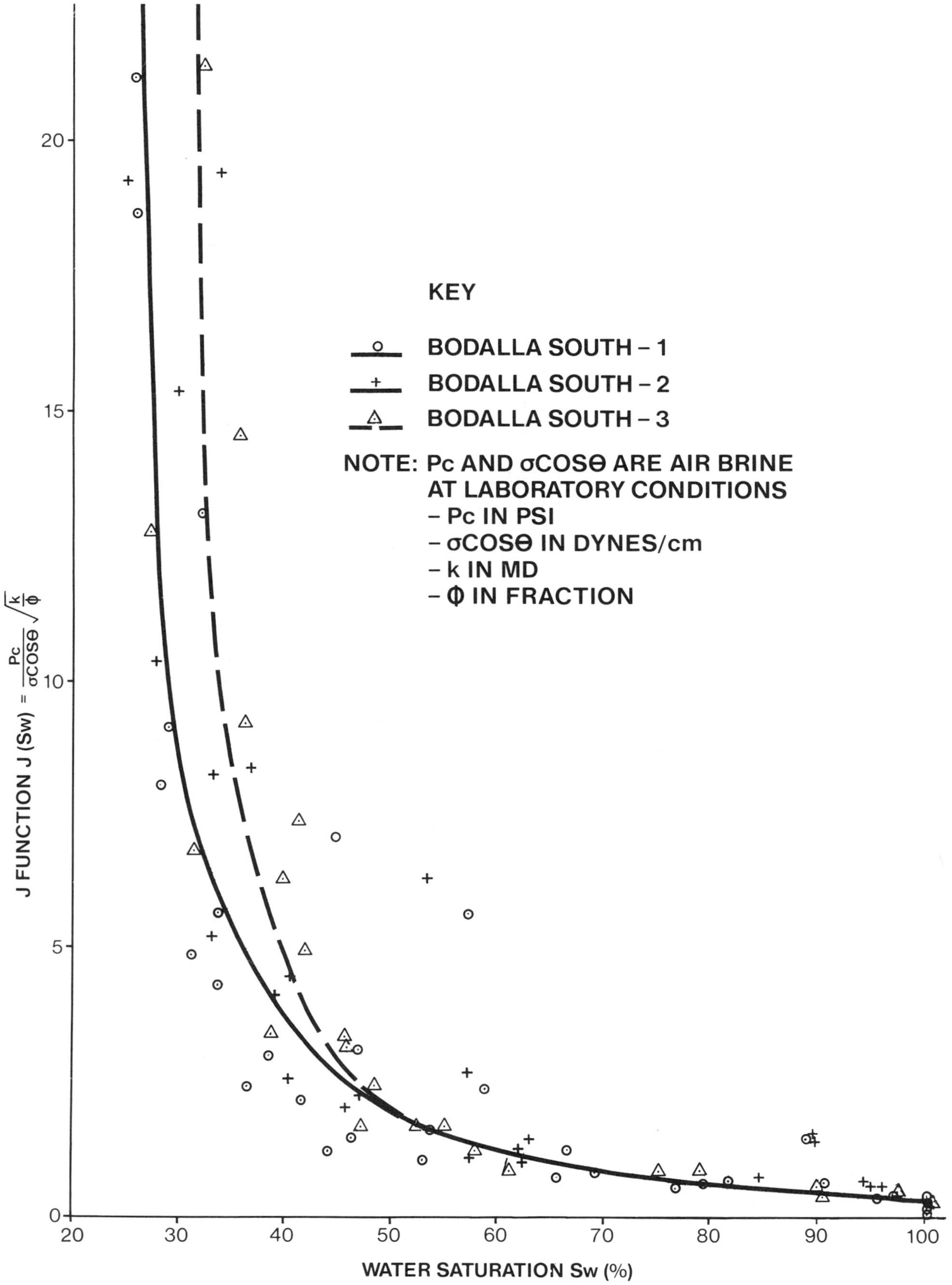
KEY
BODALLA SOUTH - 1
BODALLA SOUTH - 2
BODALLA SOUTH - 3
NOTE: Pc AND σCOSΘ ARE AIR BRINE AT LABORATORY CONDITIONS
- Pc IN PSI
- σCOSΘ IN DYNES/cm
- k IN MD
- Φ IN FRACTION
J FUNCTION J (Sw) = Pc/σCOSΘ √(k/Φ)
WATER SATURATION Sw (%)
0
5
10
15
20
20
30
40
50
60
70
80
90
100

Figure 16. "*J*" function versus water saturation plot for Bodalla South No.'s 1, 2, and 3.

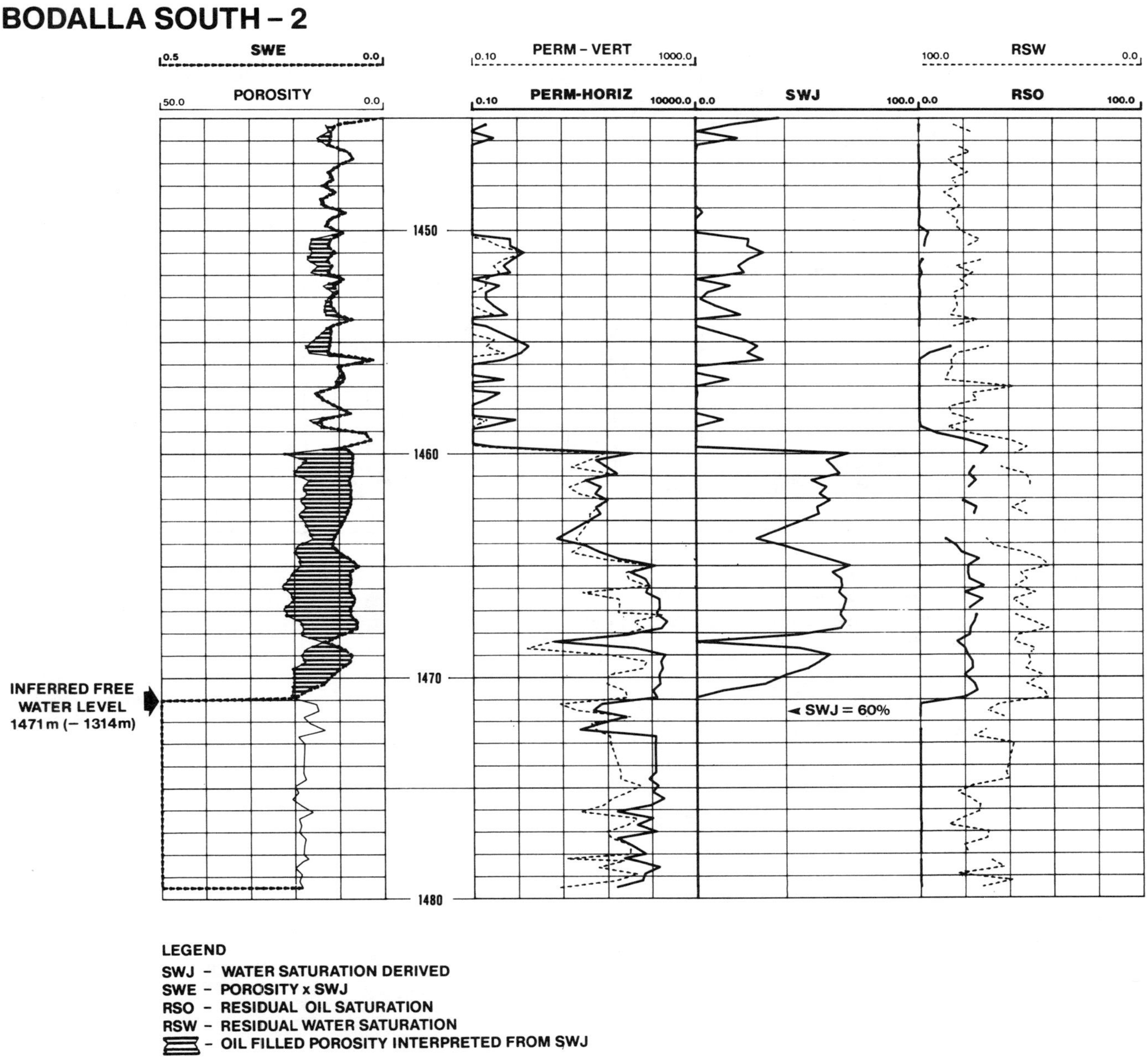
BODALLA SOUTH - 2
0.5 SWE 0.0
50.0 POROSITY 0.0
0.10 PERM - VERT 1000.0
0.10 PERM-HORIZ 10000.0
0.0 SWJ 100.0
100.0 RSW 0.0
0.0 RSO 100.0
1450
1460
1470
1480
INFERRED FREE WATER LEVEL 1471 m (– 1314m)
SWJ = 60%
LEGEND
SWJ - WATER SATURATION DERIVED
SWE - POROSITY x SWJ
RSO - RESIDUAL OIL SATURATION
RSW - RESIDUAL WATER SATURATION
- OIL FILLED POROSITY INTERPRETED FROM SWJ

Figure 17. "*J*" function analyses for Bodalla South No. 2, Hutton Sandstone.

The oils are light and sweet with high gravities—49° API in the Windorah Formation and 47° API in the Hutton Sandstone. They yield high proportions of diesel on distillation. Both oils have a relatively low pour point of 6°C.

Both crudes contain very few alkanes in the C_1 to C_6 range and the bulk of the oil is normal alkanes in the C_7 to C_{18} range (see Figure 18A and B). This accounts for the high gravities and low pour point. The absence of the C_1 to C_6 compounds explains the extremely low gas-oil ratio, measured at 5 standard ft^3/STB for the Hutton oil and 2 standard ft^3/STB for the Windorah Formation oil.

From PVT analyses, reservoir pressures of 2295 psia (15,824 kPa) and 2486 psia (17,141 kPa) were measured for the Hutton and Windorah Formation, respectively, at temperatures of 101°C (214 °F) and 103°C (218°F). Formation volume factors are 1.062 and 1.052 RB/STB, respectively, and viscosities of 1.223 cp and 0.758 cp, respectively, were measured.

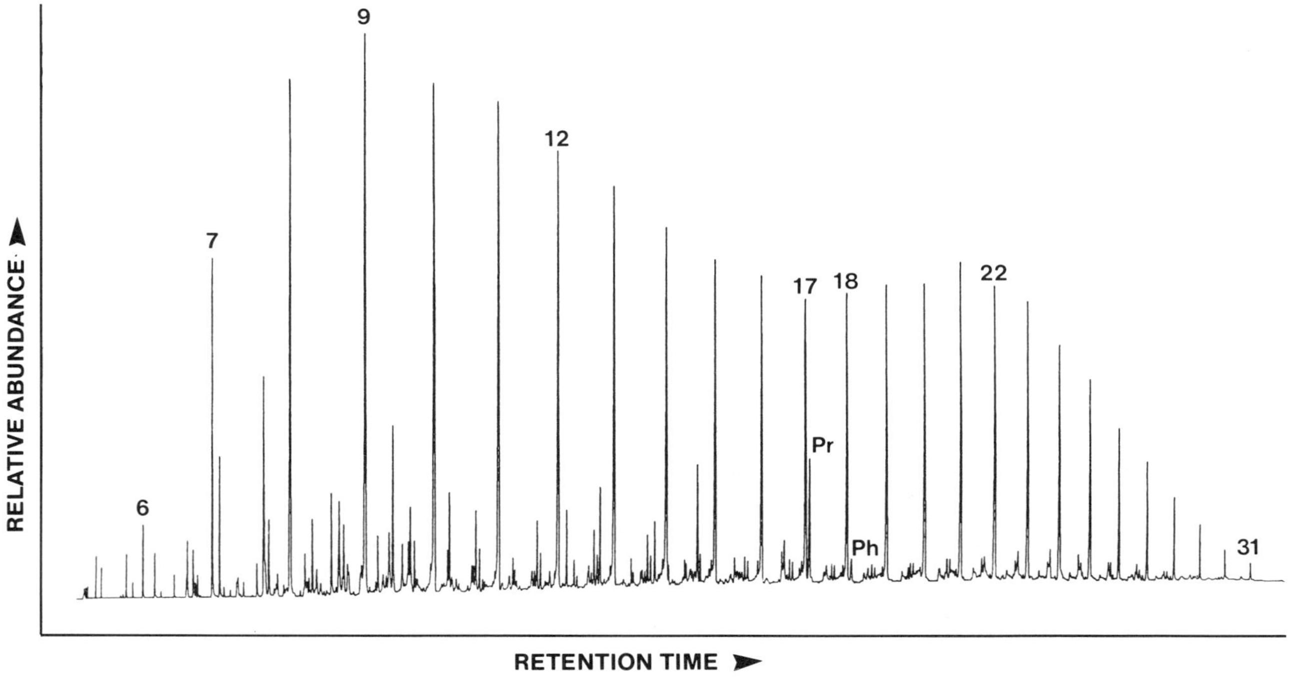

RELATIVE ABUNDANCE
9
12
7
17
18
22
Pr
6
Ph
31
RETENTION TIME

Figure 18A. Typical whole oil gas chromatogram of oil sample from the Hutton Sandstone at Bodalla South. (Sampled from Bodalla South No. 1.)

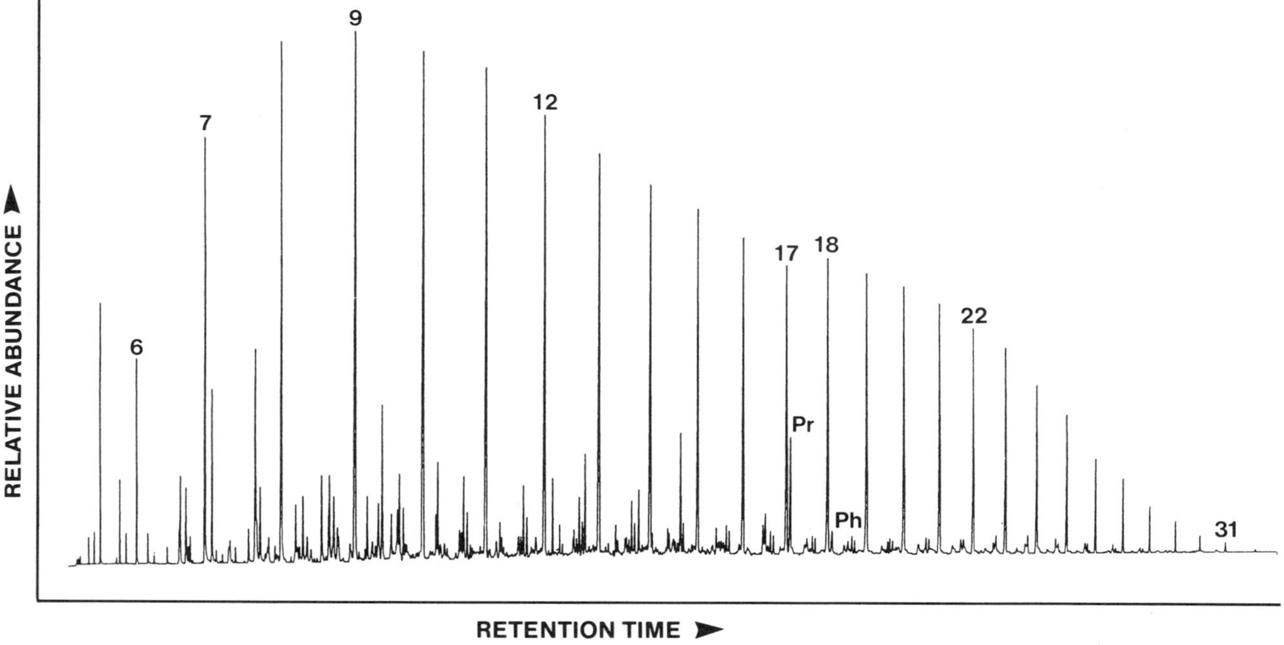

RELATIVE ABUNDANCE
9
12
7
17
18
22
6
Pr
Ph
31
RETENTION TIME

Figure 18B. Typical whole oil gas chromatogram of oil sample from the Windorah Formation at Bodalla South. (Sampled from Bodalla South No. 1.)

EXPLORATION AND DEVELOPMENT CONCEPTS

The variability in sand distribution brought about by the dominantly fluvial environment of deposition of the Hutton Sandstone results in generally poor predictability. Added to this are the low relief of the structural closures and the apparent lack of a significant seismic reflection from the top of the reservoir porosity. These factors together make exploration and development of the Hutton Sandstone reservoir very difficult. Three of the seven wells (the three most recently drilled) have not intersected commercial hydrocarbon-bearing sandstones primarily due to adverse stratigraphic changes in the reservoir section.

The elucidation of geological factors relating to environments of deposition and of the orientation/distribution of sands at the Hutton-Birkhead transition is crucial in the location of additional wells both within the current confines of the field and as potential step-out wells designed to locate additional sand bodies. As with other Hutton Sandstone oil fields in the area, the combination of rapid facies changes, the low relief, and the resolution limits of seismic data are significant factors that handicap exploration and development. Alternative methods such as soil geochemistry and controlled source audio-frequency magnetotellurics (CSAMT) have been employed with varying (and arguable) success.

The recording of high-quality new seismic data and the reprocessing of previously recorded data (even data acquired in the last few years) has been carried out in an attempt to extract as much stratigraphic information as possible and to refine structure maps as far as possible. Trials of seismic inversion processing show that some success may be achievable; however, at the time of writing, this work was in the early stages of evaluation.

A reservoir simulation of the Windorah Formation easily achieved an exceptionally close match between predicted and actual water breakthroughs, and on this basis the model is regarded as reliable. Further predictions show that the reservoir, being thin and consistently permeable with good bottom and top seals and strong edge-water drive, will be adequately swept with the existing wells that are structurally high in relation to the oil-water contact. The model was run until watercuts reached 95%, which represented a recovery factor of 60%. As watercuts in the structurally lower wells have increased to in excess of 95%, they have been shut in to reduce reservoir voidage. Resultant increases in production at the structurally higher wells with a diminished decline trend has resulted. Thus, the Windorah Formation reservoir requires no further development drilling. Current plans focus on artificial lift in order to maintain oil production rates.

ACKNOWLEDGMENTS

The authors wish to thank the ATP 269P (1, 3) Joint Venturers, namely Beach Petroleum N.L., Bligh Oil & Minerals N.L., Northern Michigan Exploration Co., Oil Company of Australia N.L., TCPL Resources Ltd., and Western Mining Corporation Ltd. for permission to publish this paper. The views expressed in this paper are those of the authors and not necessarily those of the Joint Venture.

REFERENCES CITED

Bowering, O.J.W., 1982, Hydrodynamics and hydrocarbon migration—a model for the Eromanga basin: APEA Journal, v. 22, pt. 1, p. 227-236.

Dolan, P., P.D. Griffiths, and S.R. Welton, 1988, The successful recovery of well productivity in the Bodalla South Oilfield: APEA Journal, v. 28, pt. 1, p. 7-18.

Duddy, I.R., 1987, Apatite fission track analysis of one sample from Bodalla South No. 2, Eromanga Basin Queensland: a report prepared for LASMO Energy Aust. Ltd. Geotrack International, 9 p.

Farley, F.R., 1967, Report on Geograph seismic survey of the Tallyabra area Queensland Authority to Prospect 99P: Unpublished Report: B.P. Petroleum Development Australia Pty. Ltd.

Finlayson, D.M., S.H. Leven, and M.A. Etheridge, 1988, Structural styles and basin evolution in Eromanga Region, Eastern Australia: American Association of Petroleum Geologists Bulletin, v. 72, n. 1, p. 33-48.

Halyburton, R.V., and A.L. Robertson, 1984, Geology of the Jackson Oilfield: APEA Journal, v. 24, pt. 1, p. 62-84.

Harland, W.B., A. Cox, P.G. Llewellyn, C.A.G. Pickton, A.G. Smith, R. Walters, and K.E. Fancett, 1982, Geologic time scale: Cambridge University Press, 1 p.

Mulready, J., 1971, Final Report Authority to Prospect 99P, Queensland: Unpublished Report, Alliance Oil Management.

Newton, C.B., 1986, The Tintaburra Oilfield: APEA Journal v. 26, pt. 1, p. 334-352.

Saxena, Ram.S., 1979, Facies models and subsurface exploration methods for deltaic sandstone reservoirs: PESA Overseas Lecture Course Notes, p. 48-57.

Senior, B.R., and J.A. Mabbutt, 1979, A proposed method of defining deeply weathered rock units based on regional geological mapping in southwest Queensland: Journal of the Geological Society of Australia, v. 26, p. 237-254.

Veevers, J.J., ed., 1984, Phanerozoic earth history of Australia: Oxford Geological Sciences Series, 418 p.

Watts, K.J., 1987, The Hutton Sandstone-Birkhead Formation Transition, ATP 269P(1), Eromanga Basin: APEA Journal, v. 27, pt. 1, p. 215-228.

Waxman, M.H., and L.J.M. Smits, 1968, Electrical conductivities in oil bearing shaly sands: SPE Journal, v. 8, n. 2, p. 107-122.

Appendix 1. Field Description

Field name .... *Bodalla South*

Ultimate recoverable reserves .... *5.4 million bbl*

Field location:

- **Country** .... *Australia*
- **State** .... *Queensland*
- **Basin/Province** .... *Eromanga*

Field discovery:

- **Year field discovered** *1984*
- **Year second pay discovered** *1984*
- **Third pay** .... *NA*

Discovery well name and general location
(i.e., Jones No. 1, Sec. 2T12NR5E; or Smith No. 1, 5 mi west of Sheridan, Wyoming):

- **First pay** .... *Bodalla South No. 1*
- **Second pay** .... *Bodalla South No. 1*
- **Third pay** .... *NA*

Discovery well operator .... *LASMO Australia*
(if more than one pay in field, list operators of discovery well in other pays)

- **Second pay** .... *LASMO Australia*
- **Third pay** .... *NA*

IP in barrels per day and/or cubic feet or cubic meters per day:

- **First pay** .... *200 BOPD*
- **Second pay** .... *220 BOPD*
- **Third pay** .... *NA*

All other zones with shows of oil and gas in the field:

Age	Formation	Type of Show
Cretaceous	*Cadna-Owie*	*Fluorescence*
Cretaceous	*Namur*	*Fluorescence*
Jurassic	*Westbourne*	*Fluorescence*
Jurassic	*Birkhead*	*Fluorescence*
Permian	*Toolachee*	*Fluorescence*
Permian	*Patchawarra*	*Fluorescence*

Geologic concept leading to discovery and method or methods used to delineate prospect, e.g., surface geology, subsurface geology, seeps, magnetic data, gravity data, seismic data, seismic refraction, nontechnical:

Searching for paleoclosures along elongate Tertiary anticlines. Closures located by seismic data.

Structure:

Province/basin type (see St. John, Bally, and Klemme, 1984) .... *Eromanga, intracratonic*

Tectonic history

Four major tectonic episodes account for the structure of the Bodalla South field: Early to Middle Carboniferous, Late Permian, Middle Triassic, middle Tertiary. All episodes were crustal shortening events, occurring as reactivations over older basement trends. Extensive regional erosion followed the Carboniferous, Triassic, and Tertiary episodes.

Regional structure

The Bodalla South oil field is contained within an anticlinal structure on the western, upthrown flank of the northwest-southeast-trending Tallyabra fault.

Local structure

Asymmetric anticline, steeply dipping eastern side, shallower dips to the west.

Trap

Trap type(s)

Anticlinal trap with multiple pays; minor pinchout trapping may also be present.

Basin stratigraphy (major stratigraphic intervals from surface to deepest penetration in field):

Age	Formation	Depth to Top
Early-Late Cretaceous	*Winton*	*Surface*
Early Cretaceous	*Mackunda*	*150 m*
Early Cretaceous	*Allaru Mudstone*	*200 m*
Early Cretaceous	*Toolebuc*	*800 m*
Early Cretaceous	*Wallumbilla*	*820 m*
Early Cretaceous	*Cadna-Owie Hooray Sandstone*	*1065 m*
Early Cretaceous	*Murta Member*	*1145 m*
Early Cretaceous	*Namur Member*	*1170 m*
Late Jurassic	*Westbourne*	*1265 m*
Late Jurassic	*Adori Sandstone*	*1380 m*
Middle-Late Jurassic	*Birkhead*	*1390 m*
Middle Jurassic	*Hutton Sandstone*	*1450 m*
Early Jurassic	*Windorah Formation*	*1590 m*
Triassic	*Nappamerri*	*1630 m*
Late Permian	*Toolachee*	*1655 m*
Late Carboniferous-Early Permian	*Patchawarra*	*1675 m*

Location of well in field .... *NA*

Reservoir characteristics:

Number of reservoirs *2*

Formations .... *Hutton Sandstone, Windorah Formation ("basal Jurassic")*

Ages .... *Hutton, Middle Jurassic; Windorah, Early Jurassic*

Depths to tops of reservoirs .... *Hutton, 1450 m; Windorah, 1590 m*

Gross thickness (top to bottom of producing interval) .... *Hutton, 12 m; Windorah, 7.5 m*

Net thickness—total thickness of producing zones

Average .... *Hutton, 5.4 m; Windorah, 3.6 m*

Maximum .... *Hutton, 8.8 m; Windorah, 4.8 m*

Average ....

Maximum ....

Lithology .... *Quartzose to lithic sandstones generally well sorted*

Porosity type .... *Intergranular*

Average porosity .... *Hutton, 18%; Windorah, 17%*

Average permeability .... *Hutton, 675 md; Windorah, 1606 md*

Seals:

Upper

Formation, fault, or other feature .... *Hutton, Birkhead Formation; Windorah, basal Jurassic shale unit*

Lithology .... *Hutton, lithic sandstone; Windorah, shale*

Lateral

Formation, fault, or other feature .... *Hutton, Windorah, rollover*

Lithology .... *NA*

Source:

Formation and age .... *Either Jurassic or Permian shales*

Lithology .... *Jurassic shale; Permian coal*

Average total organic carbon (TOC) .... *Jurassic, 1%; Permian, 20%*

Maximum TOC .... *Jurassic, 5.0%; Permian, 70%*
Kerogen type (I, II, or III) .... *II and II-III*
Vitrinite reflectance (maturation) .... *R_0 = 0.6 (Jurassic)-0.8 (Permian)*
Time of hydrocarbon expulsion .... *Late Cretaceous-Tertiary*
Present depth to top of source .... *~1800 m*
Thickness .... *Various, generally greater than 20 m*
Potential yield .... *Unknown*

Appendix 2. Production Data

Field name .... *Bodalla South*

Field size:

Proved acres .... *Hutton, 860; Windorah, 346*
Number of wells all years *7*
Current number of wells (as of year) *5*
Well spacing .... *Various*
Ultimate recoverable .... *Hutton, 1.9 million bbl; Windorah, 3.2 million bbl*
Cumulative production (to 12/88) .... *Hutton, 0.85 million bbl; Windorah, 2.56 million bbl*
Annual production (1988) .... *Hutton, 0.25 million bbl; Windorah, 0.40 million bbl*
Present decline rate .... *Less than 100 bbl per day per month*
Initial decline rate .... *NA*
Overall decline rate .... *NA*
Annual water production .... *Hutton, 0.60 million bbl; Windorah, 0.48 million bbl*
In place, total reserves .... *10.3 million bbl*
In place, per acre-foot .... *Hutton, 780; Windorah, 710 bbl*
Primary recovery .... *5.1 million bbl*
Secondary recovery .... *NA*
Enhanced recovery .... *NA*
Cumulative water production .... *2.14 million bbl*

Drilling and casing practices:

Amount of surface casing set .... *200 m*
Casing program

16-in. conductor to 10 m; 9⅝-in. R3 LT and C K-55 or N-20 to 200 m; 7-in. R3 LT and C L-80 or N-80 from surface.

Drilling mud .... *Floculated gel/Lime spud mud, then KCL brine, then KCL polymer*
Bit program .... *HTC-X3A, J11, J22, DB, FV*
High pressure zones .... *Nil*

Completion practices:

Interval(s) perforated .... *Varies from 2 to 7 m*
Well treatment .... *NA*

Formation evaluation:

Logging suites .... *BHCS-GR, HDT, LDT, CNL, DLL, MSFL, SP*
Testing practices .... *Generally open-hole DST off bottom*
Mud logging techniques .... *Standard; hotwire, chromatograph drill rate, pump pressure, etc.*

Oil characteristics:

Type .... *Paraffinic*
(Tissot and Welte Classification in "Petroleum Formation and Occurrence," 1984, Springer-Verlag, p. 419)
API gravity .... *Hutton, 47.3°; Windorah, 48.6°*

Base .... *NA*
Initial GOR .... *Hutton, 5.0 standard ft³/STB; Windorah, 2.0 standard ft³/STB*
Sulfur, wt% .... *Hutton, Windorah, 0.01*
Viscosity, SUS .... *Hutton, 1.223 cp; Windorah, 0.758 cp*
Pour point .... *Hutton, 6°C; Windorah, 6°C*
Gas-oil distillate .... *NA*

Field characteristics:

Average elevation .... *200 m*
Initial pressure .... *Hutton, 2295 psi; Windorah, 2486 psi*
Present pressure .... *Hutton, 2295 psi; Windorah, 2486 psi*
Pressure gradient .... *Hutton, 0.491 psi/ft; Windorah, 0.484 psi/ft*
Temperature .... *Hutton, 101°C; Windorah, 103°C*
Geothermal gradient .... *4.7°C/100 m*
Drive .... *Aquifer*
Oil column thickness .... *Hutton, 12 m max.; Windorah, 7.5 m*
Oil-water contact .... *Hutton, 1314 ss (L.K.O.); Windorah, 1460 ss m*
Connate water .... *Hutton, 40%; Windorah, 40%*
Water salinity, TDS .... *Hutton, 1000 ppm; Windorah, 1000 ppm*
Resistivity of water .... *Hutton, 0.95 m; Windorah, 0.95 m*
Bulk volume water (%) .... *NA*

Transportation method and market for oil and gas:

Road transport to pipeline terminal.

Leidy Gas Field, Clinton and Potter Counties, Pennsylvania

JOHN A. HARPER
Pennsylvania Geological Survey
Pittsburgh, Pennsylvania

FIELD CLASSIFICATION

BASIN: Appalachian
BASIN TYPE: Foredeep
RESERVOIR ROCK TYPE: Sandstone
RESERVOIR ENVIRONMENT OF DEPOSITION: Beach/Nearshore

RESERVOIR AGE: Devonian
PETROLEUM TYPE: Gas
TRAP TYPE: Combination Structural-Diagenetic-Stratigraphic Pinchout

INTRODUCTION

"When the history of the natural gas industry is written," penned Al Ingham, former chief geologist for Peoples Natural Gas Company in Pittsburgh, "a prominent date will be January 8, 1950" (Ingham, 1954). That was the date that the Leidy Prospecting Company discovered natural gas in the Lower Devonian Ridgeley Sandstone in Clinton County, Pennsylvania, and in so doing began an exciting new period of intense oil and gas exploration and development in the Appalachian basin.

LOCATION

Leidy field is located in north-central Pennsylvania (Figure 1), in the Allegheny High Plateaus Section of the Appalachian Plateaus physiographic province. It lies in parts of five townships in Clinton and Potter counties, and consists of six pools, one in the Upper Devonian Lock Haven Formation and five in the Ridgeley Sandstone. These pools occur on both flanks of the faulted Wellsboro anticline, a structure that trends northeast to southwest for about 153 km (95 mi) through north-central Pennsylvania. At Leidy, the surface structure indicates closure on the anticline (the Leidy Dome on Figure 1) of several hundred feet. Although gas was discovered in the Ridgeley Sandstone in 1950, much of the field was exhausted by 1959 and most of the reservoirs have been used for storage since then (Figure 2). Leidy is one of two important gas fields located in northern Clinton and southern Potter counties. East Fork-Wharton field, with the Ridgeley reservoir called Wharton pool (Figure 1), is situated on the Marshlands anticline about 11 km (7 mi) northwest of Leidy. Wharton pool, which was discovered in 1933 and converted to storage in 1968, occurs as a combination structural and stratigraphic trap-controlled reservoir. The northern edge of the field coincides with a stratigraphic and porosity pinchout in the Ridgeley Sandstone (Figure 1). Wharton pool had ultimate reserves of 36 bcf. Leidy field, which still produces some gas from one pool, has ultimate recoverable reserves in the Ridgeley Sandstone estimated at 175 bcf.

HISTORY

Discovery

Contrary to popular belief, Leidy field existed nearly a hundred years before the discovery of gas in the Ridgeley Sandstone in 1950. Discovered in 1864, Leidy field had a sporadic history at best, begun by completion of a well drilled on the William Sansom farm just northeast of the village of Hammersley Fork, Clinton County. No record of the interval penetrated, or of the gas volume, remains from this first well (Johnson, 1923), but the flow must have been sufficient at least for domestic purposes. A second well, called the Ox Bow well by Chance (1880), was drilled in 1878 to a depth of almost 548 m (1800 ft) in the Lock Haven Formation (formerly called Chemung). The operator reported some small shows of oil, but within the last 244 m (800 ft) of drilling enough gas was encountered to fire a boiler. The flurry of oil activity in Venango County and other areas in western Pennsylvania apparently spurred this drilling.

Post-Discovery

Over the next 25 years the Clinton Natural Gas & Oil Company drilled approximately 15 wells in the Lock Haven Formation in what is now called Hammersley Fork pool (Figure 2). Some of these wells had relatively good flows of gas. For example, Johnson (1923) reported that the first two wells drilled by the company produced over 600 MCFGD; four of them had a combined output of 4000 MCFGD. Johnson called this play in the Lock Haven Formation the Kettle Creek field, but in the excitement that ensued from the discovery of gas in the Ridgeley Sandstone in 1950 that name seems to have been lost.

The Ridgeley Sandstone (called the "Oriskany" by

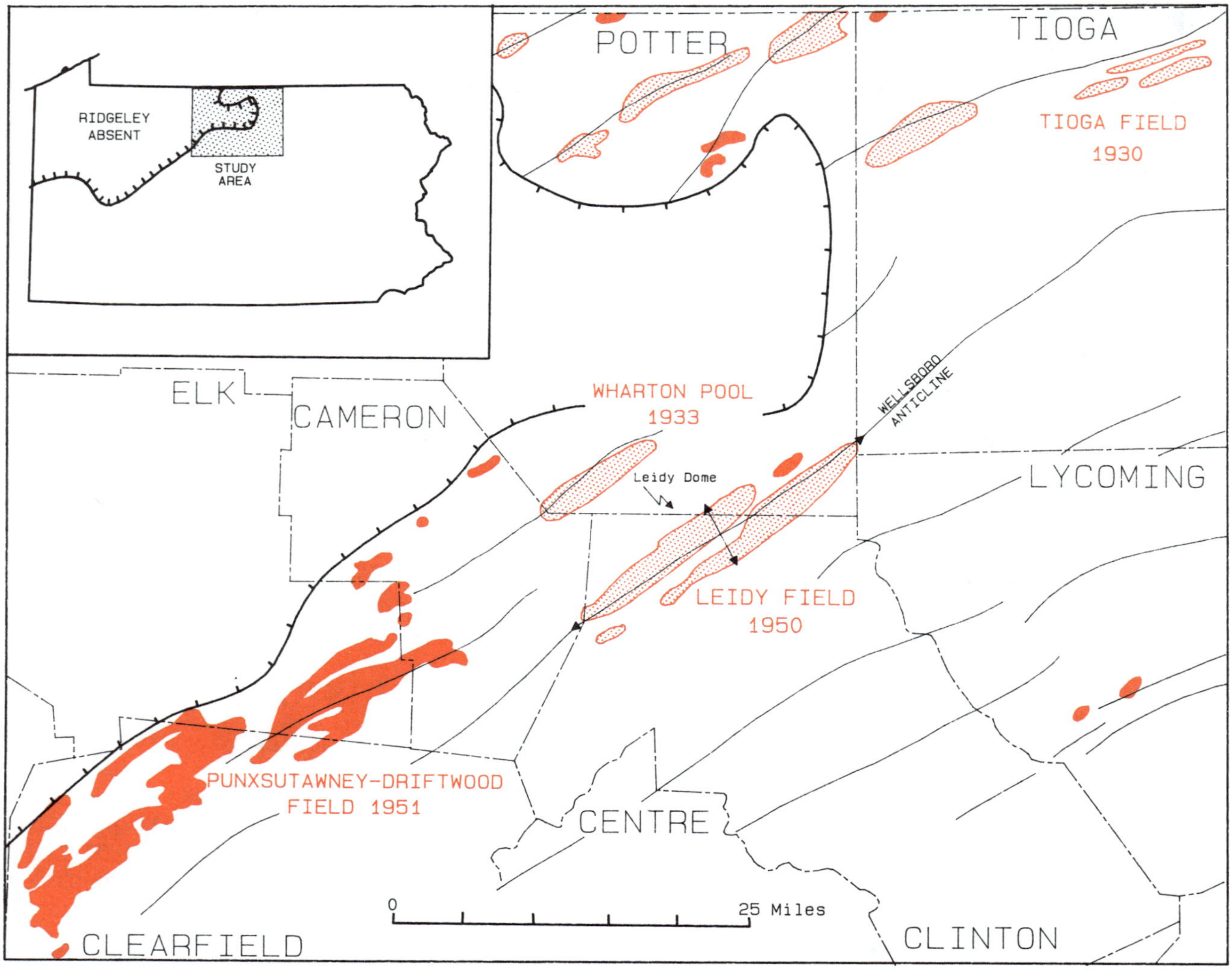

RIDGELEY ABSENT
STUDY AREA
POTTER
TIOGA
TIOGA FIELD 1930
ELK
CAMERON
WHARTON POOL 1933
WELLSBORO ANTICLINE
Leidy Dome
LYCOMING
LEIDY FIELD 1950
PUNXSUTAWNEY-DRIFTWOOD FIELD 1951
CENTRE
0
25 Miles
CLEARFIELD
CLINTON

Figure 1. Map showing location of Leidy field and other fields and pools in Ridgeley Sandstone in north-central Pennsylvania. Stippled areas indicate gas storage in the Ridgeley. Data from Abel and Heyman (1981) and Harper et al. (1982).

drillers and some authors) was considered to be prohibitively deep throughout most of western and north-central Pennsylvania until about 1930. At that time only 36 wells in all of Pennsylvania had penetrated to the depth of the Ridgeley, and only three of those had produced modest amounts of gas (Fettke, 1950). In 1930, natural gas was discovered in the Wayne-Dundee field in Schuyler County, south-central New York. The discovery well of the field had an open flow of 6000 MCFGD and a shut-in well-head pressure of 5033 kPa (730 psi) (Kreidler, 1953), and spurred activity throughout the Appalachian basin. A few months later, Tioga field was discovered in Tioga County, Pennsylvania (Figure 1). The East Penn Development Co. #1 L. B. Palmer well had a natural open flow of 22,000 MCFGD and a reservoir pressure of 11,480 kPa (1665 psi). Over the next 10 years Pennsylvania's operators discovered 10 new Ridgeley fields in Potter and adjacent Tioga counties, and by 1949 over 550 "deep" wells had been drilled in north-central and western Pennsylvania in the search for new Ridgeley reserves (Fettke, 1950).

Despite stories to the contrary by popular magazine and newspaper writers (e.g., Davidson, 1951), many industry and government geologists thought the Leidy area had high potential for hydrocarbon exploration in the Ridgeley. New York State Natural Gas Corp. (now part of Consolidated Natural Gas Corp.) had acquired leases on about 4452 ha. (11,000 ac) on Wellsboro anticline, several years prior to the discovery of Leidy field (Ingham, 1954). But two particular circumstances kept

Figure 2. Map of Leidy field, showing pool outlines and locations of wells drilled as Ridgeley Sandstone tests. Numbers indicate specific wells referred to in text, including (1) Leidy Prospecting Company #1 Dorcie Calhoun, (2) New York State Natural Gas Corporation #1 Finnefrock, (3) New York State Natural Gas Corporation #1 Pennsylvania Tract 45, and (4) Phillips Petroleum Company #6 Pennsylvania Tract 81.

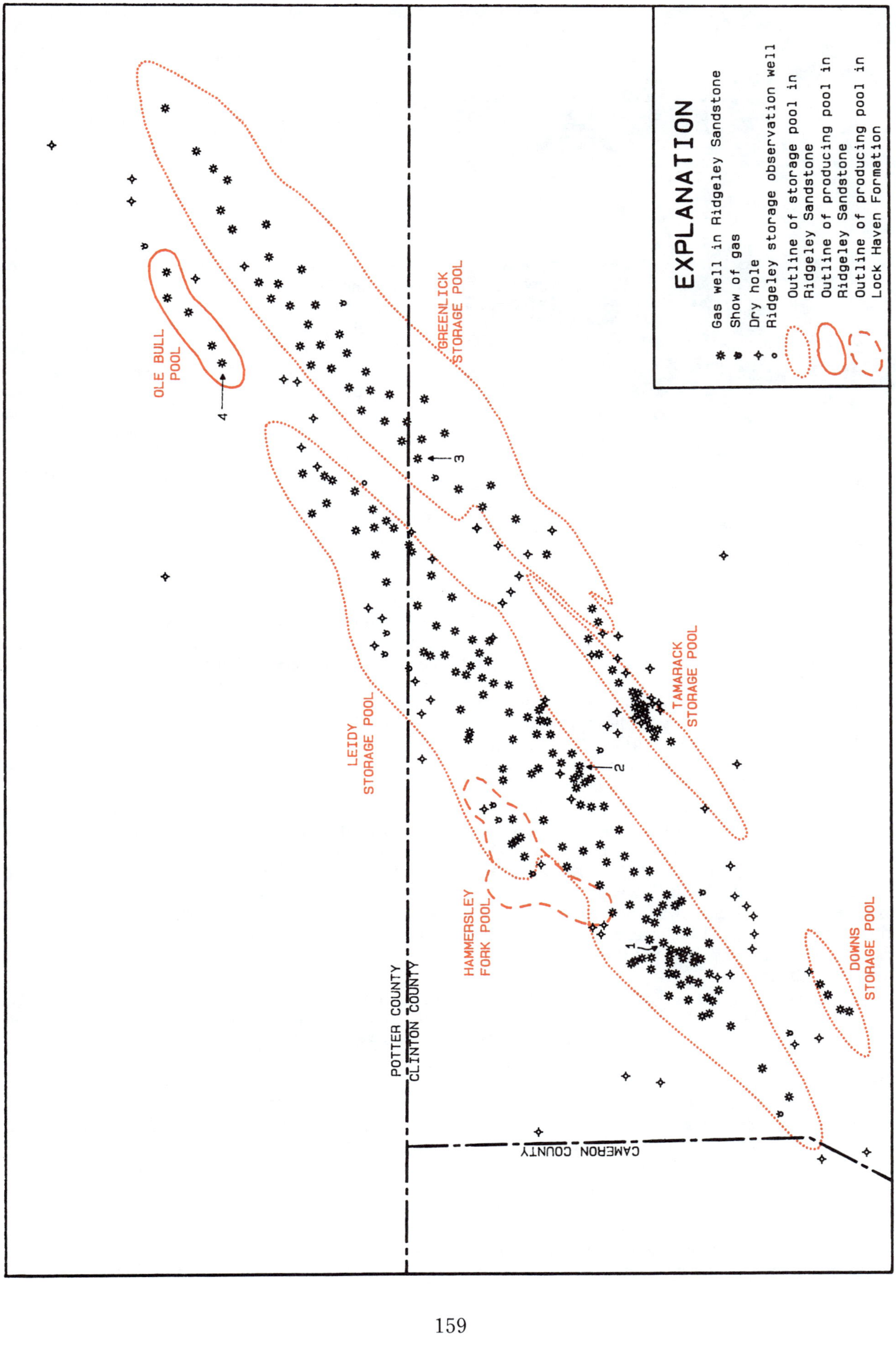
EXPLANATION
Gas well in Ridgeley Sandstone
Show of gas
Dry hole
Ridgeley storage observation well
Outline of storage pool in Ridgeley Sandstone
Outline of producing pool in Ridgeley Sandstone
Outline of producing pool in Lock Haven Formation
OLE BULL POOL
GREENLICK STORAGE POOL
LEIDY STORAGE POOL
TAMARACK STORAGE POOL
HAMMERSLEY FORK POOL
DOWNS STORAGE POOL
POTTER COUNTY
CLINTON COUNTY
CAMERON COUNTY
1
2
3
4

Figure 3. Photograph of Dorcie Calhoun (right) talking with drilling contractor Sam Jack outside a cable-tool rig used to drill to the Ridgeley Sandstone in Leidy field (photo reproduced from Philadelphia Inquirer magazine, November 12, 1950).

the successful drilling of the prospect from becoming a reality until 1950. First, it had been known since the 1930s that, as a result of pinch-out to the north and west, the Ridgeley Sandstone was absent in a large area of northern Pennsylvania and southern New York (Figure 1). The northern boundary of production in nearby Wharton pool is governed by this pinchout. Secondly, the sandstone in all of the wells drilled on the Wellsboro anticline in Tioga County to the northeast of Leidy had been too tightly cemented to produce more than a show of gas. Therefore, New York State Natural Gas considered the anticline in the Leidy Township area too risky to explore at that time. Such was not the case with some of the farmers and small businessmen who lived and worked in the area, however, and the tale of their discovery is worth retelling.

As the story goes, Dorcie Calhoun (Figure 3) had known there was gas on the family farm near Leidy, Clinton County, since he was a boy. He had even seen it bubbling out of nearby Kettle Creek, where he liked to go fishing. Dorcie spent years telling folks in the area that there was gas under Kettle Creek and the Calhoun farm. Very few people listened until he managed to convince one prominent businessman to finance him in his venture. This "stamp of approval" from a well-respected member of the community, plus the knowledge that New York State Natural Gas thought enough of the area to lease large tracts of land, suddenly made the venture itself more respectable, and brought other potential investors in the area to Dorcie with their share of the stake. After forming the Leidy Prospecting Company, Dorcie bought an old dilapidated cable-tool rig used for drilling shallow oil wells in Bradford, Pennsylvania. With the help of some of the investors and the man he'd hired to supervise operation of the rig, he hauled it back to Leidy. It was raining when they arrived, and the lead truck became mired in mud less than halfway up the hillside to the site that had been chosen for the first well. When it became apparent that the trucks would not move, Dorcie simply decided to drill on the spot (Figure 2).

The old cable-tool rig spent more time in repair than it did in drilling. The framework had to be rewelded, wooden pieces were replaced by iron or steel ones, sections had to be brought in from outside the area, and additional guy wires were attached to keep the thing from falling over. By the time the well reached total depth, the rig had been turned into a fairly heavy piece of equipment. Still, most of the knowledgeable people who saw it were amazed that it was able to drill to the 609 m (2000 ft) for which it had been originally designed, let alone a depth of 1707 m (5600 ft).

Suddenly, on Sunday afternoon, January 8, 1950, the drillers heard a rumble deep in the ground and began hauling the tools out of the hole faster than the bull wheel could take them in. Realizing the meaning of this phenomenon, the experienced rig hands scattered and hid in any convenient spot. The cable had broken, leaving the bit in the hole, but the gas was under enough pressure to squeeze past the obstruction and still fling the cable out of the hole so that it almost demolished the rig. Total depth was 1725 m (5659 ft), the open flow was estimated to be 15,000 MCFGD, and the shut-in well-head pressure was estimated to be 28,960 kPa (4200 psi). Dorcie Calhoun and his corporation of farmers and small businessmen had struck it rich, and at the same time had ushered in a new era of hydrocarbon prospecting in the Appalachian basin.

Between 1950 and 1960, approximately 300 wells were drilled along the Wellsboro anticline in the vicinity of Leidy. All of the early drilling was done with cable-tool rigs. This drilling activity discovered five fault-block pools in the Ridgeley Sandstone, including the aforementioned Leidy (Figure 2). Tamarack pool (originally called South Leidy field), situated on the southeastern flank of the anticline, was discovered February 25, 1952, with the successful completion of the McGuire et al. #1 Jents well. This well had an open flow of 60,000 MCFGD and a shut-in well-head pressure of 4137 kPa (600 psi) after only two hours. Downs pool was discovered May 18, 1954, in the Downs Oil & Gas Co. #1 Downs well. This pool, which is also situated on the southeastern flank of the anticline, is directly south of Leidy pool and southwest of Tamarack. The discovery well had a reported open flow of 6048 MCFGD and a shut-in well-head pressure of 11,763 kPa (1706 psi) in 19 hours. Greenlick pool, on the southeastern flank, was discovered January 25, 1955, with the completion of the New York State Natural Gas #7 Pennsylvania Tract 16 well in Stewardson Township, Potter County. This well had an open flow of 35,500 MCFGD and a a shut-in well-head pressure of 29,235 kPa (4,240 psi) in 39 hours. The final pool in the field, Ole Bull, was discovered on January 9, 1959, when the Phillips Petroleum #1 Pennsylvania Tract 81 well was completed with an open flow of 983 MCFGD at a shut-in well-head pressure of 30,524 kPa (4427 psi) in 24 hours. Ole Bull pool is situated on the northwestern flank of the anticline, just to the northeast of Leidy pool.

Early drilling was haphazard; there was no attempt at spacing or conservation. Tamarack pool (pre-storage), for example, consisted of 29 producing Ridgeley wells, all but eight of which occupied an area of 62.3 ha. (154 ac). This computes to an average of 3 ha. (7.5 ac) per well. Later discovery and development drilling by established companies such as New York State Natural Gas and Phillips Petroleum was more consistent with established drilling and production practices. Greenlick pool, for example, contained 40 producing Ridgeley wells, each with an average spacing of 16.2 ha. (40 ac).

Some of the wells in Leidy field were spectacular successes, such as the New York State Natural Gas #1 Finnefrock (Figure 2), which had a reported natural open flow of 145,000 MCFGD, the largest official open flow in the Appalachian basin. New York State Natural Gas completed the well in Leidy pool on February 2, 1951, at a depth of 1932 m (6339 ft), 0.9 m (3 ft) into the Ridgeley. Other wells were spectacular for reasons other than just production success. The New York State Natural Gas #1 Pennsylvania Tract 45 (Figure 2), for example, caught fire with a natural open flow originally estimated at between 150,000 and 200,000 MCFGD. When the well was finally gauged, the open flow was 124,000 MCFGD. This was the fifth well drilled in Greenlick pool. It found gas at the top of the Ridgeley on June 10, 1955, at 2008 m (6588 ft), right at the top of the sandstone, and caught fire on June 16 as workmen were trying to dislodge a dril stem from the hole. After several days of trying to get the well under control, New York State Natural Gas officials brought in "Red" Adair, the famous well-fire expert, who celebrated his 40th birthday by blasting out the fire with 127 kg (280 lb) of nitroglycerine. Adair was no stranger in the Leidy area; he had put out fires in at least two earlier wells drilled in Leidy field.

Because of the hazards of high-pressure blow outs, E. C. Ingham, the General Superintendent of New York State Natural Gas, devised a new stimulation tool that lessened the dangers of drilling to the Ridgeley in Leidy field (Ingham, 1954). The device was shaped specifically for vertically stimulating the Ridgeley; after drilling to the top of the sandstone, the charge was set in the hole and detonated with a timing device. The force of the charge was directed downward into the sandstone, and proved to be very successful.

Most of the Ridgeley pools in Leidy field produced huge quantities of gas, but they were relatively short lived. Leidy pool produced approximately 94 bcf before being converted to storage in 1959. Greenlick pool was converted to storage in 1961 after producing about 50 bcf, and Tamarack and Downs were consolidated and converted to storage in 1971 after producing a combined 11.4 bcf. At the time of this writing, only Ole Bull is still producing small quantities of gas (approximately 2500 mcf per year) from the Ridgeley Sandstone (Harper, 1986).

DISCOVERY METHOD

At the time that Leidy field was discovered in 1864, exploration for oil and gas was basically a hit-or-miss operation. Wells were drilled where someone noticed a natural oil seep, or where property could be leased to forward-thinking individuals. Several years of experience became a driller's most valuable asset. It was only after about 15 years, and the pioneering subsurface geological work of John F. Carll of the Second Pennsylvania Geological Survey, that exploration and development of hydrocarbon resources started to become a refined and definite scientific endeavor.

The exploration for gas in the Ridgeley Sandstone in north-central Pennsylvania and south-central New York originated from an accidental discovery of gas in the Tully Limestone in 1928. The Belmont Quadrangle Drilling Corp. #1 Gilbert well, in the town of Wirt, Allegany County, New York, had a natural open flow of 3500 MCFGD with a high percentage of condensate (Fettke, 1961). Described by Finn (1949) as a "freak well," the Tully discovery spurred new interest in the geology of New York state and led to the eventual discovery of numerous gas fields and pools in the Ridgeley Sandstone between New York and West Virginia. Most of the exploratory activity followed the concept of the "anticlinal theory" of oil and gas accumulation. T. Sterry Hunt (1861) had proposed the theory based on observed coincidences between oil production and mapped structures in Ontario, Canada. I. C. White (1885, 1892) revived the theory after successfully demonstrating its utility in the gas fields of West Virginia and Pennsylvania, and it was still a strongly favored concept at the time of the Ridgeley discoveries.

When the Tioga field was discovered in Pennsylvania in 1930, the Fourth Pennsylvania Geological Survey undertook to map and define the surface structures throughout north-central western Pennsylvania to benefit gas operators in their search for deep reservoirs. Of particular interest was Cathcart's report (1934) on the structures of the northern Pennsylvania plateau region and their relation to deep gas occurrences. This report showed, for the first time, the names, locations, and trends of the major surface anticlines and synclines in north-central and northeastern Pennsylvania. Cathcart's report was primarily responsible for convincing Dorcie Calhoun's investors that his venture would be more than just a long shot.

STRUCTURE

Tectonic History

Leidy field is situated in the Appalachian orogenic belt, which extends from the maritime provinces of Canada southward along the eastern edge of the United States to the coastal plain of the Gulf of Mexico. The belt consists of several northeast-striking tectonic provinces with distinct sedimentary and stratigraphic regimes, each with its own superimposed physiographic style (Thompson and Sevon, 1982). The degree of deformation increases across the belt from west to east. Three major Paleozoic orogenic events caused by plate-boundary collisions, the Taconian, Acadian and Alleghanian, created these tectonic provinces.

The Taconian orogeny occurred during the Late Ordovician and Early Silurian. The orogeny primarily affected the easternmost and northernmost rocks in the Appalachian trend, resulting in westward overthrusting and close folding of the rocks that now occupy the Piedmont physiographic province. This tectonic disturbance resulted from subduction of the Proto-Atlantic oceanic crust beneath the North American continental crust, creating a mountainous landmass, Appalachia, that became the major sediment source for the newly formed Appalachian basin throughout the Silurian and into the Devonian. Large volumes of predominantly noncalcareous clastic sediments were shed from Appalachia westward into the basin, creating the Taconic clastic wedge or Queenston delta complex. Drake and Epstein (1967) estimated the thickness of the Upper Ordovician clastic sequence in the Central Appalachians at 2987-3900 m (9800-12,800 ft), and Colton (1970) suggested that the original thickness, prior to development of a major unconformity at the end of the Ordovician, would have been appreciably greater still.

The next major tectonic event, the Acadian orogeny, began sometime during the late Early Devonian and continued through early Late Devonian (Faill, 1985). The tectonic disturbance resulted from the collision between the North American and Avalonian plates. The collision apparently occurred in a rotating fashion—rocks in the Northern Appalachians were affected first, and the deformation spread gradually southward through the Southern Appalachians. With one exception, this orogeny is presently represented along the entire Appalachian tectonic belt by a sequence of deformational, metamorphic and/or plutonic structures. Opposite the Central Appalachians, direct evidence for the orogeny is lacking. However, the Acadian clastic wedge, or Catskill delta complex, provides a large body of indirect evidence for the orogeny in this area. Acadian tectonism either recreated or rejuvenated the Taconian-developed landmass, Appalachia, and sediments eroded from it were shed westward into the Appalachian basin from the Middle Devonian to the Pennsylvanian. This clastic wedge was estimated by Dott and Batten (1976) at more than 288,000 km^3 (69,000 mi^3) in volume.

The third major orogenic event was the Alleghany orogeny, which apparently began in the Late Pennsylvanian (Van der Voo, 1979), and extended through most of the Permian. The Alleghany orogeny, caused by the collision of the North American and African plates, uplifted, deformed and eroded the clastic wedge sediments deposited during and after the earlier orogenies and spread them westward onto the craton margin. The folded and faulted Appalachian mountains in the Central and Southern Appalachians represent the most visible evidence of Alleghanian tectonic deformation, but thrusting also affected the basement by carrying the crystalline rocks of the Blue Ridge and Piedmont westward over the Paleozoic sedimentary rocks (Faill, 1985). Cook et al. (1979), Harris et al. (1981), and others have even speculated, based on seismic evidence, that the entire orogenic belt in the Central and Southern Appalachians may be allochthonous. Folding and faulting in the Appalachian Plateau province is also considered by many investigators to be the exclusive result of Alleghanian deformation, but this is an oversimplification. Based on stratigraphic evidence, it is apparent that many of the Plateau folds were already growing before the first surges of the Alleghany orogeny (see, for example, Bradley and Pepper, 1938; Piotrowski and Harper, 1979; Harper, 1987).

Appalachian Plateau Structure

Surface structures in western and north-central Pennsylvania are arcuate and trend roughly northeast to southwest. They consist of numerous, simple, low-amplitude, *en echelon* folds and a few recognizable high-angle faults; fold amplitudes decrease to the northwest. Many anticlines are narrow structures between broad, open synclines. Folds are generally asymmetrical, with the southeastern limb commonly a few degrees steeper than the northwestern limb. Domes and saddles occur along some of the anticlines (e.g., Leidy Dome in Figure 1), and Cathcart (1934) recognized that many of these occur near bends in the structures.

Shallow subsurface structures in north-central Pennsylvania are relatively simple folds whose axes generally match the axes of the surface folds. Deeper subsurface folds (i.e., Lower or Middle Devonian), on the other hand, may have axes displaced somewhat to the northwest of the surface and shallow subsurface fold axes. These deeper folds are also more complex, commonly involving large amounts of faulting (Figure 4). These structures typically consist of anticlinal flanks thrust over depressed cores. Gwinn (1964) suggested that tectonic thickening of 50 to 150% in the structurally disturbed formations occurs along the margins of the depressed crests near the lips of the thrusted flank blocks, but the thickening appears to be more apparent (due to attitude) than real.

These structures resulted mainly from detachment within the incompetent rocks of the Upper Silurian Salina Group and Tonoloway Formation, and perhaps within the Upper Ordovician Reedsville Formation (Figure 5). Basement and/or lower Paleozoic structures beneath the Plateau probably helped govern the placement of shallower structures, however. Numerous authors, including Shumaker (1974), Root (1978a, 1978b), Harper and Piotrowski (1978), Piotrowski and Harper (1979), Parrish and Lavin (1982), Lavin et al. (1982), Rodgers and Anderson (1984), Laughrey and Harper (1986), and Harper (1987), have speculated that recurrent movement along deep-seated structures since

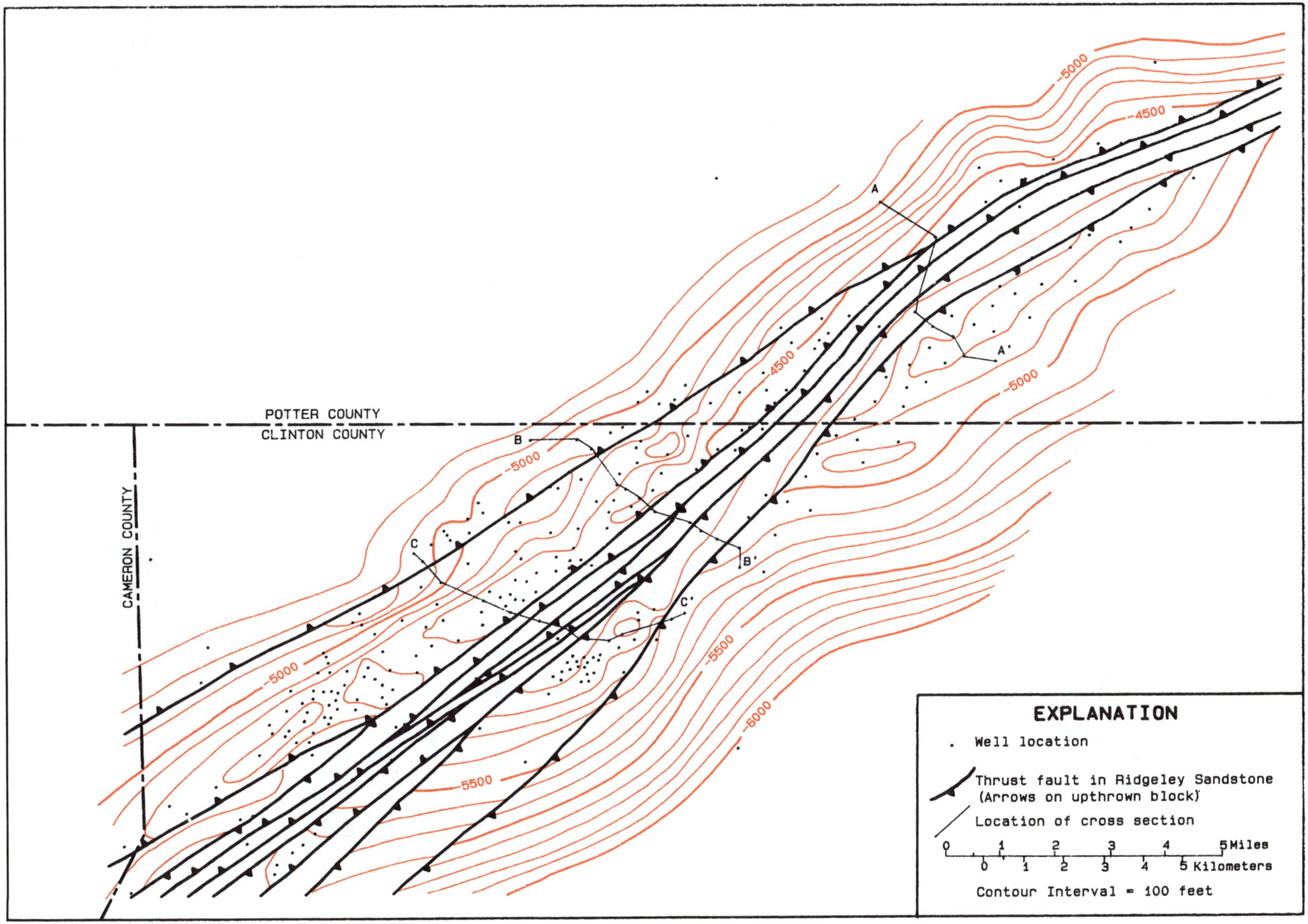

POTTER COUNTY
CLINTON COUNTY
CAMERON COUNTY
-5000
-4500
-5500
-6000
5500
A
A'
B
B'
C
C'
EXPLANATION
Well location
Thrust fault in Ridgeley Sandstone (Arrows on upthrown block)
Location of cross section
0 1 2 3 4 5 Miles
0 1 2 3 4 5 Kilometers
Contour Interval = 100 feet

Figure 4. Structure map of Ridgeley Sandstone in Leidy field, based on well data.

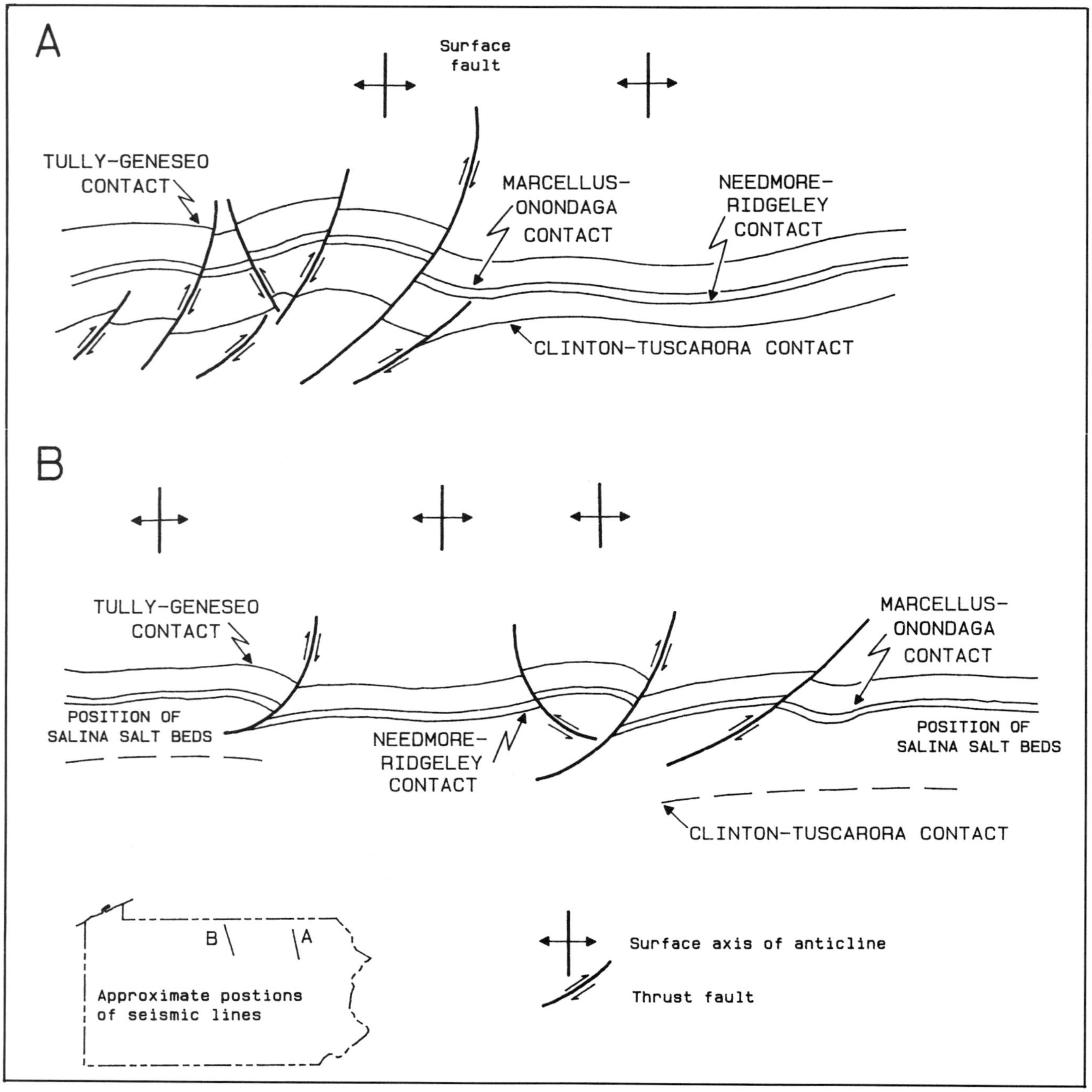
A
Surface fault
TULLY-GENESEO CONTACT
MARCELLUS-ONONDAGA CONTACT
NEEDMORE-RIDGELEY CONTACT
CLINTON-TUSCARORA CONTACT
B
TULLY-GENESEO CONTACT
POSITION OF SALINA SALT BEDS
NEEDMORE-RIDGELEY CONTACT
MARCELLUS-ONONDAGA CONTACT
POSITION OF SALINA SALT BEDS
CLINTON-TUSCARORA CONTACT
B
A
Approximate postions of seismic lines
Surface axis of anticline
Thrust fault

Figure 5. Interpretations of proprietary seismic lines in north-central and northeastern Pennsylvania. A. Section showing folding and faulting caused by detachment in Ordovician shales. B. Section showing folding and faulting caused primarily by detachment in Upper Silurian Salina Group evaporites. General direction on both sections is north (left) to south (right). Notice that most thrust faults are antithetic.

original deformation took place has affected sedimentation, folding, faulting, and mineral-resource emplacement throughout geologic time. Seismic studies indicate the presence of down-to-the-east basement faults striking northeast to southwest (parallel to regional strike) beneath the surface folds. These basement faults have been interpreted as extensional faults associated with the opening of the Proto-Atlantic ocean in the late Precambrian and early Paleozoic (Rankin, 1976). They may have controlled both the ramping of decollement thrusts from lower to higher stratigraphic levels and the development of the high-angle thrust faults that created the anticlines. Strike-normal features, or cross-strike structural discontinuities (CSDs of Wheeler, 1980), occur throughout the Plateau province. The exact nature of these features is not well known, but some probably represent the surface and shallow subsurface signatures of reactivated transcurrent basement faults, possibly Precambrian through Early Ordovician transform faults (Thomas, 1977; Lavin et al., 1982). Others may represent the signatures of tear faults within the Paleozoic cover. They seem to have acted as boundaries between independent decollement blocks during tectonic transport, as indicated by offset of structural axes (Kowalik and Gold, 1976). In Leidy field and adjacent areas, the Wellsboro anticline and Leidy Dome are directly related to salt movement in the Salina Group (Figure 5B).

There are many questions that have to be answered about the genesis of Appalachian Plateau structures. These include, but are not limited to, direction and mechanism of tectonic transport (thrusting from the southeast versus southeastward gravity sliding), method of fold construction (flow of salt into anticlinal flanks versus dissolution of salt in anticlinal cores), and timing of the detachment (Alleghanian versus pre-Alleghanian).

The Ridgeley Sandstone was recognized as a faulted reservoir very early in its development history through the use of structure contouring. Finn (1949) and Gwinn (1964) described several Ridgeley fields in Pennsylvania, New York, and West Virginia that have fault characteristics similar to that of Leidy field.

Leidy Field Structure

Leidy field structure is illustrated in Figures 4 and 6 to 8. Because of the lack of nonproprietary seismic information across the field, the structure map and cross sections are based solely on well data and surface mapping. The structure consists of a salt-cored anticline with overthrusted limbs at the level of the Ridgeley Sandstone, but becomes a simple asymmetrical fold at the surface (Figure 9). Surface structural relief is generally on the order of 152 to 244 m (500 to 800 ft), whereas structural offset at depth may be more than 457 m (1500 ft). This discrepancy is due to termination of the splay thrusts, which originated in the Salina salt beds, within the Upper Devonian shale sequence (Wiltschko and Chapple, 1977; proprietary seismic evidence). Complexity of faulting in the Ridgeley Sandstone increases from the northeast to the southwest (Figure 4), and thrust faults in the Ridgeley may be more numerous and anastomosing than illustrated. The major flank faults traverse the entire length of the field and continue northeast and southwest for an undetermined distance (probably the length of the anticline—approximately 177 km (110 mi). The exact number and distribution of faults is not known for certain, but they have created a set of independently producing pools situated on separate blocks. Except for the anomalous Downs pool, which is situated on an upthrown block between opposing thrusts near the core of the anticline, all of the Ridgeley gas production occurs in fault-produced flexures on the anticlinal flanks.

STRATIGRAPHY

The known stratigraphic sequence in the vicinity of Leidy field consists of Middle Cambrian through Pennsylvanian rocks (Figure 10) comprising clastics, carbonates, and evaporites. Most carbonate rocks occur in the Cambrian through Upper Ordovician and in the Lower Silurian through Middle Devonian. The Upper Ordovician through Lower Silurian, and the Middle Devonian through Pennsylvanian, are dominated by clastics. Evaporites commonly occur as cements in many clastic rock sequences, but dominate as bedded salts only in the Upper Silurian Salina Group.

The primary reservoir rocks in Pennsylvania and adjacent states are Upper Devonian multi-tiered marine sandstones of the Venango, Bradford, and Elk Groups (Laughrey and Harper, 1986, Table 1). These western Pennsylvania reservoirs are laterally equivalent with the Catskill and Lock Haven formations in the study area. Reservoirs of secondary importance include the low-permeability sandstones of the Lower Silurian Medina Group (the northwestern Pennsylvania equivalent of the Tuscarora Formation in Figure 10), and the shallow-shelf quartz arenites and sublitharenites of the Lower Devonian Ridgeley Sandstone. Tertiary reservoirs include, in ascending order: (1) Upper Cambrian arenaceous dolostones of the Gatesburg Formation; (2) Upper Ordovician sandstones of the Bald Eagle Formation; (3) Lower Silurian sandstones of the Tuscarora Formation; (4) Silurian dolomitized bioherm rubble of the Lockport Dolomite (equivalent to the Mifflintown Formation in Figure 10); (5) Middle through Upper Devonian organic-rich shales and siltstones (e.g., the Marcellus Formation); and (6) various Carboniferous sandstone units.

On the basis of organic content, source rocks occur in only a few of these stratigraphic sequences; the most important to oil and gas production in Pennsylvania are Upper Ordovician mudrocks and carbonates (the Utica Shale and "Trenton," "Black River," and "Chazy" carbonates in Figure 10) and the Devonian shales (C. D. Laughrey, 1987, personal communication). This latter sequence of organic-rich shales began in the Middle Devonian with deposition of the Marcellus Formation, and included much of the Upper Devonian sequence in western Pennsylvania (Piotrowski and Harper, 1979). In north-central Pennsylvania, the turbidite sequences of the Brallier Formation take the place of most of these Upper Devonian organic-rich shales. Only the Burket Shale Member of the Harrell Formation, equivalent to the Geneseo black shales of western Pennsylvania, occurs in north-central Pennsylvania.

TRAPS

Three types of traps—structural, stratigraphic, and diagenetic—affect hydrocarbon production in Leidy field. Ridgeley Sandstone reservoirs have combination structural/diagenetic traps; simple stratigraphic changes have very little to do with trapping and production in the Ridgeley in this field. Lock Haven reservoirs

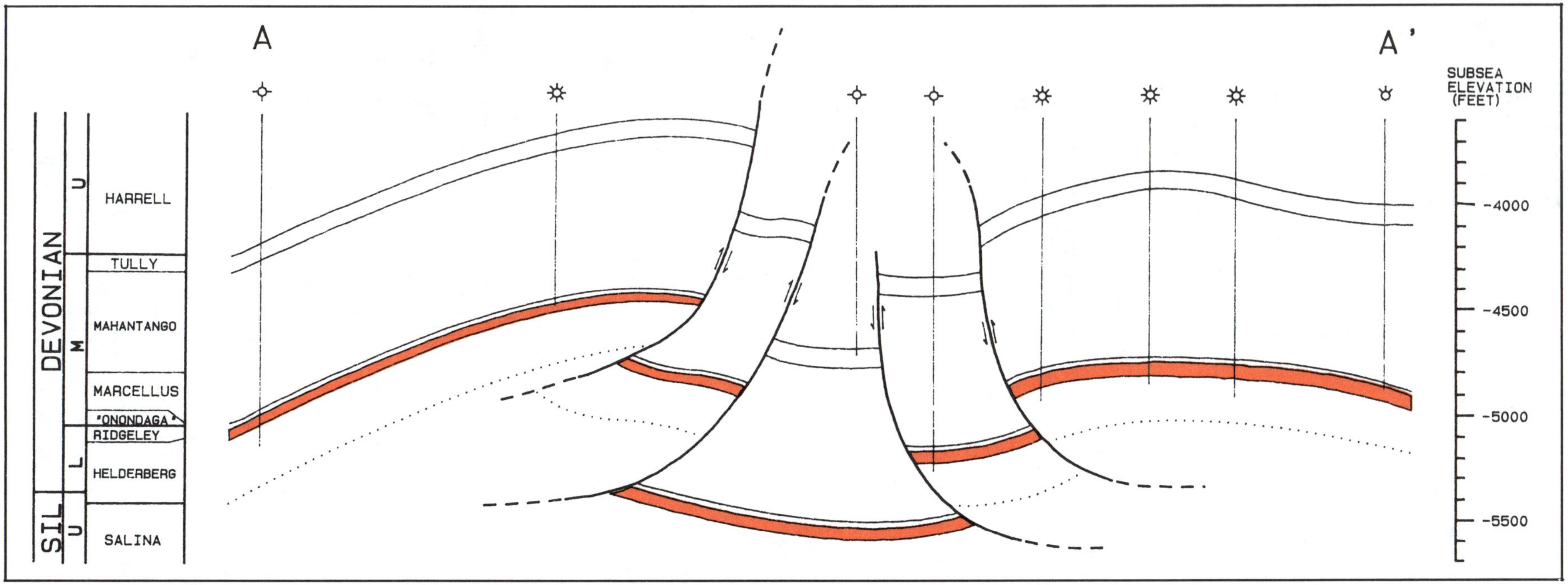

A
A'
SUBSEA ELEVATION (FEET)
-4000
-4500
-5000
-5500
DEVONIAN
SIL
U
M
L
U
HARRELL
TULLY
MAHANTANGO
MARCELLUS
'ONONDAGA'
RIDGELEY
HELDERBERG
SALINA

Figure 6. Cross section A-A′ Colored portions indicate position of Ridgeley Sandstone. See Figure 4 for location of cross section.

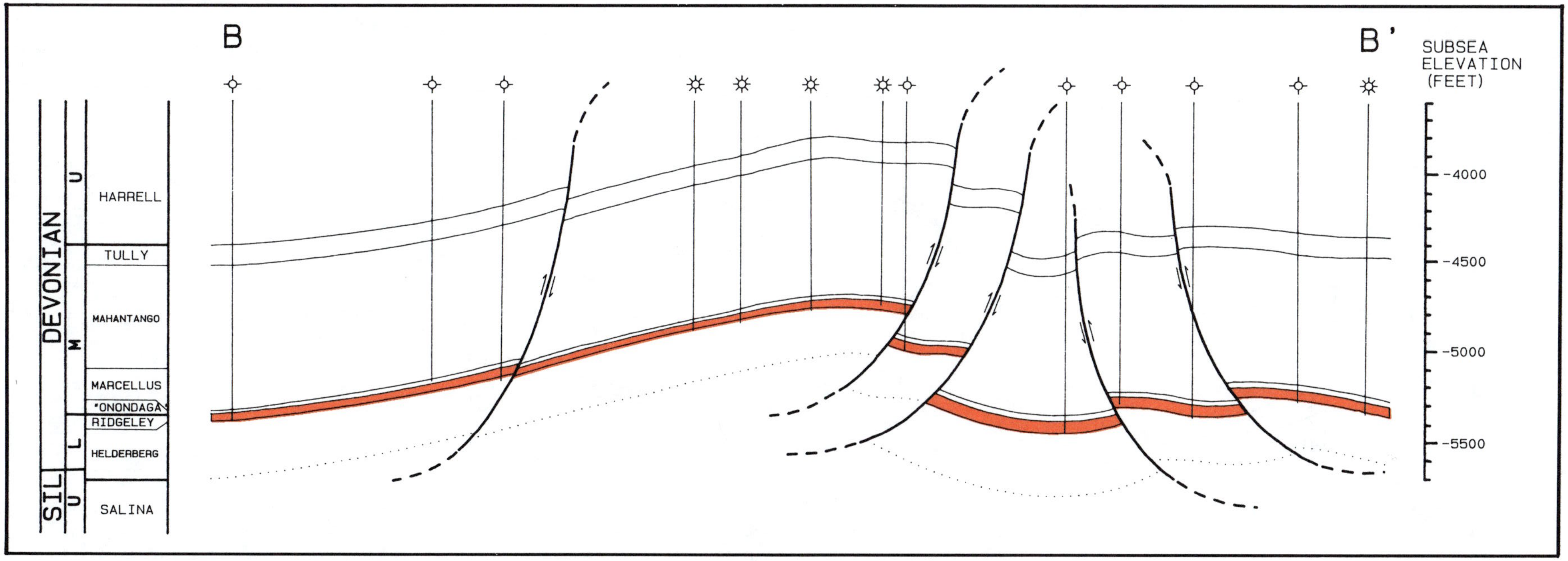
B
B'
SUBSEA
ELEVATION
(FEET)
-4000
-4500
-5000
-5500
DEVONIAN
SIL
U
M
L
U
HARRELL
TULLY
MAHANTANGO
MARCELLUS
*ONONDAGA
RIDGELEY
HELDERBERG
SALINA

Figure 7. Cross section B-B′. Colored portions indicate position of Ridgeley Sandstone. See Figure 4 for location of cross section.

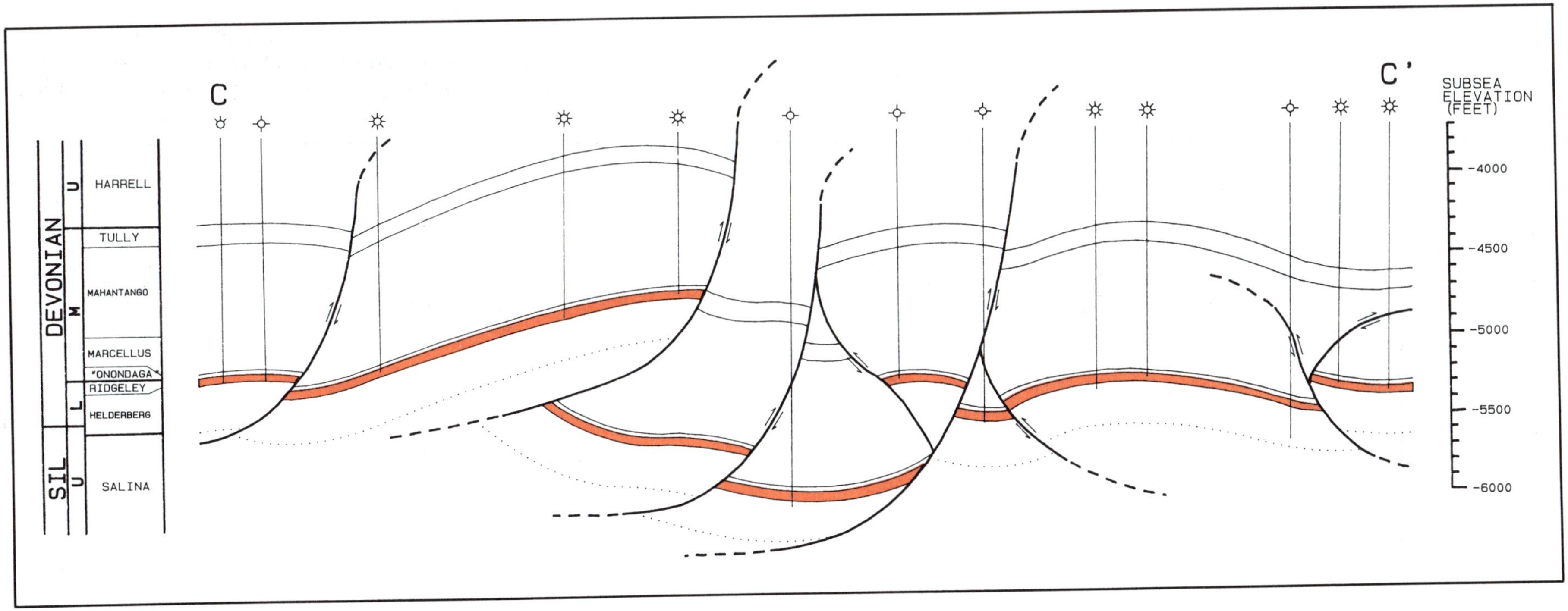
C
C'
SUBSEA
ELEVATION
(FEET)
-4000
-4500
-5000
-5500
-6000
DEVONIAN
SIL
U
M
L
U
HARRELL
TULLY
MAHANTANGO
MARCELLUS
"ONONDAGA"
RIDGELEY
HELDERBERG
SALINA

Figure 8. Cross section C-C′. Colored portions indicate position of Ridgeley Sandstone. See Figure 4 for location of cross section.

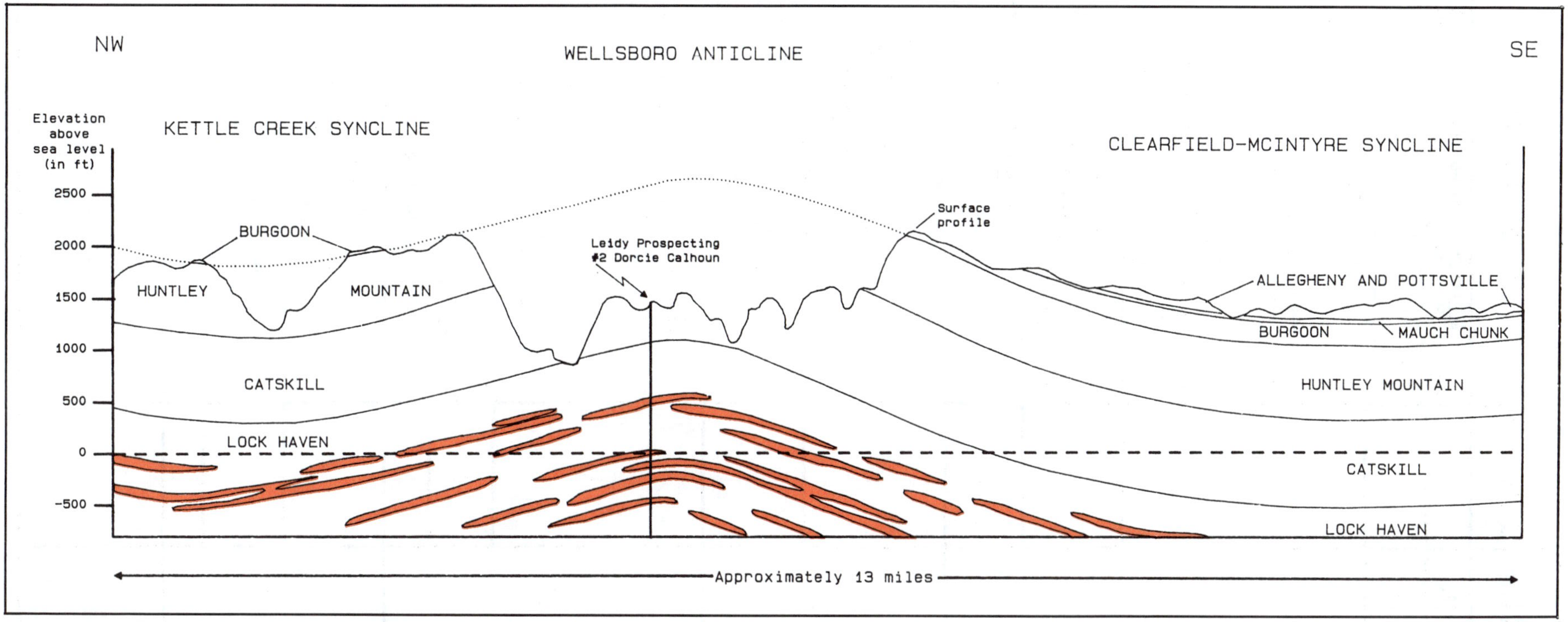
NW
WELLSBORO ANTICLINE
SE
Elevation above sea level (in ft)
2500
2000
1500
1000
500
0
-500
KETTLE CREEK SYNCLINE
CLEARFIELD-MCINTYRE SYNCLINE
BURGOON
HUNTLEY
MOUNTAIN
Leidy Prospecting #2 Dorcie Calhoun
Surface profile
ALLEGHENY AND POTTSVILLE
BURGOON
MAUCH CHUNK
CATSKILL
HUNTLEY MOUNTAIN
LOCK HAVEN
CATSKILL
LOCK HAVEN
Approximately 13 miles

Figure 9. Generalized cross section of shallow formations on Wellsboro anticline in Leidy field. Most of historical production from Lock Haven Formation occurs on northwestern flank of anticline (just left of indicated well). Modified from Ebright and Ingham (1951).

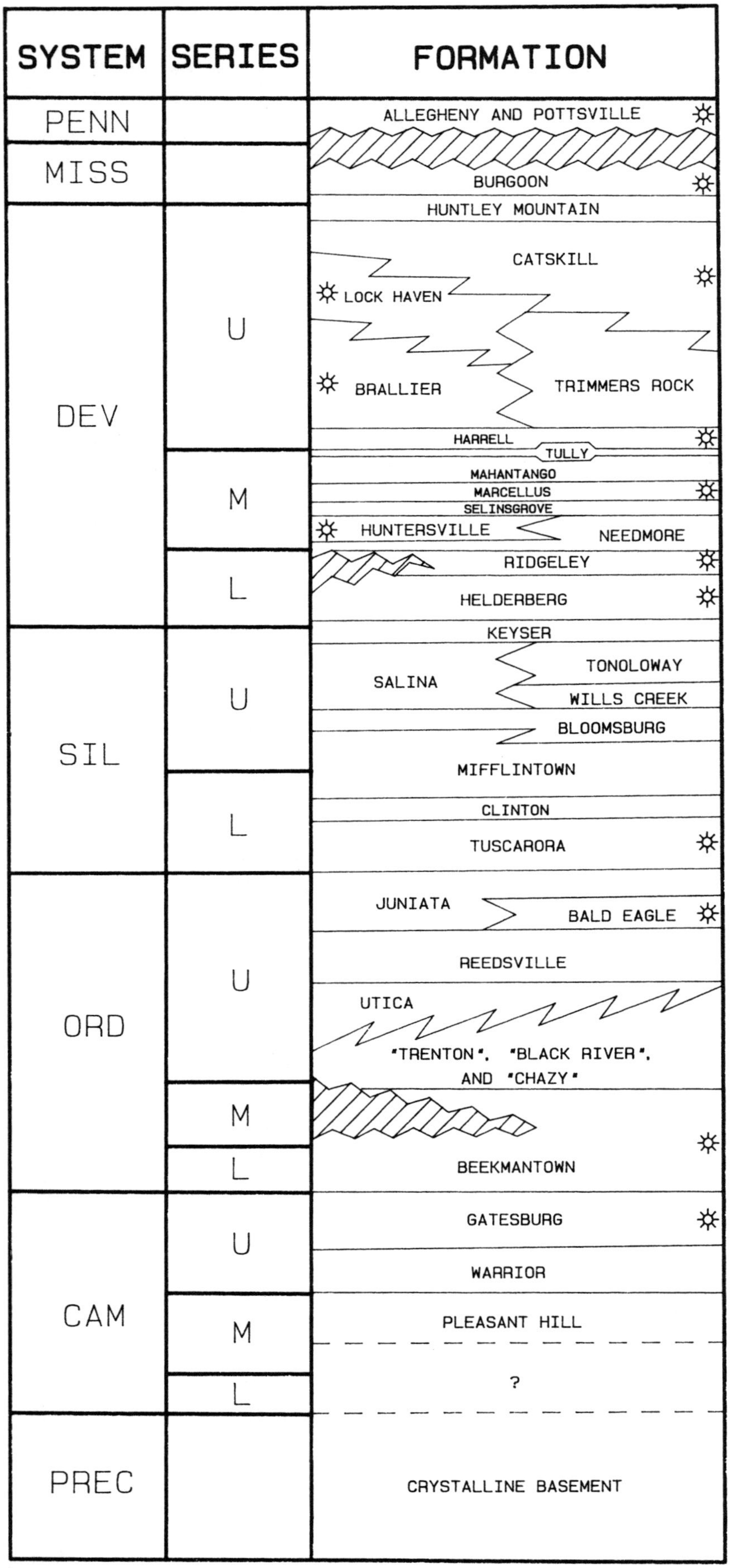
SYSTEM
SERIES
FORMATION
PENN
MISS
DEV
SIL
ORD
CAM
PREC
U
M
L
U
L
U
M
L
U
M
L
ALLEGHENY AND POTTSVILLE
BURGOON
HUNTLEY MOUNTAIN
CATSKILL
LOCK HAVEN
BRALLIER
TRIMMERS ROCK
HARRELL
TULLY
MAHANTANGO
MARCELLUS
SELINSGROVE
HUNTERSVILLE
NEEDMORE
RIDGELEY
HELDERBERG
KEYSER
SALINA
TONOLOWAY
WILLS CREEK
BLOOMSBURG
MIFFLINTOWN
CLINTON
TUSCARORA
JUNIATA
BALD EAGLE
REEDSVILLE
UTICA
"TRENTON", "BLACK RIVER", AND "CHAZY"
BEEKMANTOWN
GATESBURG
WARRIOR
PLEASANT HILL
?
CRYSTALLINE BASEMENT

are mostly combination stratigraphic/diagenetic traps, but structure probably had some influence on the best producing areas.

Structural traps in the Ridgeley Sandstone exist where flexure occurs on the thrust-faulted limbs (see Figures 6 to 8). In some cases, these flexures constitute secondary anticlines with closure of less than 30 m (100 ft) to more than 91 m (300 ft); other flexures create homoclines terminating at the faults. The upper seal of these reservoirs is the relatively impermeable Needmore Shale. Structure also helped influence diagenesis by providing fluid migration pathways in the form of numerous fractures in the brittle rocks of the section. Diagenetic effects on the structural traps consist of porosity enhancement/degradation contacts. The upper few feet of the Ridgeley Sandstone typically consists of friable sandstone, and this is commonly the zone from which the best gas production comes. At about 1.2 m (4 ft) below the top of the formation in these wells, the sandstone becomes very calcareous, and the typically rounded, frosted sand grains show signs of quartz overgrowths (Ebright and Ingham, 1951). Porosity enhancement in the upper few feet of the Ridgeley, therefore, appears to be due to dissolution of carbonate and silica cements. Given the burial history of Leidy field (see below), it is unlikely that porosity enhancement is related to the inhibition of cementation by early hydrocarbon emplacement.

Stratigraphic trapping in the Ridgeley Sandstone, also intimately associated with diagenetic changes, occurs in those areas of Pennsylvania where the sandstone wedges out northwestward toward the "no-sand area." Updip porosity loss occurs about 1.6 km (1 mi) from this pinchout (Heyman, 1969). Wharton pool, adjacent to Leidy field (Figure 1), lies near the edge of the "no-sand area" and its production is directly related to porosity entrapment near that feature. Termination of the Ridgeley Sandstone at the "no-sand area" occurs about 11.3 km (7 mi) northwest of Leidy, however, and the only stratigraphic effect is a gradual northward thinning of the sandstone.

Stratigraphic traps are more important in the Lock Haven Formation than in the Ridgeley Sandstone. Lock Haven reservoir sandstones, like those of the Bradford and Venango Groups in western Pennsylvania, are discontinuous lenses having geometries of a few tens to several thousands of square feet, separated horizontally and vertically by less permeable mudrocks. Lock Haven reservoir rocks probably have undergone as much diagenetic alteration as Bradford and Venango reservoir rocks. If so, diagenetic trapping is even more important in these rocks than is sandstone lensing. Modern production from Upper Devonian reservoir rocks in western Pennsylvania generally occurs without regard to structure, and may even occur in synclines (e.g., Laughrey, 1982). The few Upper Devonian fields and pools in north-central Pennsylvania certainly occur in association with surface anticlines (Harper et al., 1982), but this is probably due to the long-standing acceptance of the "anticlinal theory" among Appalachian basin operators, and to the relative lack of modern exploration and development of Upper Devonian targets in this area of the state. The Lock Haven production in Hammersley Fork pool occurs along the northern flank of the Leidy Dome on the Wellsboro anticline, where homoclinal dips of only a few degrees are common. Although structure may contribute to Lock Haven gas accumulation and production, it certainly is not as important as stratigraphy and diagenesis.

◄

Figure 10. Generalized stratigraphic column for Leidy field area and north-central Pennsylvania. Facies changes are generally from west to east (left to right). Gas well symbols indicate formations with production histories in the state.

RESERVOIRS

Ridgeley Sandstone

The Ridgeley Sandstone consists mostly of quartz arenites in New York and in central and western Pennsylvania (Basan et al., 1980). It is typically a medium- to fine-grained, well-sorted, grain-supported sandstone. Data from cores recovered from wells located about 1.6 km (1 mi) apart in the northeastern end of Leidy pool indicate the sandstone is light gray, medium grained, fossiliferous, and contains a small fraction of pyrite. Quartz grains are well rounded and exhibit high sphericity. They consist of an average of 95% quartz, 3% feldspar, 1% rock fragments, and 1% accessory grains and fossil shell fragments (Laughrey et al., in preparation). Data from a dry hole on the northwestern edge of Leidy pool indicate that composition includes 87% detrital silica, 11% authigenic silica, 1% feldspar, and 1% carbonate cement (Rosenfeld, 1953). Secondary porosity due to feldspar dissolution has been documented in the Ridgeley Sandstone in western and central Pennsylvania by several investigators (e.g., Basan et al., 1980; Sanders, 1982), and it is probable that feldspars accounted for a few percent more of the detrital fraction in the Ridgeley at the time of deposition than at present (C. D. Laughrey, 1987, personal communication). The sandstones may actually be more "diagenetic quartz arenites" than depositional ones (McBride, 1987) because of replacement of feldspars by authigenic silica. The Ridgeley Sandstone has many incipient to strongly developed fractures, both horizontal and vertical, along the faulted anticlines. The lower portion of the formation tends to contain abundant carbonate; the core reports, and drill-cutting descriptions from nearby wells (Fettke, 1961), refer to the sandstone in this zone as a sandy or silty limestone (Figure 11).

Basan et al. (1980) describe the Ridgeley in north-central Pennsylvania as transgressive beach-sand deposits, characterized by low-angle cross stratification to horizontal stratification, minor laminations, and scattered fossils of nearshore brachiopods. *Skolithos* trace

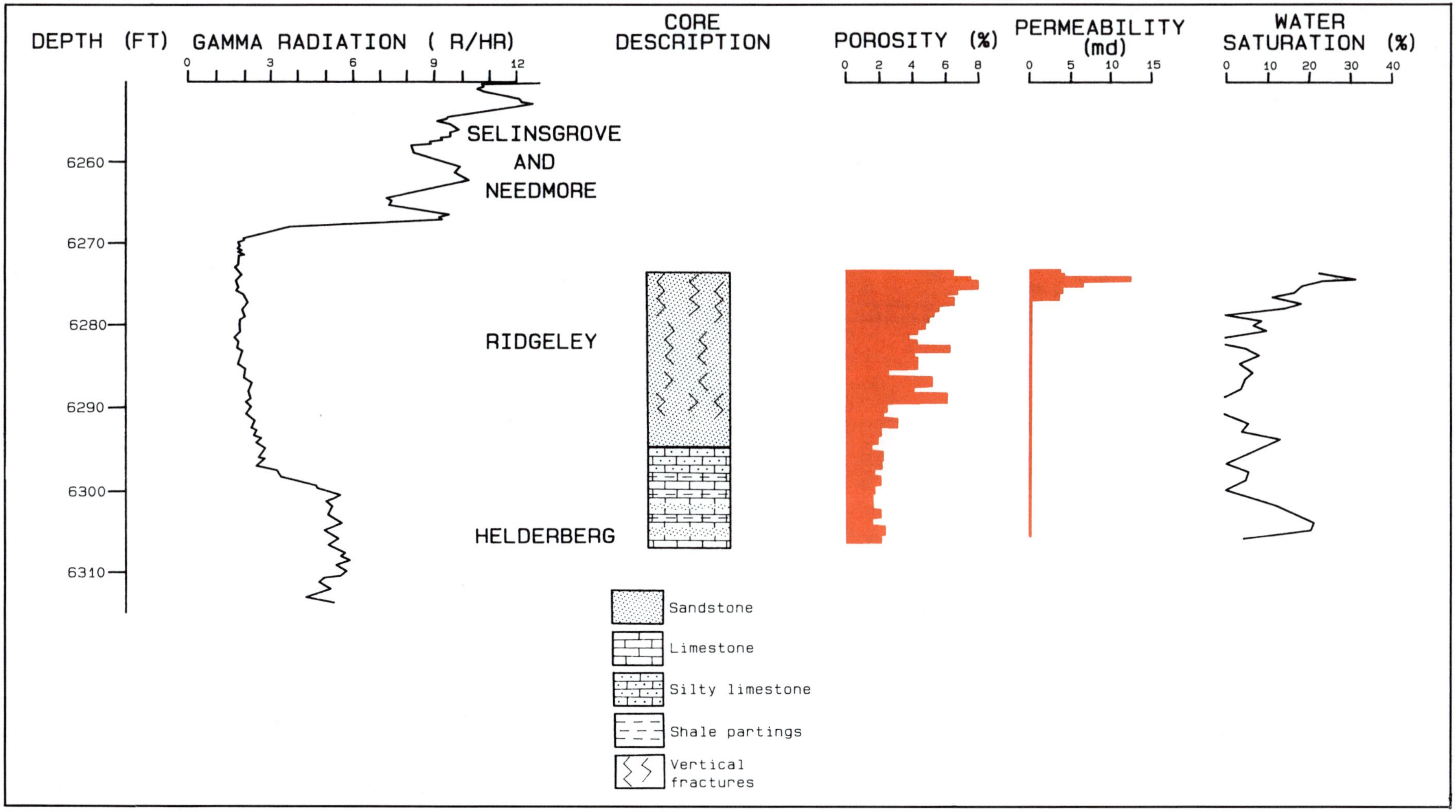
DEPTH (FT)
GAMMA RADIATION (R/HR)
0 3 6 9 12
CORE DESCRIPTION
POROSITY (%)
0 2 4 6 8
PERMEABILITY (md)
0 5 10 15
WATER SATURATION (%)
0 10 20 30 40
6260
6270
6280
6290
6300
6310
SELINSGROVE AND NEEDMORE
RIDGELEY
HELDERBERG
Sandstone
Limestone
Silty limestone
Shale partings
Vertical fractures

Figure 11. Correlation of geological and petrophysical properties of the Ridgeley Sandstone in Leidy field, based on core data from the New York State Natural Gas #6 Pennsylvania Tract 81 well in Leidy pool. Notice that porosity and permeability are greatest in the upper four or five feet, but so is water saturation. Data courtesy of Consolidated Gas Supply Corp.

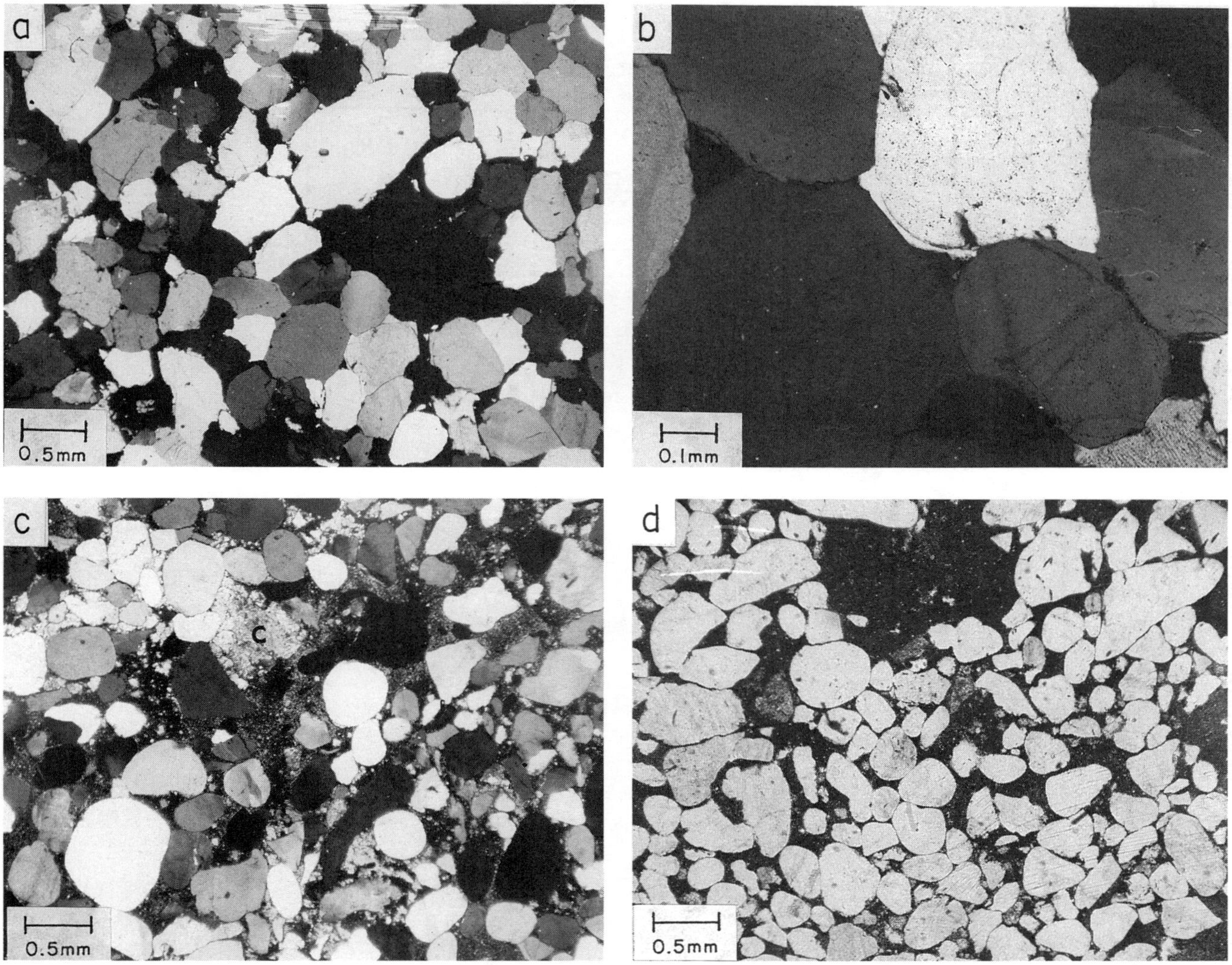
a
0.5mm
b
0.1mm
c
C
0.5mm
d
0.5mm

Figure 12. Photomicrographs of Ridgeley Sandstone from Leidy field and vicinity. (a) Silica-cemented quartz arenite, and (b) enlarged view of portion of (a), both under cross nicols. Note extensive authigenic cements coating otherwise well-rounded quartz grains. Sample, from Roeter and White #1 Harry Ludy well in Leidy pool, is particularly representative of beach-bar sand complex of Basan et al. (1980). (c) Carbonate-cemented quartz arenite, under crossed nicols. Well-rounded quartz grains, and angular calcite grains (C) replacing dissolved feldspars, in matrix of fine-grained calcite cement. (d) Different view of same sample, in plain light. From Shaw and Smith #1 Frank Harmon well near Hyner, about 20.9 km (13 mi) southeast of Leidy field.

fossils commonly occur in the cleaner sandstones, and ***Palaeophycos*** traces occur in the lower sections. Description of a core from a Leidy field well (Basan et al., 1980) shows that the Ridgeley consists of 11.9 m (39 ft) of fine-grained, dark gray, massive sandstone, bioturbated and calcareous in the lower part, with numerous unoriented brachiopod valves and natural fractures through most of the section. Minor shale partings occur about 3 m (10 ft) above the base.

Common diagenetic modifications of the Ridgeley Sandstone include development of authigenic cements, both silicate and carbonate (Figure 12), and pressure solution. Calcite and dolomite cements have replaced some detrital grains and occluded some of the pore spaces, reducing the total intergranular volume by 2 to 8%. Dissolution of these cements, and of detrital grains, created secondary porosities of as much as 20% in the upper few feet. Porosity in a typical well in Leidy field ranges from 1.39% to 7.82% (Figure 11), and enhanced porosity may occur in association with fractures. Although higher porosities generally mean better gas recovery, they have disadvantages as well in the Ridgeley—producible water saturations commonly are higher. Core analyses indicate that water saturation in some wells in Leidy field ranges from near 0% in low-porosity, low-permeability zones to over 30% in the highest-porosity zone near the top of the formation (Figure 11). In seeming contradiction to this, the log suite of a Leidy field well (Figure 13) indicates that water saturations are higher in the lower-porosity zones. This, however, is probably due to irreducible water in the formation that was not recorded during core analyses.

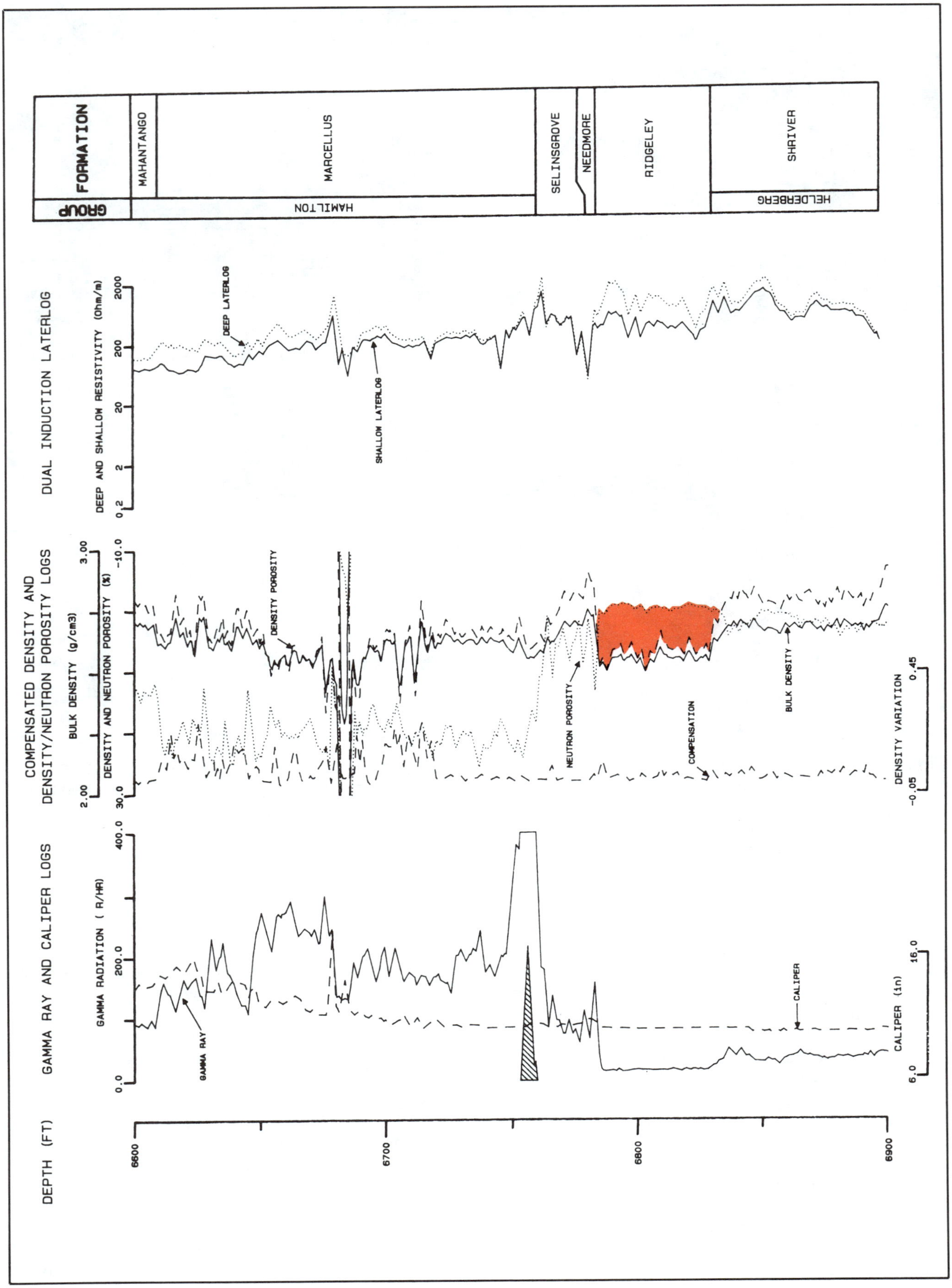
GROUP
FORMATION
MAHANTANGO
MARCELLUS
HAMILTON
SELINSGROVE
NEEDMORE
RIDGELEY
SHRIVER
HELDERBERG
DUAL INDUCTION LATERLOG
DEEP AND SHALLOW RESISTIVITY (Ohm/m)
DEEP LATERLOG
SHALLOW LATERLOG
COMPENSATED DENSITY AND
DENSITY/NEUTRON POROSITY LOGS
BULK DENSITY (g/cm3)
DENSITY AND NEUTRON POROSITY (%)
DENSITY POROSITY
NEUTRON POROSITY
COMPENSATION
BULK DENSITY
DENSITY VARIATION
GAMMA RAY AND CALIPER LOGS
GAMMA RADIATION (R/HR)
GAMMA RAY
CALIPER
CALIPER (in)
DEPTH (FT)
6600
6700
6800
6900

▲

Figure 13. Geophysical log suite of Mahantango Formation through Shriver Chert, from a typical well in Leidy field. The neutron-density crossplot (shaded) indicates that relatively high porosities exist in the Ridgeley Sandstone.

Permeabilities (to air) in the Ridgeley Sandstone range from less than 0.1 md to greater than 40 md, with lower permeabilities in tightly cemented zones and higher permeabilities in fractured or high-porosity zones. A typical well in Leidy field has permeabilities ranging from 0.03 to 13.03 md (Figure 11).

Shut-in well-head pressures in the Ridgeley Sandstone in Leidy field initially ranged from 18,341 to 28,959 kPa (2660 to 4200 psi), and were considered to be anomalously high for the depth of the reservoir, which was about 1706.9 m (5600 ft) (Ebright and Ingham, 1951). However, subsequent drilling and production decreased these pressures significantly. Using the youngest pool in the field, Ole Bull, as an example, Figures 14 and 15 illustrate pressure decline during a brief period in the early 1960s, just a few months after discovery. Figures 16 and 17 are decline curves of gas production from Ole Bull pool. Production rose sharply in early 1961 with the successful completion of the Phillips Petroleum Co. #6 Pennsylvania Tract 81 well (Figure 2, well #4) on December 30, 1960. This well had an after-treatment open flow of 7770 MCFGD and an initial shut-in well-head pressure of 22,305 kPa (3235 psi) in 7 days. Combined with the other four wells in the pool, production reached a peak of 233,578 mcf in January 1961. Production in the pool in 1961 amounted to 1,881,061 mcf. Except for Ole Bull, all Ridgeley reservoirs in Leidy field are now in storage (Figure 2). Table 1 lists the pertinent data for the two largest storage pools, Leidy and Greenlick.

Figure 18 shows the variations in thickness of the Ridgeley in Leidy field. Higher thickness values are probably due to structural, rather than stratigraphic, thickening. Although the Ridgeley averages 9.9 m (32.6 ft) in gross thickness in Leidy field, the best pay zone is considerably thinner. It is generally limited to the higher-porosity zones near the top of the sandstone

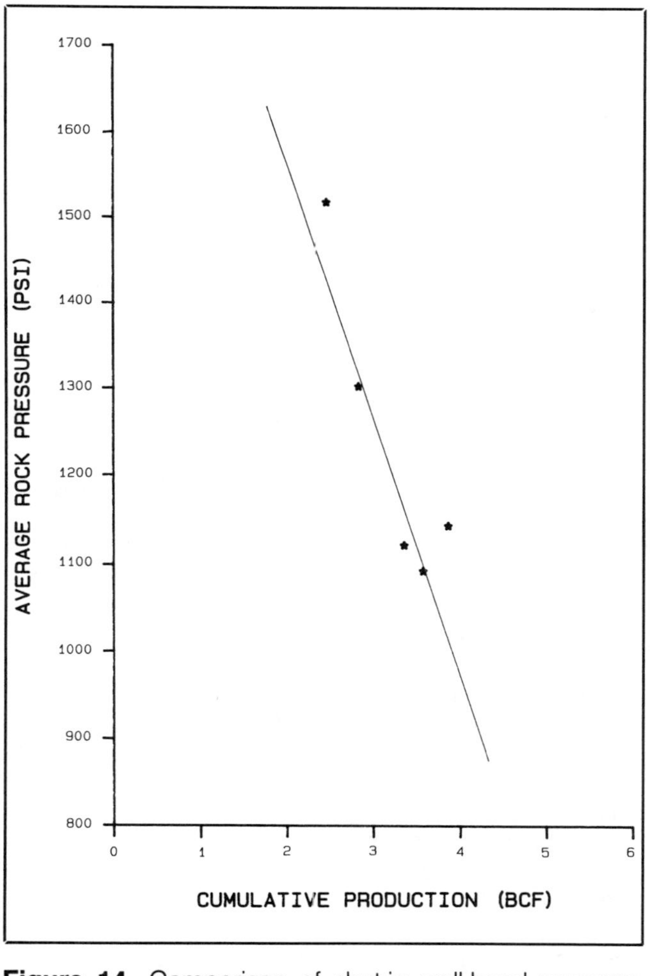
AVERAGE ROCK PRESSURE (PSI)
1700
1600
1500
1400
1300
1200
1100
1000
900
800
0
1
2
3
4
5
6
CUMULATIVE PRODUCTION (BCF)

Figure 14. Comparison of shut-in well-head pressure decline with cumulative production, Ole Bull pool, Leidy field. Straight line is the predicted decline curve. Data courtesy of Consolidated Gas Supply Corp.

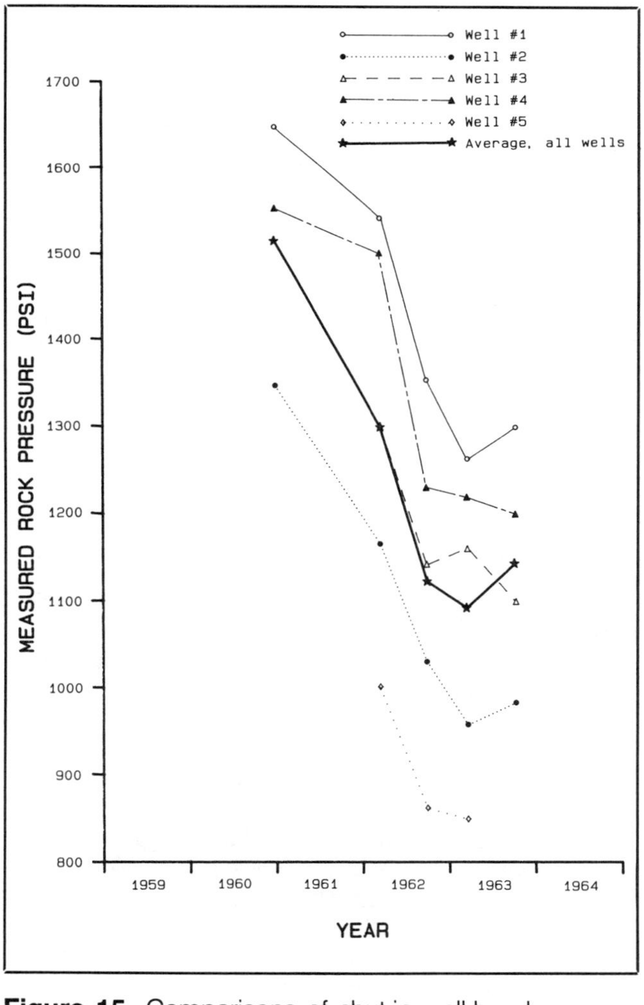
Well #1
Well #2
Well #3
Well #4
Well #5
Average, all wells
MEASURED ROCK PRESSURE (PSI)
1700
1600
1500
1400
1300
1200
1100
1000
900
800
1959
1960
1961
1962
1963
1964
YEAR

Figure 15. Comparisons of shut-in well-head pressure decline in five producing wells in Ole Bull pool over the first three years of pool production.

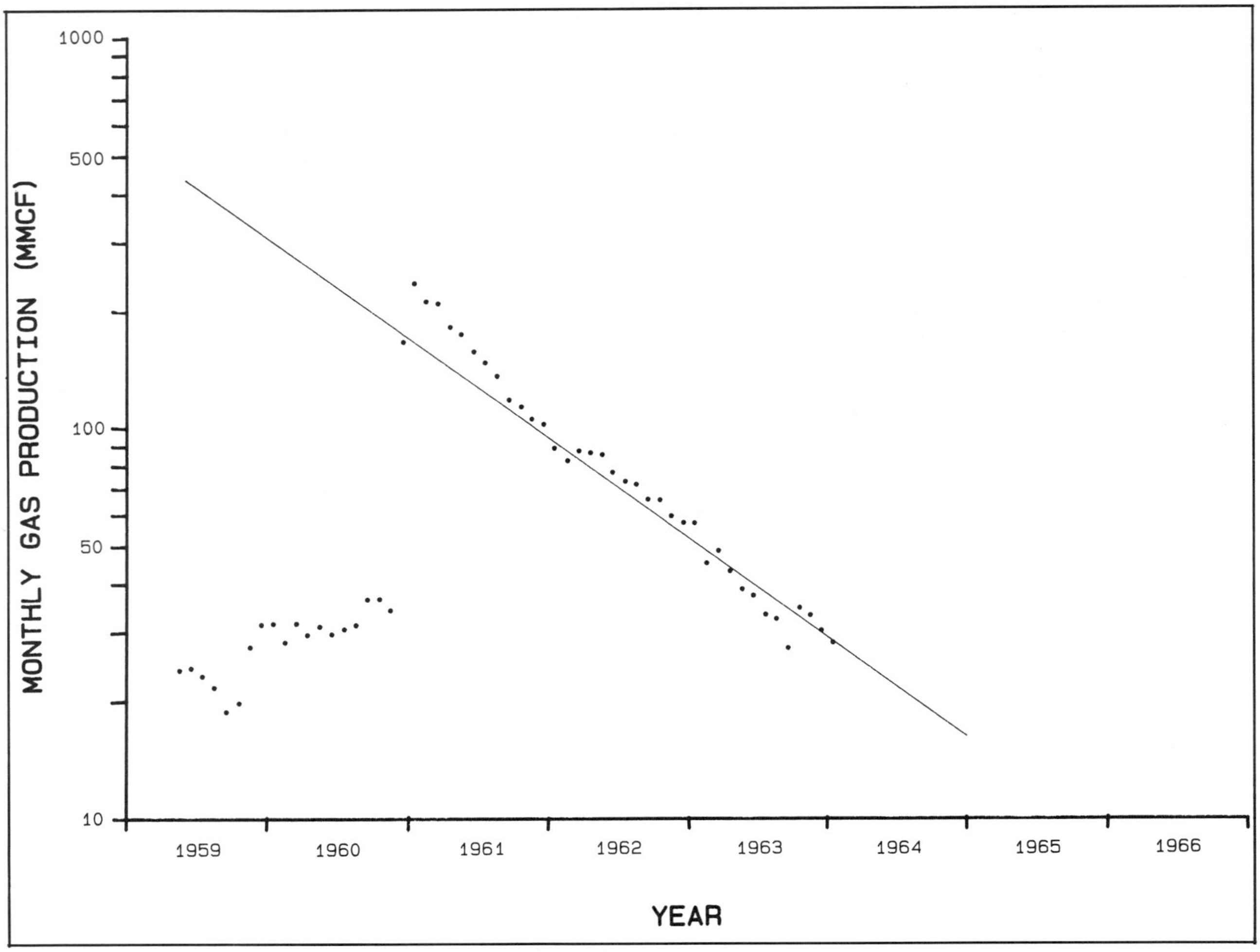

1000
500
100
50
10
MONTHLY GAS PRODUCTION (MMCF)
1959
1960
1961
1962
1963
1964
1965
1966
YEAR

Figure 16. Monthly decline curve for all wells in Ole Bull pool from May 1959 (date of first withdrawal) to January 1964. Straight line is ideal decline curve based on initial reserves of 5 bcf. Data courtesy of Consolidated Gas Supply Corp.

(Ebright and Ingham, 1951) (Figure 11). Modern well-completion technology has helped increase the pay zone, however, so that now most of the formation will produce some gas.

Ridgeley natural gas primarily is dry gas, composed of greater than 95% methane (Table 2). Heavier hydrocarbons and nonhydrocarbon gases typically are very minor constituents. Oil does not occur in the Ridgeley Sandstone in Pennsylvania.

Lock Haven Formation

Not much is known of the specific geologic and engineering characteristics of the Lock Haven Formation because it has only recently been recognized as a potentially good reservoir. Other than a few minor pools consisting of between two and 20 wells located sporadically throughout north-central Pennsylvania, Lock Haven reservoirs were practically unheard of until the 1980s. Discovery and development of large volumes of gas in the Lock Haven Formation in Council Run field in Centre County, south of Clinton County, recently has spurred a flurry of activity in this and adjacent areas, with drilling proceeding generally in a northeasterly direction. Standard geological and engineering studies, such as log interpretation and mapping, supplemented with some seismic and geochemical work, constitute the common techniques used to develop this reservoir.

No cores of the Lock Haven have been taken in Leidy field or any of the surrounding areas. Fettke (1961) described drill cuttings from the Lock Haven in Wharton Township, Potter County about 11.3 km (7 mi) northwest of Leidy field, but his descriptions of the interbedded sandstones, siltstones, and shales, is too general to be of much use. Although some Lock Haven strata crop out in the valley of Kettle Creek, which runs along the northwestern edge of Leidy field (Ebright and Ingham, 1951), most of the data pertaining to the lithology of the formation come from the outcrop belt to the south.

In the type locality in southern Clinton County, about 48.3 km (30 mi) southeast of Leidy field, the outcrop of

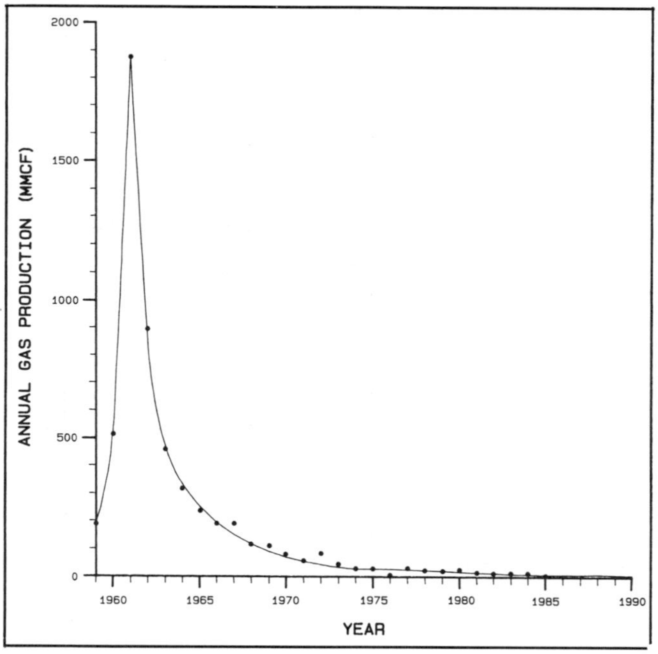
ANNUAL GAS PRODUCTION (MMCF)
2000
1500
1000
500
0
1960
1965
1970
1975
1980
1985
1990
YEAR

Figure 17. Production decline from all wells in Ole Bull pool. Data from files of Pennsylvania Geological Survey.

the Lock Haven consists of interbedded, predominantly marine sandstones, siltstones, shales, and mudrocks, and some conglomeratic beds 1181 m (3876 ft) thick (approximately 701 m or 2300 ft in Leidy field). The rocks are represented by 22.5% sandstone, 28.1% siltstone, 49.1% shale, and 0.2% conglomeratic sandstone (Faill and Wells, 1977). These percentages vary from outcrop to outcrop, and undoubtedly also vary from well to well. Thin coquinite lenses are present but not common. Zones of hematitic sandstone, siltstones, and shales (probably the Mansfield iron ore of Rogers, 1858) occur near the top of the formation. The sandstones tend to be discontinuous, thin to medium bedded, very fine grained, siliceous, locally calcareous, locally hematitic, and fossiliferous. The siltstones are more persistent than the sandstones; they are thin to medium bedded, siliceous, and contain thin shale partings. The conglomerates, such as those shown in Figure 19, consist of well-sorted, medium- to coarse-grained sandstones supporting well-rounded quartz pebbles generally less than 1 cm (1/3 in.) in diameter. Both upper and lower formational contacts are gradational, based on color changes or percentages of coarse to fine clastics. Based on local and regional studies of the Upper Devonian in Pennsylvania (Kelley and Wagner, 1972; Piotrowski and Harper, 1979), the Lock Haven Formation represents the same depositional facies as the Bradford and Elk groups in western Pennsylvania. These formations consist of near-shore marine sequences (beach, bar and subaqueous channel deposits)

Table 1. Reservoir characteristics of selected Leidy field storage pools from Consolidated Gas Transmission Corp.).

Storage Reservoir	Year Activated	Acreage	Input/Out Wells	Pressure Control Wells	Compressor Station Horsepower	Pipeline Size (Max/Min)	Cushion Gas (MMCF)		Working Gas (MMCF)		Unused Capacity (MMcf)	Designed Maximum Deliverability (MMcf)	Maximum Storage Pressure (psig)
							Native	Injected	Native	Injected			
Leidy*	1959	44,407	75	50	25,800	20″/2″	—	52,022	—	61,201	—	1224	4200
Greenlick	1961	20,160	34	1	13,600	16″/2″	3570	23,460	—	27,030	—	612	4240

*Includes Tamarack Storage pool.

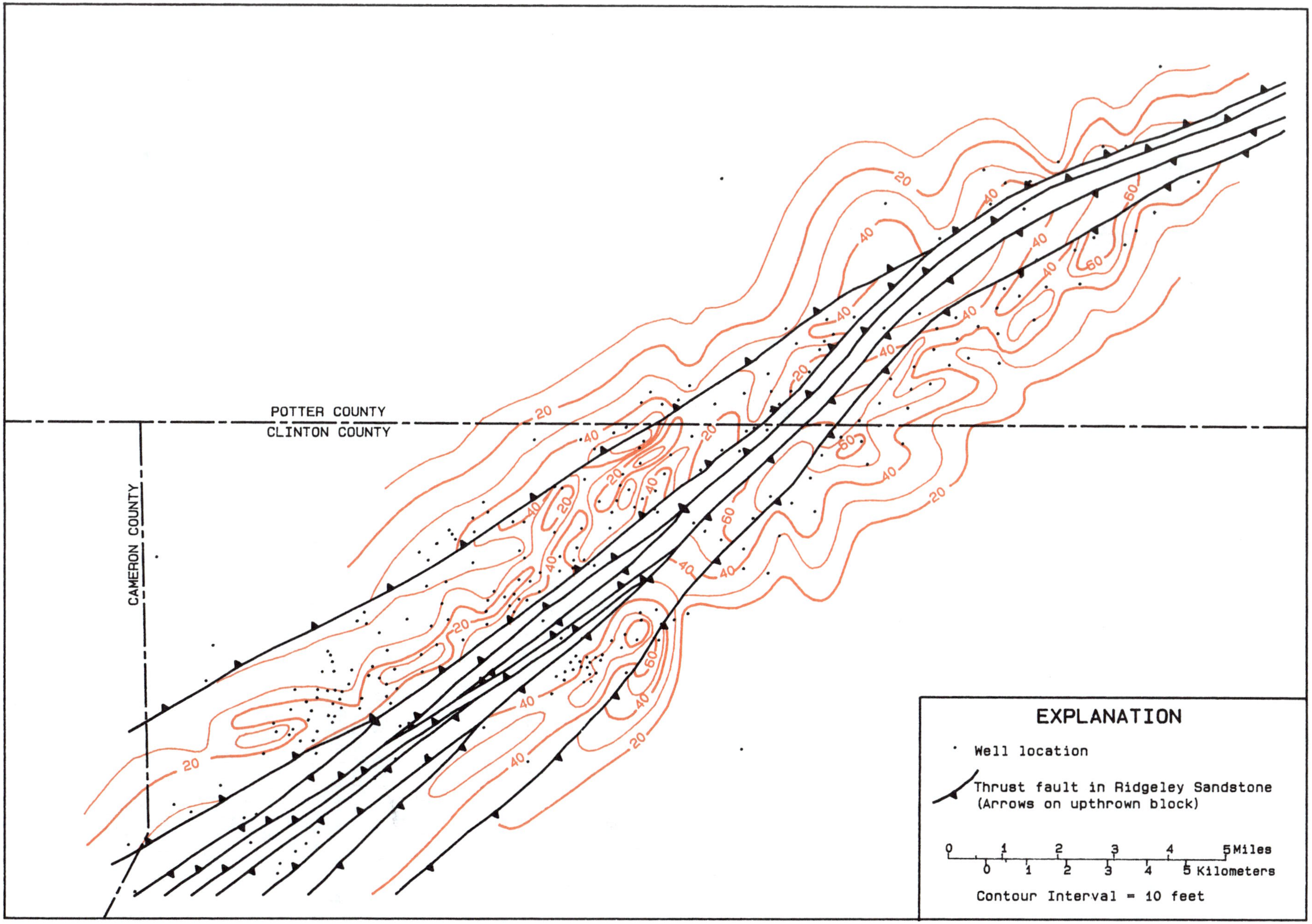
POTTER COUNTY
CLINTON COUNTY
CAMERON COUNTY
20
40
60
EXPLANATION
Well location
Thrust fault in Ridgeley Sandstone (Arrows on upthrown block)
0 1 2 3 4 5 Miles
0 1 2 3 4 5 Kilometers
Contour Interval = 10 feet

Figure 18. Isopach map of Ridgeley Sandstone in Leidy field.

a

b

Figure 19. Photographs of outcrops of Lock Haven Formation in Tioga County, Pennsylvania, northeast of Leidy field. (a) Outcrop showing coarser-grained sandstones (lighter-colored rock at geologist's hand) in sequence of interbedded mudstones, siltstones and sandstones. (b) Close-up of portion of light-colored rock from (a). Quartz-pebble conglomerates such as this are scattered throughout the formation in north-central Pennsylvania, and may be responsible for what hydrocarbon production has occurred in the formation. Pen for scale measures 12.7 cm (5 in.).

Table 2. Analysis of gas from Ridgeley Sandstone, Clinton County, Pennsylvania. BTU value — 1045, specific gravity — 0.583 (from Moore and Sigler, 1987, sample number 10316).

Component	Mole Percent
Methane	95.5
Ethane	2.9
Propane	0.5
N-Butane	0.1
Isobutane	Trace
N-Pentane	0.1
Isopentane	Trace
Cyclopentane	Trace
Hexanes plus	0.1
Nitrogen	0.5
Oxygen	Trace
Argon	Trace
Hydrogen	0.0
Hydrogen sulfide	0.0
Carbon dioxide	0.2
Helium	0.01

to slope turbidites (Bayles, 1949; Wolfe, 1963; Piotrowski and Harper, 1979; Laughrey and Harper, 1986).

Upper Devonian sandstones and siltstones include sublitharenites, subarkoses, quartz arenites, quartz wackes, lithic wackes, and arkosic wackes in order of dominance (see Pettijohn et al., 1973, for information on the classification of sandstones used here). Monocrystalline and polycrystalline quartz grains account for greater than 60% of the detrital fraction in most specimens. Other constituents include feldspars (generally less than 5% on average), rock fragments (about 6%), micas (about 6%), minor accessory minerals such as zircon and tourmaline, fossil fragments, and allochems. Cements and matrix, commonly consisting of clay minerals, carbonates, silica, hematite, anhydrite, and feldspar, may account for almost 16% of the composition. See Appendix 1 for a summary of these data. Laughrey et al. (in preparation) describes the diagenetic changes that occurred in Upper Devonian reservoir sandstones in Pennsylvania and, apparently, throughout the Appalachian basin. In general, compaction, cementation, and replacement processes reduced intergranular volume in the sandstones and siltstones. Some cementation helped preserve relict primary porosity, however, and clay mineral authigenesis created microporosity. Secondary porosity resulted from the direct dissolution of cements and unstable grains; secondary porosity also was created when replacement minerals were later dissolved. There also is a scattered amount of fracture porosity present in these rocks (Laughrey and Harper, 1986).

Average porosities in the Upper Devonian rocks range from 14.3% in the oil district of northwestern Pennsylvania (Fettke, 1938; McGlade, 1964; Lytle, 1965; Dixon, 1974) to 8.5% in the gas-producing belt throughout the rest of western and north-central Pennsylvania (Kimmel and Fulton, 1983). Figure 20 shows a log suite for the Lock Haven Formation in a typical well in Leidy field.

Permeabilities vary widely from region to region and reservoir to reservoir. On the average, permeabilities in the oil district range from 0.1 md to more than 33 md (some reservoir permeabilities are measured in darcys), and in the gas belt the average reported *in situ* permeability of Bradford Group sandstones is 0.0031 md (Laughrey and Harper, 1986). Areas of significantly higher permeabilities do occur in these rocks (e.g., Laughrey, 1982), but they are relatively insignificant in terms of the volume of existing and potential reservoirs in the Upper Devonian of the Appalachian basin. Measured rock pressures, including areas adjacent to Leidy field, generally range from 3447 to 10,342 kPa (500 to 1500 psi).

As with the Ridgeley, Lock Haven natural gas is primarily dry gas, composed of greater than 95% methane and with considerably lesser amounts of heavier hydrocarbons (Table 3). Although oil is not commonly found in producible amounts in the Lock Haven Formation, it is not unknown. One of the more interesting oil fields in Pennsylvania, the Gaines field 41.8 km (26 mi) northeast of Leidy, had estimated reserves of over 2.2 million bbl from two reservoir sands in the Lock Haven (Lytle, 1950). The Ox Bow well described by Chance (1880) in what is now Leidy field had shows of oil at depths less than 243.8 m (800 ft). A recently completed well in Bradford County, just east of Tioga, reported the first new producible oil reservoir in north-central Pennsylvania since the Gaines field was discovered in 1898. Unfortunately, there still is very little information available for this phenomenon.

Table 3. Analysis of gas from Lock Haven Formation, Centre County, Pennsylvania. BTU value — 1038, specific gravity — 0.584 (from Moore and Sigler, 1987, sample number 17173).

Component	Mole Percent
Methane	95.1
Ethane	2.5
Propane	0.3
N-Butane	0.1
Isobutane	Trace
N-Pentane	0.1
Isopentane	0.0
Cyclopentane	0.1
Hexanes plus	0.2
Nitrogen	1.2
Oxygen	0.1
Argon	Trace
Hydrogen	0.1
Hydrogen sulfide	0.0
Carbon dioxide	Trace
Helium	0.15

SOURCE

Although most of the organic-rich shales deposited in the Appalachian basin during the Paleozoic have the potential to produce hydrocarbons, the most probable sources of natural gas in the Lock Haven and Ridgeley reservoirs in Leidy field are the Devonian shales, particularly the Middle Devonian Marcellus Formation. Fracturing of the Lower and Middle Devonian strata probably acted as open pathways for hydrocarbon-generating fluids migrating horizontally and vertically through the section. Timing of this fluid migration, and the maturation of the organic material, is difficult to establish accurately, but it probably occurred at, or after, the end of the Paleozoic. Overburden or dewatering pressures may have forced the hydrocarbon-rich fluids downward into the Ridgeley as well as upward into the Lock Haven and other reservoirs. Oliver (1986), however, has proposed an interesting alternative hypothesis that also could provide a satisfactory explanation for both maturation and hydrocarbon emplacement, particularly in formations many miles from likely source beds. When the North American and African plates converged and collided during the Paleozoic, overthrusting from the east buried the North American continental margin sediments, expelling fluids westward toward the continental interior. "Some fluids are expelled into permeable beds of the foreland basin and platform...;[they] carry heat, minerals, and petroleum, or the ingredients for petroleum, into the foreland, and perhaps well into the midcontinent" (Oliver, 1986). Data concerning abnormal pore pressures and tectonic jointing in the Devonian shales (Engelder and Oertel, 1985) complement this hypothesis.

Very few geochemical data exist outside of company files for the Marcellus Formation, or for any other potential source bed, in north-central Pennsylvania.

Figure 20. Geophysical log suite of a portion of the Lock Haven Formation from a typical well in Leidy field. Percent sandstone curve is based on computer log. The neutron-density crossplot (shaded) indicates only a few scattered sandstones have porosities sufficient to be potential reservoirs.

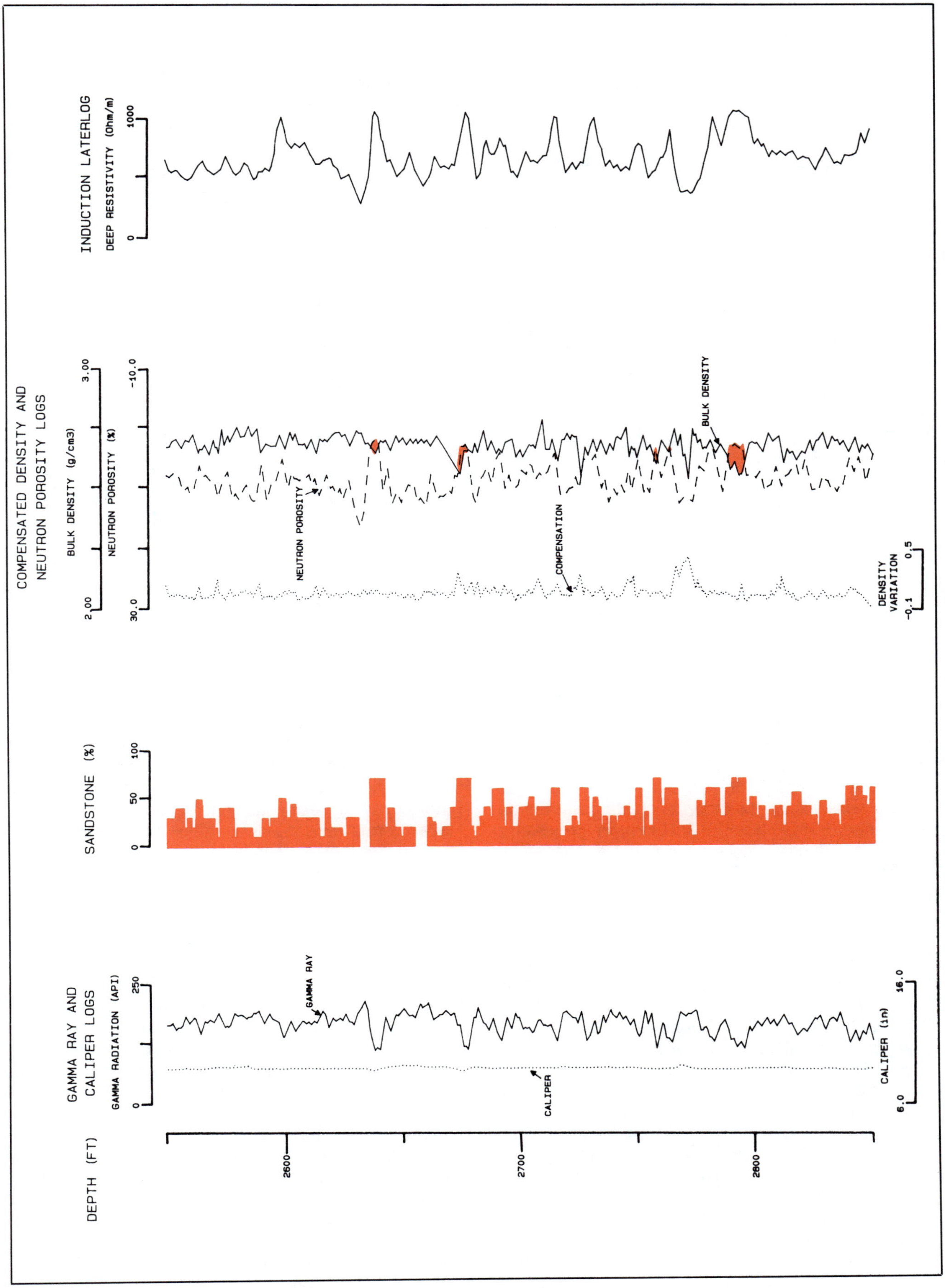

INDUCTION LATERLOG
DEEP RESISTIVITY (Ohm/m)
0
1000
COMPENSATED DENSITY AND
NEUTRON POROSITY LOGS
BULK DENSITY (g/cm3)
2.00
3.00
NEUTRON POROSITY (%)
30.0
-10.0
NEUTRON POROSITY
COMPENSATION
BULK DENSITY
DENSITY
VARIATION
-0.1
0.5
SANDSTONE (%)
0
50
100
GAMMA RAY AND
CALIPER LOGS
GAMMA RADIATION (API)
0
250
GAMMA RAY
CALIPER
CALIPER (in)
6.0
16.0
DEPTH (FT)
2600
2700
2800

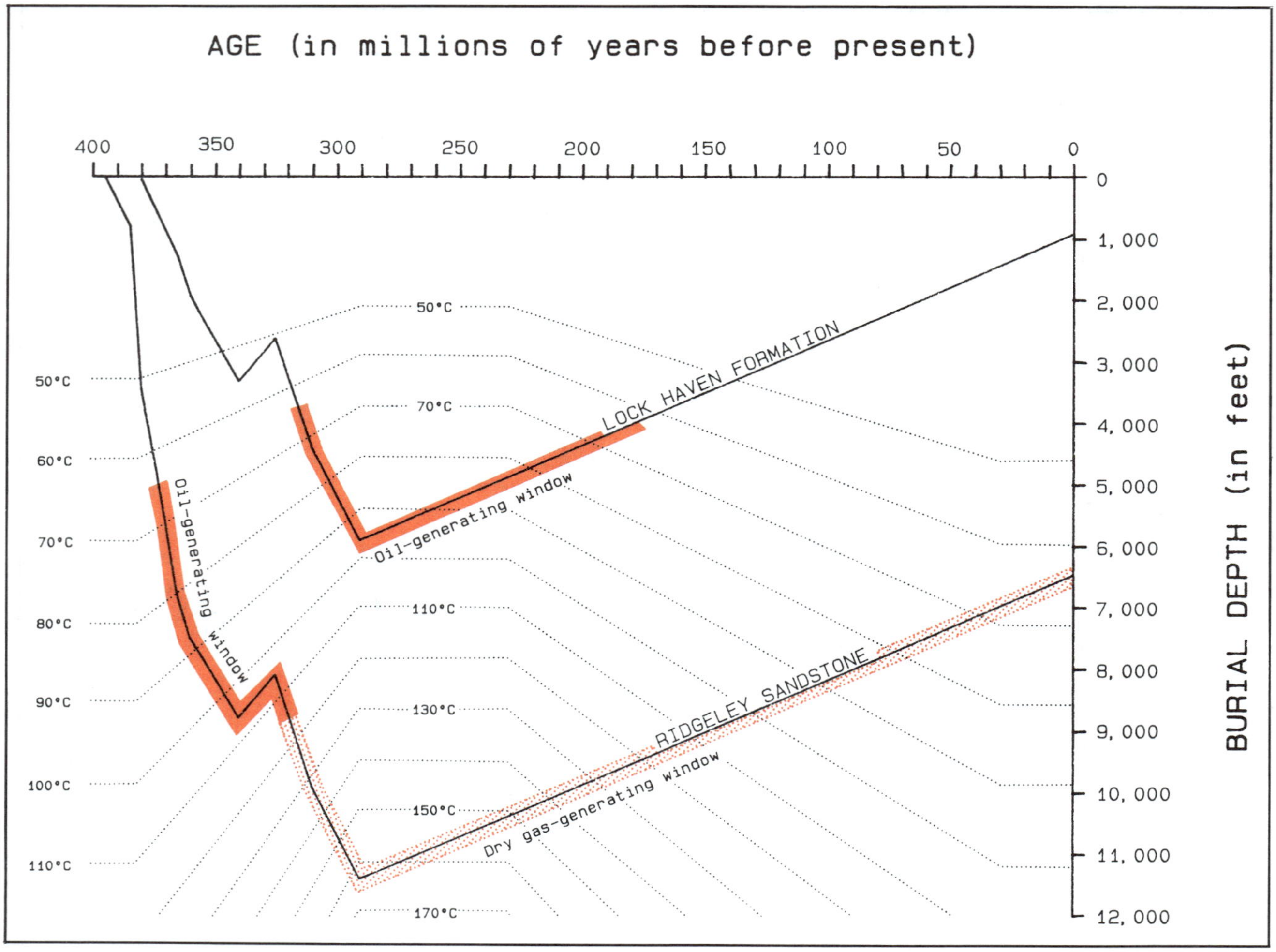

AGE (in millions of years before present)
400
350
300
250
200
150
100
50
0
0
1, 000
2, 000
3, 000
4, 000
5, 000
6, 000
7, 000
8, 000
9, 000
10, 000
11, 000
12, 000
BURIAL DEPTH (in feet)
50°C
60°C
70°C
80°C
90°C
100°C
110°C
50°C
70°C
110°C
130°C
150°C
170°C
LOCK HAVEN FORMATION
Oil-generating window
Oil-generating window
RIDGELEY SANDSTONE
Dry gas-generating window

Figure 21. Burial depth curve for Leidy field. The curve assumes: (1) an initial geothermal gradient during the Early Devonian of 25°C/km (14°F/1000 ft) and a surface temperature of 25°C (77°F); (2) an increase in the geothermal gradient where the organic-rich Marcellus Formation passed 50°C (122°F) and began generating hydrocarbons, thereby increasing pore pressures in the rock (Tillman and Barnes, 1983); and (3) a maximum geothermal gradient of 40°C/km (104°F/1000 ft) during the Alleghany orogeny due to dewatering of the shales. The Ridgeley Sandstone and associated source beds (Marcellus Formation about 12.2 m [40 ft] above the top of the Ridgeley) are in the dry gas-generating window. The Lock Haven Formation is still within the oil-generating window. Oil shows and production from Lock Haven sandstones may be due to sources within the formation.

Some total organic carbon (TOC), vitrinite reflectance (R_o), and gas volume data are available for broad areas of the Appalachian basin (Basan et al., 1980; Streib, 1981), and for Pennsylvania in particular (Pennsylvania Geological Survey files). In McKean County, Pennsylvania, about 80.5 km (50 mi) northwest of Leidy field, Marcellus TOC averages 5.78%, mean R_o is 1.26, and gas volume measures 0.014 m^3/m^3 (0.51 ft^3/ft^3. Basan et al. (1980) indicated that TOC values for the Marcellus throughout the basin range from 1.04 to 9.63%, averaging 3.68%. They rated the average source quality of the Marcellus as very good. Carbon values increase from east to west across the Appalachian basin, indicating increased thermal maturity to the east. Marcellus organic matter most commonly consist of type I kerogens, but type II kerogens occur in some areas influenced by higher terrigenous input.

Burial History

Figure 21, illustrating the burial history curve for Leidy field, assumes at least one major change in the geothermal gradient during the geologic record. The higher temperatures necessary for thermal alteration in the hydrocarbon-generating window in the Appalachian basin typically are regarded as being due solely to depth of burial (e.g., Basin et al., 1980). As such, source rocks must be buried to improbable depths in order to gain the particular geochemical traits they exhibit. Based on vitrinite reflectance of some shale partings in the Ridgeley Sandstone, Basan et al. (1980) determined that the Ridgeley in Leidy field had a maximum burial depth of 5974 m (19,600 ft) and that maximum erosion of the overburden was 3932 m (12,900 ft). Having such depths to work with, many authors (e.g., Friedman and

Sanders, 1982; Lakatos and Miller, 1983) assume that the present geothermal gradient is a valid basis for computing burial depth of ancient sediments in the Appalachian basin. The present geothermal gradient in the Leidy field area is about 25 °C/km (72.5 °F/mi) and the surface temperature is about 11 °C (51.8 °F) (Maurath and Eckstein, 1981).

The maximum burial depth calculated by Basan et al. (1980) is unrealistic. Based on a plot of the interval transit time of the Devonian shales as determined from sonic logs (using the method employed by Magara, 1976), the approximate amount of overburden in the Leidy field area should have been around 3505.2 m (11,500 ft) (Figure 22). This means that approximately 1524 m (5000 ft) of post-Early Pennsylvanian-age strata have been removed by erosion.

Although it is possible that the geothermal gradient has not changed over geologic time, it is more likely that temperatures were significantly higher in the late Paleozoic (during the Alleghany orogeny), masking the actual overburden stresses throughout the Appalachian basin. If the Ridgeley were buried to 3505.2 m (11,500 ft), as proposed by Figure 21, a higher gradient at some time in the geologic past would be necessary. When the burial and thermal data shown in Figure 21 are used, Lopatin modeling of the time-temperature history of the Ridgeley and adjacent strata yields the thermal maturities recorded for the Leidy field area. Tillman and Barnes (1983) determined that the geothermal gradient in the northern Appalachian basin should have been 40 °C/km (155 °F/mi) at the time of dewatering of the shales, based on mineralogic evidence. Figure 21 assumes this 40 °C/km gradient, but the gradient may have been somewhat higher in the Leidy field area because of structural complications. The figure also assumes that the gradient began increasing when the organic-rich shales of the Marcellus Formation reached a burial temperature of 50 °C (122 °F), the lower limit of the oil-generating window.

EXPLORATION CONCEPTS

Discovery and development of the Ridgeley Sandstone reservoirs in Leidy field in 1950 initiated one of the most intense periods of leasing and drilling in Appalachian basin history by renewing an almost dormant interest in the deeper structural plays in western and central Pennsylvania (Ingham, 1954). Operators in Pennsylvania drilled 718 Ridgeley wells during the next four years, of which 413 succeeded in finding gas (Fettke, 1956). The largest Ridgeley play in Pennsylvania, the giant Punxsutawney-Driftwood field centered in Clearfield County (Figure 1), was discovered in 1951 and kept drillers busy into the 1960s. This field had a cumulative production of 384,511,914 MCFG at the end of 1985 from reservoirs having combination structural-stratigraphic-diagenetic traps associated with the "no-sand area" and the Chestnut Ridge anticline, a southern equivalent of the Wellsboro anticline in Clinton County (Abel and Heyman, 1981, plates 2 and 3).

Because of the variety and complexity of factors responsible for the development of Ridgeley Sandstone fields and pools in the Appalachian basin, no single exploratory tool or concept would make it easier to explore for such reservoirs throughout the basin today than 30 to 35 years ago. New Ridgeley reservoirs are still not easy to find, despite the fact that the search for Leidy-type reservoirs is more scientific. The importance of serendipity in the discovery of Ridgeley reservoirs in the early 1950s is always understated. In fact, without it Leidy might still be undiscovered. New York State Natural Gas was ready to stake a well near Leidy when Dorcie Calhoun began drilling; the company held back waiting to see what would occur, rather than risk the capital on a contemporaneous well (Paul Garrett, 1988, personal communication). If it had not been raining that day, Dorcie Calhoun's lead equipment truck probably would have gotten to the top of the mountain. The well would have been a dry hole (it might never have reached the Ridgeley) because it would have been drilled into the rifted core of the anticline. At that point, Leidy Prospecting Company would have failed and New York State Natural Gas would have seen Leidy as a slim prospect for future development; the deep drilling boom of the 1950s might have been delayed or might never have occurred.

Geophysical prospecting will not delineate productive zones in all areas, nor will analyses of porosity or production trends, nor will standard geological studies. Diecchio (1985) pointed out that stratigraphic traps, such as updip porosity changes, occur along the basin margin where the Ridgeley Sandstone was eroded (or not deposited), and that structural traps are more common in the central and eastern parts of the basin. In northern Pennsylvania, there sometimes occurs a fortuitous combination of these two factors (e.g., Wharton pool). Between these two broad zones of excellent gas recovery lies an area in which the structures are relatively subdued, even at depth, and intergranular porosity has not been greatly developed in the Ridgeley Sandstone. This zone has only minor Ridgeley production (e.g., Glyde pool in Washington County, southwestern Pennsylvania—the sole well in the pool produced a cumulative 155.8 MMCFG in the 25 years after it was drilled, and most of that production is from the Upper Devonian and Lower Mississippian). Where deep fields exist in this zone, production generally comes more readily from the overlying Huntersville Chert. Diecchio (1985) suggested that the more brittle Huntersville either failed to act as an effective cap rock for the Ridgeley Sandstone or that the chert serves as the better reservoir in this zone.

If the Ridgeley reservoirs of Leidy field were still undiscovered today, they could be explored for by obtaining an understanding of the three basic processes that affect reservoir development in north-central Pennsylvania. The first of these is the importance of Salina salt structures on development of such reservoirs in the area. An exploration geologist could easily prospect for structural traps in northern Clinton and southern Potter counties by using a grid of seismic lines. At least the largest faults show up as distinct reflector dis-

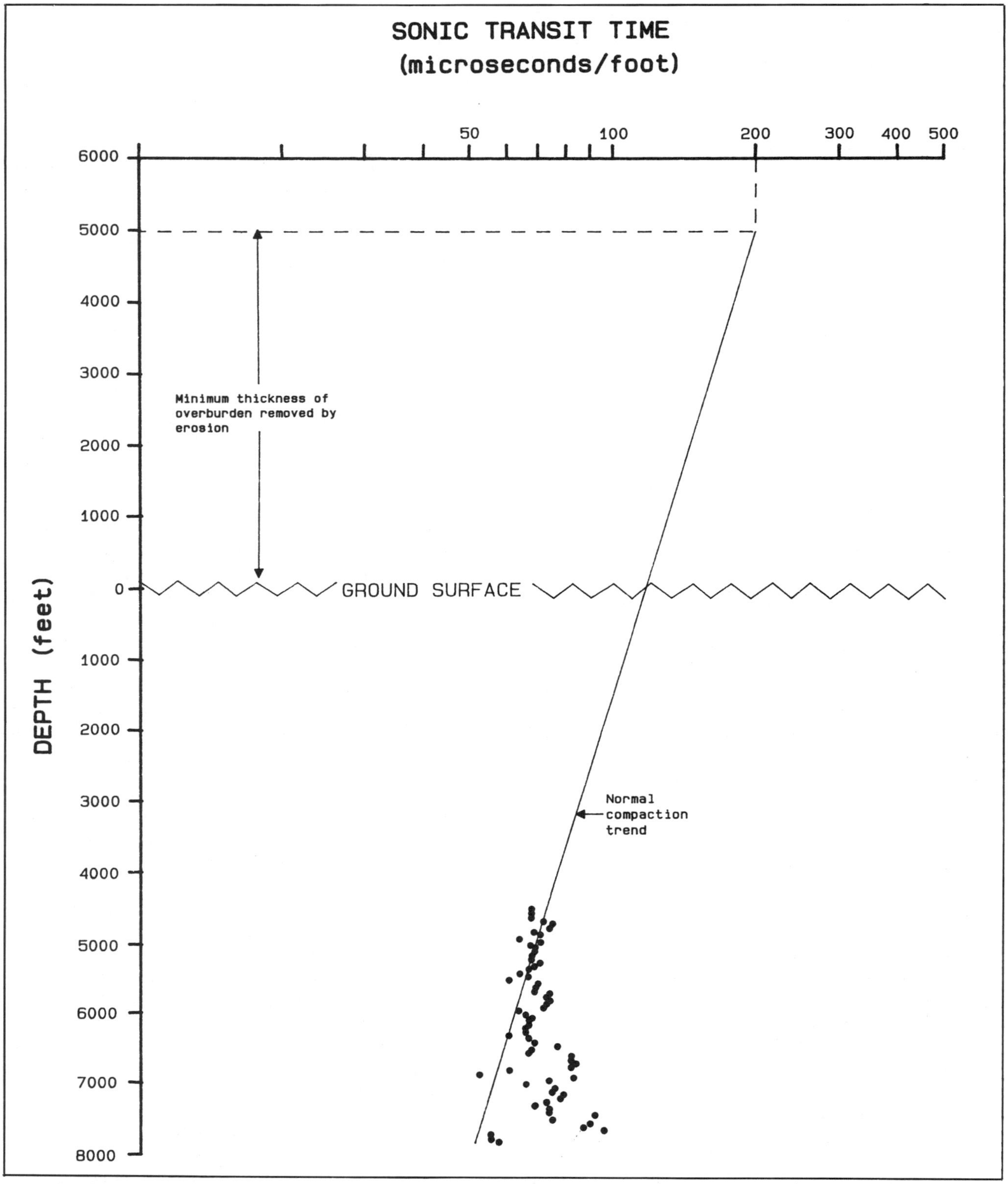
SONIC TRANSIT TIME
(microseconds/foot)
50
100
200
300
400
500
6000
5000
4000
3000
2000
1000
0
1000
2000
3000
4000
5000
6000
7000
8000
DEPTH (feet)
Minimum thickness of overburden removed by erosion
GROUND SURFACE
Normal compaction trend

Figure 22. Plot of shale transit time versus depth for the Leidy field area. Intersection of the normal compaction trend with the 200 μs/ft value indicates that a minimum of 1524 m (5000 ft) of rock were removed after maximum burial. Notice that the Marcellus Formation shales in the lower part of the diagram are characteristically undercompacted.

turbances in such surveys (Figure 5). Second, it is necessary to develop an understanding of the relationship of production to such structures—as stated above, production in Leidy field is limited to flexures on the outer fault blocks. This is also true of other fields in north-central Pennsylvania. Third, analysis of diagenetic trends within the Ridgeley in north-central Pennsylvania would soon establish the relationship of porosity development to the "no-sand area." The farther one drills from this area, the less likely one is to find significant porosity of any sort. Several dozen wells drilled in northeastern Pennsylvania over the last 35 years, including a recently completed Ridgeley dry hole in Bradford County 80.5 km (50 mi) northeast of Leidy field, have encountered significant thicknesses of Ridgeley Sandstone (10.9 m/36 ft) of sandstone in the recent well). However, despite the presence of fracturing induced by salt-generated structures, the sandstone typically is cemented tightly in this area.

Although other Ridgeley pools like those in Leidy field may be waiting to be discovered in north-central Pennsylvania, it is unlikely. Large stretches of the major anticlines within 32.2 km (20 mi) of the "no-sand area" have been drilled with no success.

There still is a good chance of developing the Lock Haven Formation in this area, however. Fractured and diagenetically altered Lock Haven sandstones in the vicinity of Leidy field have been shown to be good reservoirs, not only for natural gas but for oil as well. The Lock Haven should be test drilled throughout north-central and northeastern Pennsylvania to determine the best areas for exploiting this relatively untapped resource. There may be little chance of finding additional Leidy-type Ridgeley Sandstone reservoirs, but Lock Haven reservoirs such as Leidy's Hammersley Fork pool hold strong hope for future recoverable reserves in this area.

ACKNOWLEDGMENTS

Some of the information used in this paper was graciously provided by Consolidated Gas Transmission Corporation in Charleston, West Virginia. I thank Ronald L. Walden and Glenn Knepper for their support. Christopher D. Laughrey, Pennsylvania Geological Survey, provided assistance with the sections on reservoir geochemistry. Helpful comments on the initial draft of the manuscript were made by Christopher D. Laughrey, Thomas M. Berg, and Donald M. Hoskins of the Pennsylvania Geological Survey.

REFERENCES CITED

Abel, K. D., and L. Heyman, 1981, The Oriskany Sandstone in the subsurface of Pennsylvania: Pennsylvania Geological Survey, 4th ser., Mineral Resources Report 81, 9 p.

Basan, P. B., D. L. Kissling, K. D. Hemsley, and others, 1980, Geological study and reservoir evaluation of Early Devonian formations of the Appalachians: Robertson Research (U.S) Inc., 263 p.

Bayles, R. E., 1949, Subsurface Upper Devonian sections in southwestern Pennsylvania: American Association of Petroleum Geologists Bulletin, v. 33, p. 1682-1703.

Bradley, W. H., and J. F. Pepper, 1938, Structure and gas possibilities of the Oriskany Sandstone in Steuben, Yates, and parts of the adjacent counties, New York: U.S. Geological Survey Bulletin 899-A, 68 p.

Cathcart, S. H., 1934, Geologic structure in the plateaus region of northern Pennsylvania and its relation to the occurrence of gas in the Oriskany sand: Pennsylvania Geological Survey, 4th ser., Progress Report 108, 24 p.

Chance, H. M., 1880, The geology of Clinton County: Second Geological Survey of Pennsylvania, v. G4, Part 1, p. 1-78.

Colton, G. W., 1970, The Appalachian basin—its depositional sequences and their geologic relationships, *in* G. W. Fisher et al., eds., Studies of Appalachian geology: Central and Southern: New York, Interscience Pub., p. 5-47.

Cook, F. A., D. S. Albaugh, L. D. Brown et al., 1979, Thin-skinned tectonics in the crystalline southern Appalachians—COCORP seismic-reflection profiling of the Blue Ridge and Piedmont: Geology, v. 7, p. 563-567.

Davidson, W., 1951, Dorcie Calhoun struck it rich: Colliers, January 13, 1951, p. 16, 17, 69-71.

Diecchio, R. J., 1985, Regional controls of gas accumulation in Oriskany Sandstone, central Appalachian basin: American Association of Petroleum Geologists Bulletin, v. 69, p. 722-732.

Dixon, W. H., 1974, Reservoir geology of the Sage farm, Bradford oil field: American Association of Petroleum Geologists Eastern Section, Field Trip Guidebook, sec. III, 22 p.

Dott, R. H., and R. L. Batten, 1976, Evolution of the earth, 2d ed.: New York, McGraw-Hill, 504 p.

Drake, A. A., and J. B. Epstein, 1967, The Martinsburg Formation (Middle and Upper Ordovician) in the Delaware Valley, Pennsylvania-New Jersey: U.S. Geological Survey Bulletin 1244-H, p. H1-H16.

Ebright, J. R., and A. I. Ingham, 1951, Geology of the Leidy gas field and adjacent areas, Clinton County, Pennsylvania: Pennsylvania Geological Survey, 4th ser., Mineral Resources Report 34, 35 p.

Engelder, T., and G. Oertel, 1985, Correlation between abnormal pore pressure and tectonic jointing in the Devonian Catskill Delta: Geology, v. 13, p. 863-866.

Faill, R. T., 1985, The Acadian orogeny and the Catskill Delta, *in* D. L. Woodrow and W. D. Sevon, eds., The Catskill Delta: Geological Society of America Special Paper 201, p. 15-37.

Faill, R. T., and R. B. Wells, 1977, Bedrock geology and mineral resources of the Salladasburg and Cogan Station quadrangles, Lycoming County, Pennsylvania: Pennsylvania Geological Survey, 4th ser., Atlas 133cd, 44 p.

Fettke, C. R., 1938, Bradford oil field, Pennsylvania and New York: Pennsylvania Geological Survey, 4th ser., Mineral Resources Report 21, 454 p.

Fettke, C. R., 1950, Summarized record of deep wells in Pennsylvania: Pennsylvania Geological Survey, 4th ser., Mineral Resources Report 31, 148 p.

Fettke, C. R., 1956, Summarized record of deep wells in Pennsylvania, 1950 to 1954: Pennsylvania Geological Survey, 4th ser., Mineral Resources Report 39, 114 p.

Fettke, 1961, Well-sample descriptions in northwestern Pennsylvania and adjacent states: Pennsylvania Geological Survey, 4th ser., Mineral Resources Report 40, 691 p.

Finn, F. H., 1949, Geology and occurrence of natural gas in Oriskany Sandstone in Pennsylvania and New York: American Association of Petroleum Geologists Bulletin, v. 33, p. 303-335.

Friedman, G. M., and J. E. Sanders, 1982, Time-temperature-burial significance of Devonian anthracite implies former great (~6.5 km) depth of burial of Catskill Mountains, New York: Geology, v. 10, p. 93-96.

Gwinn, V. E., 1964, Thin-skinned tectonics in the Plateau and northwestern Valley and Ridge Provinces of the Central Appalachians: Geological Society of America Bulletin, v. 75, p. 863-900.

Harper, J. A., 1986, Oil and gas developments in Pennsylvania in 1985: Pennsylvania Geological Survey, 4th ser., Mineral Resources Report 199, 112 p.

Harper, J. A., 1987, Effects of pre-Alleghanian tectonics on Late Devonian and Early Mississippian rocks in southwestern Pennsylvania (abs.): Geological Society of American Abstracts with Programs, v. 19, no. 1, p. 17.

Harper, J. A., C. D. Laughrey, and W. S. Lytle, 1982, Oil and gas fields of Pennsylvania: Pennsylvania Geological Survey, 4th ser., Map #3.

Harper, J. A., and R. G. Piotrowski, 1978, Stratigraphy, extent, gas production, and future gas potential of the Devonian organic-rich

shales in Pennsylvania: Preprints, Second Eastern Gas Shales Symposium, METC/SP-78/6, v. 1, p. 310-329, Morgantown, WV.
Harris, L. D., A. G. Harris, W. DeWitt, and K. C. Bayer, 1981, Evaluation of southern eastern overthrust belt beneath Blue Ridge-Piedmont thrust: American Association of Petroleum Geologists Bulletin, v. 65, p. 2491-2497.
Heyman, L., 1969, Geology of the Elk Run gas pool, Jefferson County, Pennsylvania: Pennsylvania Geological Survey, 4th ser., Mineral Resources Report 59, 18 p.
Hunt, T. S., 1861, Bitumens and mineral oils: Montreal Gazette, March 1, 1961.
Ingham, A. I., 1954, The geology and development of Leidy-South Leidy gas fields, Clinton County, Pennsylvania: Producers Monthly, v. 18, no. 10, p. 22-28.
Johnson, M. E., 1923, Gas in Leidy Township, Clinton County, Pennsylvania: Pennsylvania Geological Survey, 4th ser., Mineral Resources Report 78, 7 p.
Kelley, D. R., and W. R. Wagner, 1972, Surface to Middle Devonian (Onondagan) stratigraphy, Part I (STOMDES): Pennsylvania Geological Survey, Open-File Report 1, Pittsburgh, 15 p.
Kimmel, S. L., and P. F. Fulton, 1983, Results of pressure transient well testing in Appalachian gas reservoirs: Society of Petroleum Engineers, SPE 12303, p. 59-65.
Kowalik, W. S., and D. P. Gold, 1976, The use of Landsat-1 imagery in mapping lineaments in Pennsylvania: Proceedings of the First International Conference on the New Basement Tectonics (Salt Lake City, 1974): Utah Geological Association Publication 5, p. 236-249.
Kreidler, W. L., 1953, History, geology and future possibilities of gas and oil in New York State: New York State Museum Circular 33, 58 p.
Lakatos, S., and D. S. Miller, 1983, Fission-track analysis of apatite and zircon defines a burial depth of 4 to 7 km for lowermost Upper Devonian, Catskill Mountains, New York: Geology, v. 11, p. 103-104.
Laughrey, C. D., 1982, High-potential gas production and fracture-controlled porosity in Upper Devonian Kane "sand," central-western Pennsylvania: American Association of Petroleum Geologists Bulletin, v. 66, p. 477-482.
Laughrey, C. D., and J. A. Harper, 1986, Comparisons of Upper Devonian and Lower Silurian tight formations in Pennsylvania—geological and engineering characteristics, *in* C. W. Spencer and R. F. Mast, eds., Geology of tight gas reservoirs: American Association of Petroleum Geologists Studies in Geology 24, p. 9-43.
Laughrey, C. D., R. M. Harper, and A. K. Markowski, in preparation, Oil and gas reservoir rocks of Pennsylvania: Pennsylvania Geological Survey, 4th ser., Mineral Resources Report.
Lavin, P. M., D. L. Chaffin, and W. F. Davis, 1982, Major lineaments and the Lake Erie-Maryland crustal block: Tectonics, v. 1, p. 431-440.
Lytle, W. S., 1950, Crude oil reserves in Pennsylvania: Pennsylvania Geological Survey, 4th ser., Mineral Resources Report 32, 256 p.
Lytle, W. S., 1965, Oil and gas geology of the Warren quadrangle, Pennsylvania: Pennsylvania Geological Survey, 4th ser., Mineral Resources Report 52, 84 p.
Magara, K., 1976, Thickness of removed sedimentary rocks, paleopore pressure, and paleotemperature, southwestern part of Western Canada basin: American Association of Petroleum Geologists Bulletin, v. 60, p. 555-564.
Maurath, G., and Y. Eckstein, 1981, Heat flow and heat production in northwestern Pennsylvania: Geothermal Resources Council Transactions, v. 5, p. 103-106.
McBride, E. F., 1987, Diagenesis of the Maxon Sandstone (Early Cretaceous), Marathon region, Texas—a diagenetic quartzarenite: Journal of Sedimentary Petrology, v. 57, p. 98-107.
McGlade, W. G., 1964, Oil and gas geology of the Youngsville quadrangle, Pennsylvania: Pennsylvania Geological Survey, 4th ser., Mineral Resources Report 53, 56 p.
Moore, B. J., and S. Sigler, 1987, Analyses of natural gas, 1917-85: U.S. Bureau of Mines Information Circular 9129, 1197 p.
Oliver, J., 1986, Fluids expelled tectonically from orogenic belts: Their role in hydrocarbon migration and other geologic phenomena: Geology, v. 14, p. 99-102.
Parrish, J. B., and P. M. Lavin, 1982, Tectonic model for kimberlite emplacement in the Appalachian Plateau of Pennsylvania: Geology, v. 10, p. 344-347.
Pettijohn, F. J., P. E. Potter, and R. Siever, 1973, Sand and sandstones: New York, Springer-Verlag, 618 p.
Piotrowski, R. G., and J. A. Harper, 1979, Black shale and sandstone facies of the Devonian "Catskill" clastic wedge in the subsurface of western Pennsylvania: Morgantown Energy Technology Center, EGSP Series no. 13, 40 p., Morgantown, WV.
Rankin, D. W., 1976, Appalachian salients and recesses: Late Precambrian continental breakup and the opening of the Iapetus Ocean: Journal of Geophysical Research, v. 81, p. 5605-5619.
Rodgers, M. R., and T. H. Anderson, 1984, Tyrone-Mt. Union cross-strike lineament of Pennsylvania: a major Paleozoic basement fracture and uplift boundary: American Association of Petroleum Geologists Bulletin, v. 68, p. 92-105.
Rogers, H. D., 1958, The geology of Pennsylvania: Pennsylvania Geological Survey, 1st ser. (Philadelphia), 2 v., 1401 p.
Root, S. I., 1978a, Possible recurrent basement faulting, Pennsylvania: Part 1, geologic framework: Pennsylvania Geological Survey, Open-File Report (Harrisburg), 23 p.
Root, S. I., 1978b, Possible recurrent basement faulting, Pennsylvania: Part 2, economic geology: Pennsylvania Geological Survey, Open-File Report (Harrisburg), 19 p.
Rosenfeld, M. A., 1953, Petrographic variation in the Oriskany "Sandstone Complex": Unpublished Ph.D. dissertation, Pennsylvania State University, 220 p.
Sanders, A. W., 1982, Oriskany matrix porosity in southwest-central Pennsylvania (abs.): West Virginia Geological and Economic Survey, Circular C-26, p. 50.
Shumaker, R. C., 1974, Influence of Saline salt on structure in New York-Pennsylvania: Discussion: American Association of Petroleum Geologists Bulletin, v. 58, p. 543-544.
Streib, D. L., 1981, Distribution of gas, organic carbon, and vitrinite reflectance in the eastern Devonian gas shales and their relationship to the geologic framework: Morgantown Energy Technology Center, DOE/MC/08216-1276, Morgantown, WV, 262 p.
Thomas, W. A., 1977, Evolution of the Appalachian-Ouachita salients and recesses from reentrants and promentories in the continental margin: American Journal of Science, v. 277, p. 1233-1278.
Thompson, A. M., and W. D. Sevon, 1982, Excursion 19B: Comparative sedimentology of Paleozoic clastic wedges in the central Appalachians, U.S.A.: Field Excursion Guide Book, Eleventh International Congress on Sedimentology (Hamilton, Ontario), August 22-27, 1982, 136 p.
Tillman, J. E., and H. L. Barnes, 1983, Deciphering fracturing and fluid migration histories in northern Appalachian basin: American Association of Petroleum Geologists Bulletin, v. 67, p. 692-705.
Van der Voo, R., 1979, Age of the Alleghenian folding in the central Appalachians: Geology, v. 7, p. 297-298.
Wheeler, R. L., 1980, Cross-strike structural discontinuities—possible exploration tool for natural gas in Appalachian overthrust belt: American Association of Petroleum Geologists Bulletin, v. 64, p. 2166-2178
White, I. C., 1885, The geology of natural gas: Science, v. 5, June 26, 1885, p. 43-44.
White, I. C., 1892, The Mannington oil field and the history of its development: Geological Society of America Bulletin, v. 3, p. 187-216.
Wiltschko, D. V., and W. M. Chapple, 1977, Flow of weak rocks in Appalachian Plateau folds: American Association of Petroleum Geologists Bulletin, v. 61, p. 653-670.
Wolfe, R. T., 1963, The correlation of Upper Devonian Chemung sands in west-central Pennsylvania, north of Pittsburgh, *in* V. C. Shepps, ed., Symposium on Middle and Upper Devonian stratigraphy of Pennsylvania and adjacent states: Pennsylvania Geological Survey, 4th ser., General Geology Report 39, p. 241-257.

Appendix 1. Field Description

Field name .... *Leidy field*

Ultimate recoverable reserves

Field location:

- **Country** .... *U.S.A.*
- **State** .... *Pennsylvania*
- **Basin/Province** .... *Appalachian*

Field discovery:

- **Year field discovered** *1864*
- **Year second pay discovered** *1950*
- **Third pay**

Discovery well name and general location
(i.e., Jones No. 1, Sec. 2T12NR5E; or Smith No. 1, 5 mi west of Sheridan, Wyoming):

- **First pay** .... *William Sansom No. 1, about 4 mi north of Leidy, Clinton Co., Pennsylvania*
- **Second pay** .... *Dorcie Calhoun No. 1, Leidy, Clinton Co., Pennsylvania*
- **Third pay**

Discovery well operator .... *William Sansom*
(if more than one pay in field, list operators of discovery well in other pays)

- **Second pay** .... *Leidy Prospecting Co.*
- **Third pay**

IP in barrels per day and/or cubic feet or cubic meters per day:

- **First pay** .... *Unknown*
- **Second pay** .... *15 MMCFGD*
- **Third pay**

All other zones with shows of oil and gas in the field:

Age	Formation	Type of Show
Middle Devonian	*Marcellus Formation*	*Gas*
Middle Devonian	*Selinsgrove Limestone*	*Gas*

Geologic concept leading to discovery and method or methods used to delineate prospect, e.g., surface geology, subsurface geology, seeps, magnetic data, gravity data, seismic data, seismic refraction, nontechnical:

Original discoveries were made with a nontechnical approach; subsequent exploration and development were based largely on the classic "anticlinal theory" of T. Sterry Hunt and I. C. White.

Structure:

Province/basin type (see St. John, Bally, and Klemme, 1984)

Appalachian/Continental Multicycle Basin; Bally and Snelson, 221; Klemme, IIA

Tectonic history

Appalachian basin affected by at least four episodes of regional orogenic activity between Ordovician and Triassic. Anticlines began forming during Early Devonian in response to decollement slippage and faulting in Late Silurian Salina Group salt beds.

Regional structure

Leidy Dome situated on Wellsboro anticline, occurs between Marshlands anticline to northeast and Allegheny structural front to southwest. Salt-faulted anticline at depth divides field into five separate pool blocks.

Local structure

Surface closure on Leidy Dome approximately 61 m (200 ft). Dome is approximately 24 × 5 km (15 × 3 mi).

Trap

Trap type(s) .... *Diagenetic (porosity pinchout to northwest); structural (block-faulted); stratigraphic pinchout to northwest*

Basin stratigraphy (major stratigraphic intervals from surface to deepest penetration in field):

Age	Formation	Depth to Top in m (ft)
Mississippian-Devonian	*Huntley Mountain Formation*	*0 (0)*
Upper Devonian	*Catskill Formation*	*61 (200)*
Upper Devonian	*Lock Haven Formation*	*366 (1200)*
Upper Devonian	*Brallier Formation*	*1067 (3500)*
Upper Devonian	*Harrell Formation*	*1494 (4900)*
Middle Devonian	*Tully Limestone*	*1676 (5500)*
Middle Devonian	*Hamilton Group*	*1707 (5600)*
Middle Devonian	*Onondaga Group*	*1920 (6300)*
Lower Devonian	*Ridgeley Sandstone*	*1928 (6325)*
Devonian-Silurian	*Helderberg Group*	*1939 (6360)*
Upper Silurian	*Salina Group*	*2012 (6600)*
Upper Silurian	*Wills Creek Formation*	*2591 (8500)*
Lower Silurian	*McKenzie Formation*	*2804 (9200)*
Lower Silurian	*Clinton Group*	*2896 (9500)*
Lower Silurian	*Tuscarora Formation*	*3048 (10,000)*
Upper Ordovician	*Juniata Formation*	*3109 (10,200)*
Upper Ordovician	*Bald Eagle Formation*	*3414 (11,200)*
Upper Ordovician	*Reedsville Formation*	*3780 (12,400)*
Upper Ordovician	*Utica Shale*	*3962 (13,000)*
Upper Ordovician	*"Trenton-Black River"*	*4084 (13,400)*
Upper Ordovician	*Beekmantown Group*	*4481 (14,700)*
Upper Cambrian	*Gatesburg Formation*	*4938 (16,200)*
Upper Cambrian	*Warrior Formation*	*5425 (17,800)*
Middle Cambrian	*Pleasant Hill Formation*	*5578 (18,300)*

Location of well in field

Reservoir characteristics:

Number of reservoirs *2*

Formations .... *Ridgeley Sandstone; Lock Haven Formation*

Ages .... *Early Devonian and Late Devonian, respectively*

Depths to tops of reservoirs .... *Ridgeley, 1943 m (6376 ft); Lock Haven, 549 m (1800 ft)*

Gross thickness (top to bottom of producing interval) .... *Ridgeley, avg. about 10 m (33 ft); Lock Haven, avg. about 701 m (2300 ft)*

Net thickness—total thickness of producing zones

Average .... *Ridgeley, ~4.6 m (15 ft); Lock Haven, ~8 m (25 ft)*

Maximum

Average

Maximum

Lithology

Ridgeley: Well-sorted, rounded, frosted, silica- and carbonate-cemented quartzose sandstone
Lock Haven: interbedded gray sandstones, siltstones, and shales

Porosity type .... *Intergranular, fracture*

Average porosity .... *8.34% (nonfracture)*

Average permeability .... *10.9 md (nonfracture)*

Seals:

Upper

Formation, fault, or other feature .... *Onondaga Group*

Lithology .... *Black shale and argillaceous limestone*

Lateral

Formation, fault, or other feature .... *Faults, structural closure, and stratigraphic pinchout to the northwest*

Lithology .... *NA*

Source:

Formation and age .... *Middle Devonian Marcellus Formation (Hamilton Group)*

Lithology .... *Organic-rich black shale*

Average total organic carbon (TOC) .... *~3.68%*

Maximum TOC .... *~9.60%*

Kerogen type (I, II, or III) .... *I*

Vitrinite reflectance (maturation) .... $R_o = 1.50$

Time of hydrocarbon expulsion .... *Late Permian*

Present depth to top of source .... *6150 ft (1875 m)*

Thickness .... *150 ft (46 m)*

Potential yield .... *NA*

Appendix 2. Production Data

Field name .... *Leidy field*

Field size:

Proved acres .... *9761 ha (24,120 ac)*

Number of wells all years .... *~300*

Current number of wells

Well spacing .... *None*

Ultimate recoverable .... *175 bcf*

Cumulative production (to end of 1985) .... *160 bcf*

Annual production .... *~16 bcf*

Present decline rate .... *NA*

Initial decline rate

Overall decline rate

Annual water production

In place, total reserves

In place, per acre-foot

Primary recovery

Secondary recovery

Enhanced recovery

Cumulative water production

Drilling and casing practices:

Amount of surface casing set .... *20 to 40 ft (6–12 m)*

Casing program .... *13⅜-in. surface casing; 244–274 m (800–900 ft) of 9⅝-in.; 7-in. to top of Onondaga Group (about 1920 m or 6300 ft)*

Drilling mud .... *NA*

Bit program .... *6–12 m (10–40 ft) with 13-in.; 244–274 m (800–900 ft) with 11½-in.; 8½-in. to top of Onondaga Group (about 1920 m or 6300 ft)*

High pressure zones .... *Within Hamilton Group and Ridgeley Sandstone; several well fires resulted*

Completion practices:

Interval(s) perforated .... *Modern wells only, Ridgeley Sandstone*

Well treatment .... *Early wells shot where necessary; new charge devised as a result of Ridgeley drilling. Modern wells hydrofracked in Ridgeley interval*

Formation evaluation:

Logging suites .... *Older wells not logged until after conversion to storage; modern wells logged with gamma ray, neutron, density, temperature, induction, and, sometimes, velocity*

Testing practices

Mud logging techniques

Oil characteristics:

Type .... *NA*

(Tissot and Welte Classification in "Petroleum Formation and Occurrence," 1984, Springer-Verlag, p. 419)

API gravity .... *NA*

Base .... *NA*

Initial GOR .... *All dry gas*

Sulfur, wt% .... *NA*

Viscosity, SUS .... *NA*

Pour point .... *NA*

Gas-oil distillate .... *NA*

Field characteristics:

Average elevation .... *1564 ft (477 m)*

Initial pressure .... *4200 psi (609 kPa)*

Present pressure (max storage pressure) .... *4200 psi (609 kPa)*

Pressure gradient

Temperature .... *50°C*

Geothermal gradient .... *25°C/100 m*

Drive

Oil column thickness

Oil-water contact

Connate water

Water salinity, TDS

Resistivity of water

Bulk volume water (%)

Transportation method and market for oil and gas:

Gas pipelines: 6-, 8-, 12-, 16-, 20-, 24-, and 26-in. lines, at present.

Taglu Field

PETER B. TSANG
Esso Resources Canada Limited
Calgary, Alberta, Canada

FIELD CLASSIFICATION

BASIN: Beaufort–Mackenzie
BASIN TYPE: Divergent Margin (Deltaic)
RESERVOIR ROCK TYPE: Sandstone
RESERVOIR ENVIRONMENT OF DEPOSITION: Deltaic
RESERVOIR AGE: Eocene
PETROLEUM TYPE: Gas
TRAP TYPE: Faulted Domal Anticline

LOCATION

Taglu field, located on Richards Island in the Mackenzie Delta of Canada's Northwest Territories, is the largest gas field in the Beaufort–Mackenzie basin (Figure 1). It is 190 mi (306 km) north of the Arctic Circle and 6 mi (10 km) south of the Arctic ocean. Inuvik, the closest town, is 75 mi (121 km) southeast.

Estimated recoverable reserves of the Taglu field are 3.2 tcf (91×10^9 m^3) of gas.

HISTORY

Pre-Discovery

Imperial Oil Limited first became active in the Northwest Territories in 1919, exploring claims about oil seeps on the Mackenzie River. This led to the discovery of the Norman Wells reef oil field in 1920. However, subsequent drilling around Norman Wells was unsuccessful.

In early 1961, Imperial conducted surface geological work in the Peel plateau area northwest of Norman Wells (Figure 1). In 1963, Imperial acquired a four million ac (2 million ha.) block of land for its Lower Paleozoic reef prospects. Seven wells were later drilled without success.

Minor exploration, including two shallow tests on the Tuktoyaktuk Peninsula, occurred during the early 1960s. However, significant exploration in the Beaufort area did not commence until 1964, when permits were issued for most of the onshore and shallow water acreage.

The first exploratory well in the Mackenzie delta, B.A.-Shell-IOE Reindeer D-27 was drilled in 1965 (FTD 12,668 ft; 3862 m) on the flank of a large domal structure 20 mi (32 km) southeast of Taglu. A thick Tertiary–Cretaceous deltaic to marine sandstone-shale sequence was penetrated with minor gas shows (Lerand, 1973).

The first Beaufort discovery was made in early 1970. The Imperial Oil Ltd. Atkinson H-25 well flowed 24° API gravity oil from Lower Cretaceous sandstones at 5700 ft (1738 m) on the Tuktoyaktuk Peninsula, east of the South delta (Figure 2). Thirteen other wells were drilled in 1970 without success. The second oil discovery did not come until 1971, again on the Tuktoyaktuk Peninsula. The discovery well, IOE Mayogiak J-17, flowed 32° API gravity oil from a Tertiary sandstone reservoir at 3800 ft (1158 m) and 33° API gravity oil from a porous Middle Devonian carbonate section at 9375 ft (2858 m). Almost simultaneously, the Taglu gas field was discovered in the delta itself.

Discovery

The discovery well, Imperial Oil Ltd.'s IOE Taglu G-33, was the sixteenth well drilled by Imperial in the region. It spudded on 13 April 1971 and rig released on 18 August 1971. The well was drilled to a total depth of 9825 ft (2995 m) in the Eocene. Gas was tested from seven sands within an interval of 8150 ft to 9800 ft (2484 m to 2987 m). An additional thirteen sands were interpreted to be gas-bearing. Flow rates from the tests varied from 2 to 28.7 MCFGD.

BEAUFORT MACKENZIE BASIN
LOCATION MAP

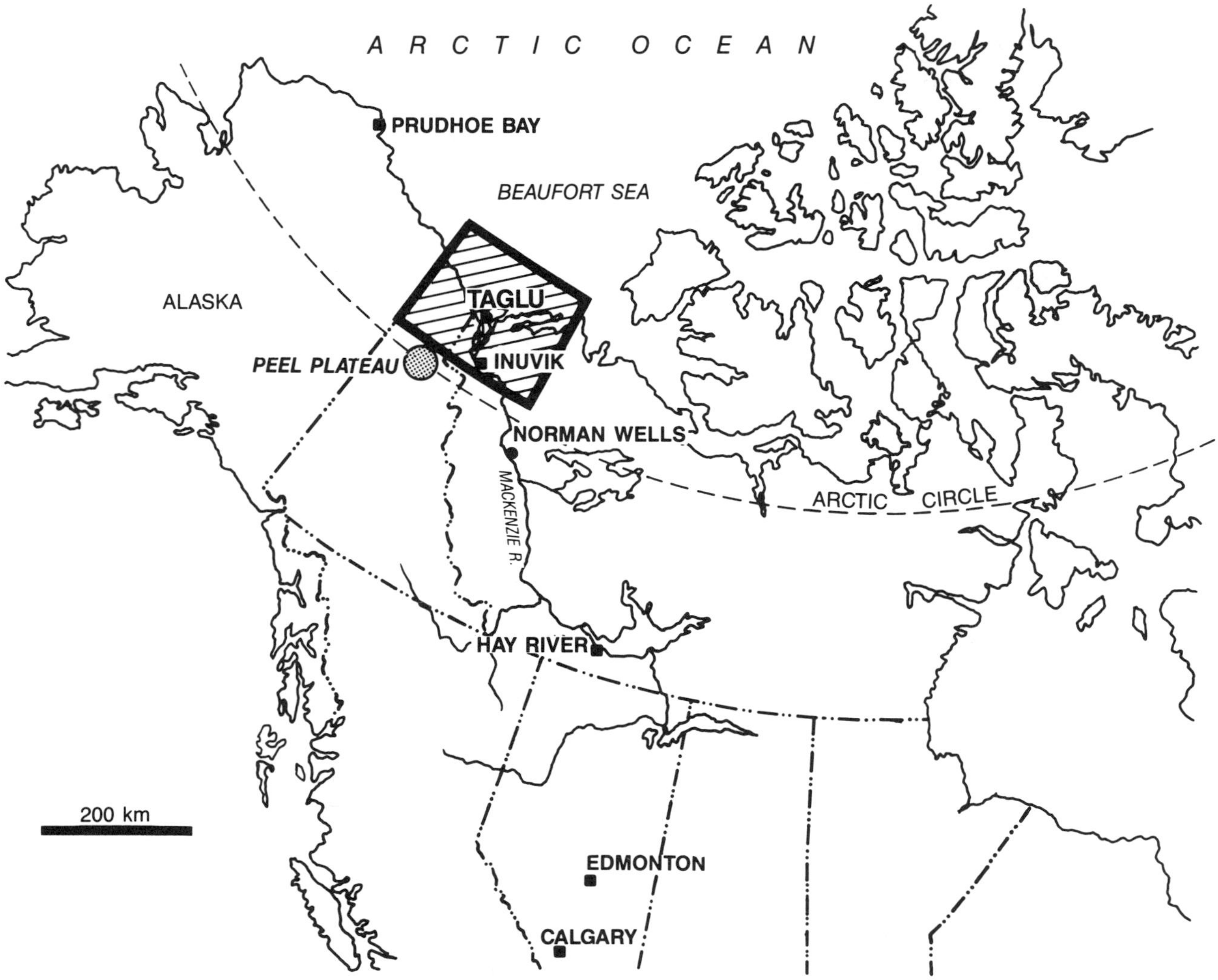
ARCTIC OCEAN
PRUDHOE BAY
BEAUFORT SEA
ALASKA
TAGLU
PEEL PLATEAU
INUVIK
NORMAN WELLS
MACKENZIE R.
ARCTIC CIRCLE
HAY RIVER
200 km
EDMONTON
CALGARY

Figure 1. Location of Taglu field.

Post-Discovery

By 1975, five wells had been drilled in the field with estimated gas reserves in excess of 3 tcf. Two other wildcats drilled later—IOE Taglu D-55, which was drilled on the downthrown side of the Taglu northern field bounding fault, and Esso Home et al. Taglu West H-06, which was drilled on a separate structure north of the field—were unsuccessful.

Over 200 wells have been drilled in the Beaufort basin since 1965, resulting in 49 oil and/or gas discoveries. In its 1988 study (Dixon et al., 1988), the Geological Survey of Canada reported that discovered reserves in the Mackenzie delta and Beaufort Sea were 11.65 tfc of gas and 1.7 billion bbl of oil. The major fields discovered up to year end 1987 were Atkinson (1970), Taglu (1971), Parsons Lake (1972), Niglintgak (1973), Adgo (1974), Kopanoar (1979), Tarsiut (1980), Issungnak (1980), and Amauligak (1984) (Figure 3).

As of December 1988, Taglu remains the largest undeveloped onshore gas field in Canada. A three-dimensional seismic survey, expected to be completed in 1988, is being conducted over the field. The survey covers about 77 mi^2 (200 km^2) and was designed to yield more accurate and detailed data for further delineation drilling and developmental study.

A gas export application was filed in 1988 with the National Energy Board of Canada. Taglu, the largest gas field in the application, represents one of the "anchor" fields that may lead to the first commercial gas production from the Beaufort Sea.

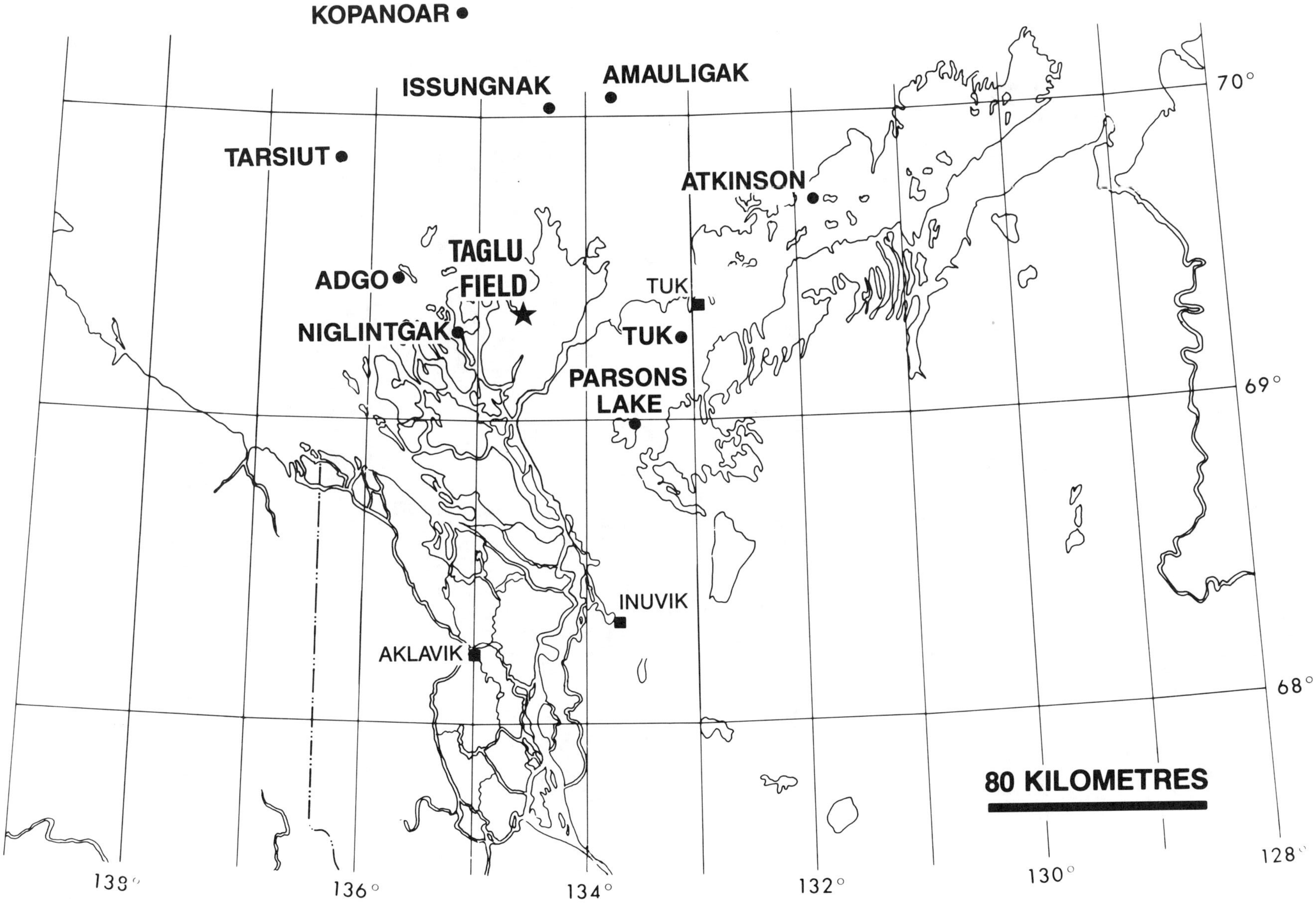
KOPANOAR
ISSUNGNAK
AMAULIGAK
TARSIUT
ATKINSON
TAGLU FIELD
ADGO
NIGLINTGAK
TUK
TUK
PARSONS LAKE
INUVIK
AKLAVIK
80 KILOMETRES
70°
69°
68°
136°
134°
132°
130°
128°

Figure 2. Four subareas based on basin configuration, structural style, and plays.

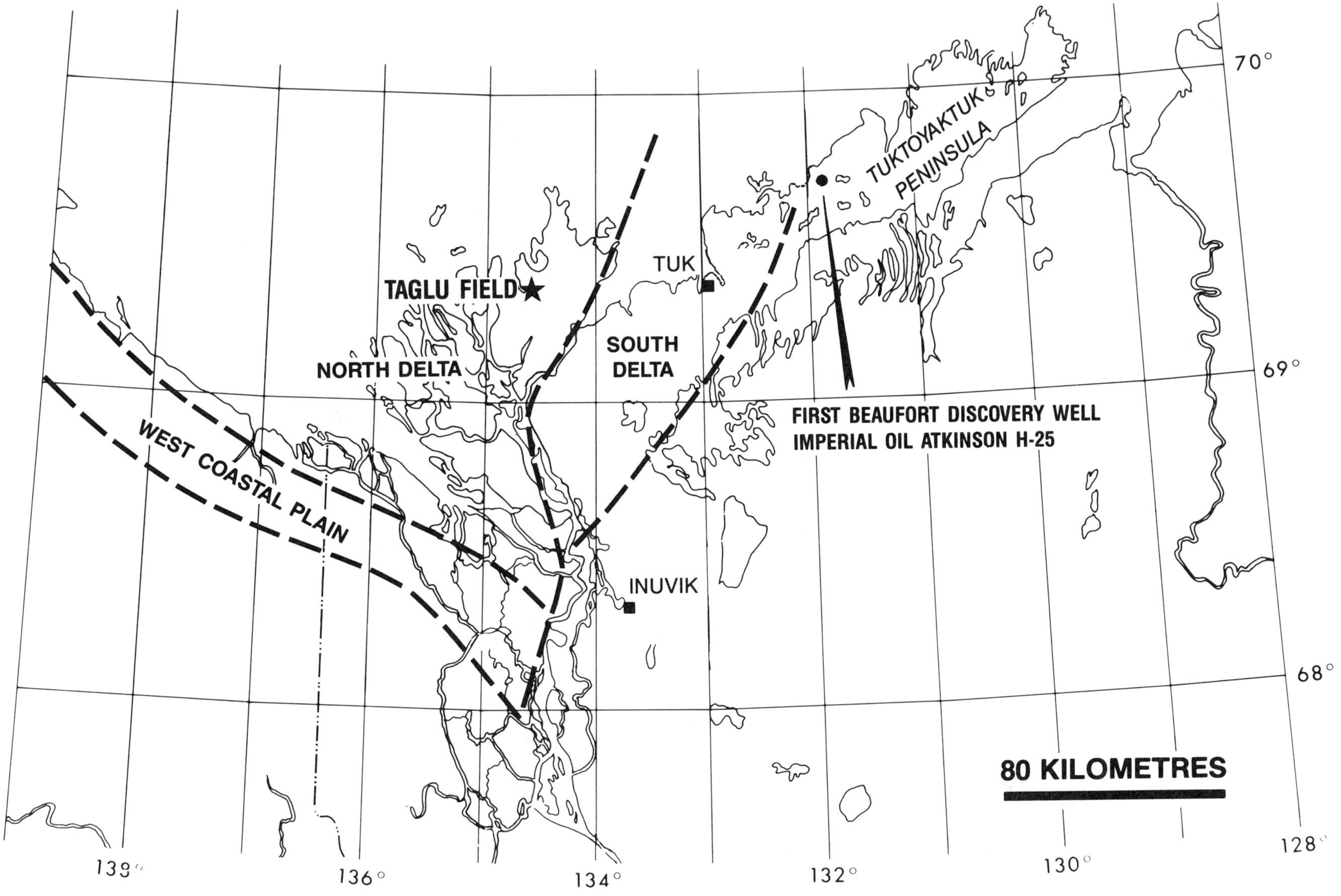
70°
69°
68°
139°
136°
134°
132°
130°
128°
TUKTOYAKTUK
PENINSULA
TUK
TAGLU FIELD
SOUTH
DELTA
NORTH DELTA
FIRST BEAUFORT DISCOVERY WELL
IMPERIAL OIL ATKINSON H-25
WEST COASTAL PLAIN
INUVIK
80 KILOMETRES

Figure 3. Beaufort basin and major discoveries.

DISCOVERY METHOD

The Taglu field was discovered using a combination of a geological play concept and seismic interpretation of the structural trap.

Initial exploration activity in the Northwest Territories had been confined to an extension of the Western Canada basin, looking for reefs in the Devonian similar to those found in the fields of Alberta. In the mid-1960s a new play started to emerge that led to the exploration of the Beaufort Sea.

Early regional geologic studies indicated that a young sedimentary basin existed in the Mackenzie delta area. Extrapolation of surface geology from the uplifted mountain areas around the Beaufort-Mackenzie basin indicated several sequences of potential reservoir sandstones and conglomerates up to 500 ft (152 m) thick, which could possibly extend into the basin. This discovery coincided with a worldwide interest in coastal geological basins such as in the North Sea, Australia, and Africa.

Because of these positive factors, Imperial Oil Ltd. acquired 8.6 million ac (3.5 million ha.) of leases in the Beaufort-Mackenzie area in 1964, and an additional 1 million ac (0.4 million ha.) in 1965.

The Beaufort-Mackenzie basin was divided at the time by Imperial into four subareas based on basin configuration, structural style, and plays. The four areas identified were the North Delta, the West Coastal Plain, the Tuktoyaktuk Peninsula, and the South Delta (Figure 2).

Early discoveries in the Tuktoyaktuk Peninsula were in sandstone reservoirs beneath the mid-Cretaceous unconformity. However, the North Delta area, where the Taglu field is located, was viewed as a unique structural province with high potential for hydrocarbon accumulations, particularly in the Tertiary sections. The main structures interpreted at the time in the delta were large domes and faulted anticlines of diapiric shale origin and large down-to-basin faults and associated anticlines. The reservoir play was identified as a deltaic clastic wedge, composed of wedge-top facies transgressing from Jurassic to present day and prograding into the basin. The lows flanking these structures were considered deep enough for kerogen to mature and generate hydrocarbons.

At this time, more seismic data were acquired in the North Delta area. Seismic processing and interpretation were hampered by variations in permafrost thickness in the region. Permafrost variations change travel times and ray paths and make multifold seismic data difficult to stack (Bowerman and Coffman, 1975). Despite the difficulties, the Taglu prospect was mapped using seismic as one of the largest anticlines with potential reservoirs in the Tertiary and Cretaceous deltaic sandstones. The Taglu G-33 wildcat location was sited at the crest of the structure to test the prospect.

TECTONIC SETTING *

The Taglu field is located in the Mackenzie delta of the Beaufort-Mackenzie basin. This delta has been described as a typical Tertiary-Holocene delta, comparable to the Mississippi and Niger deltas (Bruce and Parker, 1975; Curtis, 1986). Such deltas are characterized by diapiric structures, growth faults, and associated roll-over structures. They tend to be gas-prone, and liquid hydrocarbons are normally intermediate to high gravity and low in sulfur.

The Mackenzie delta fits these general observations. Although it lacks the salt diapirs seen in the Mississippi and Niger deltas, shale diapirs are common in the offshore Beaufort and fill an almost analogous structural role.

The Beaufort-Mackenzie basin has an area of approximately 20,000 mi^2 (51,800 km^2). The sedimentation is dominated by a thick upper Cretaceous to Holocene deltaic wedge (Figure 4). The Tertiary section exceeds 30,000 ft (9144 m) at the thickest part of the basin.

The basin is located at the intersection of three major tectonic elements: to the east and southeast, the cratonic platform and its sedimentary cover (Northern Interior platform); to the west and southwest, a complex system of fold and thrust belts and strike-slip fault zones of primarily Cretaceous and Tertiary age (Northern Cordillera and British Mountains); to the north, beneath the Beaufort Sea and Canada Deep, an oceanic basin is interpreted to have opened in Early Cretaceous time (Figure 5).

From Paleozoic to the earliest Mesozoic time, the area was dominated by a land mass to the north that provided sediment to the basin.

The Beaufort-Mackenzie basin developed to its present day configuration in two distinct tectonic stages.

During the first stage, in Early Cretaceous time, a major structural and topographical inversion occurred that was related to the opening of the southern part of the Canada Deep by sea-floor spreading and to the development of the Brooks range of northeastern Alaska and northern Yukon.

The opening of the oceanic basin resulted in the formation of a depocenter in the southeast corner of the ocean, which was connected by a shallow seaway, i.e., a paleontological connector, to the Alberta basin farther south. The mountain-building processes provided the source of clastics and drainage system required to fill the depocenter. These two events combined to initiate the formation of the Beaufort-Mackenzie basin.

---

* Information from Hawkings and Hatelid (1975) and B. L. Collot, personal communication.

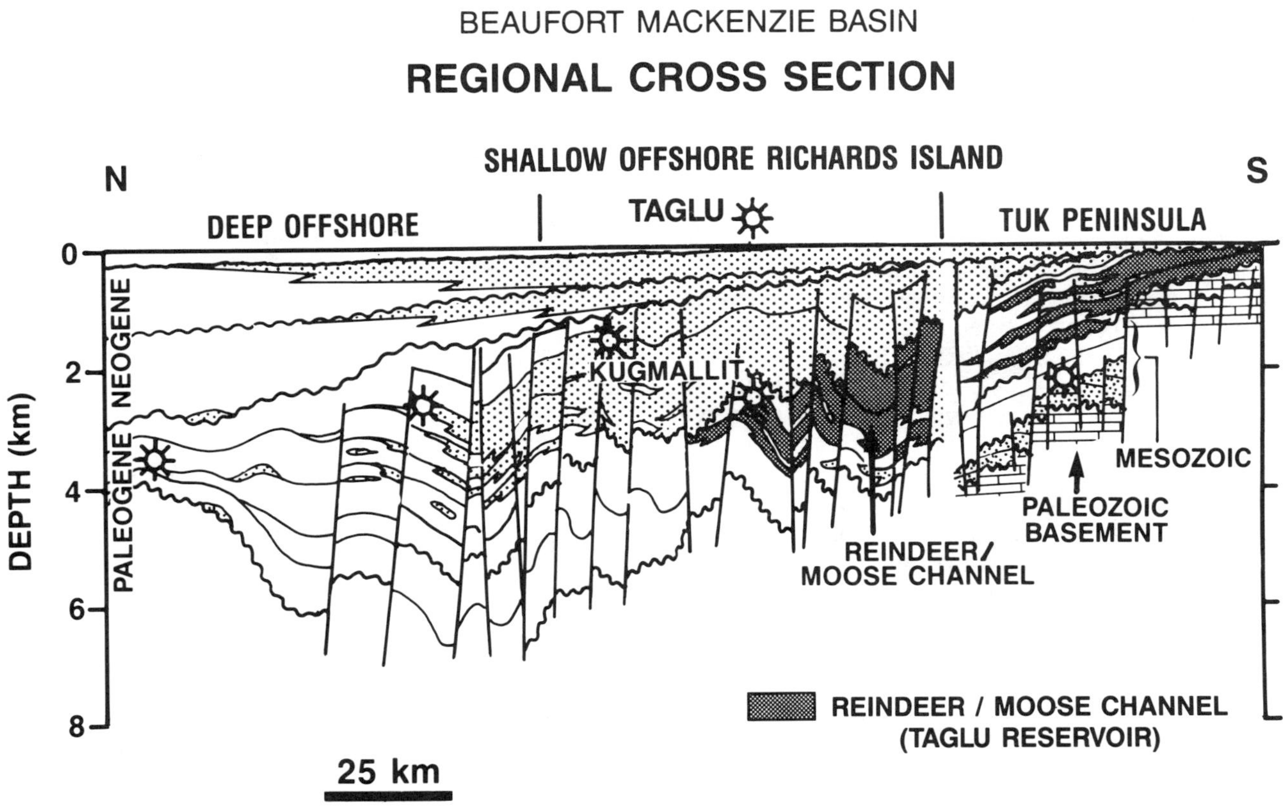
BEAUFORT MACKENZIE BASIN
REGIONAL CROSS SECTION
SHALLOW OFFSHORE RICHARDS ISLAND
N
S
DEEP OFFSHORE
TAGLU
TUK PENINSULA
0
2
4
6
8
DEPTH (km)
PALEOGENE
NEOGENE
KUGMALLIT
MESOZOIC
PALEOZOIC
BASEMENT
REINDEER/
MOOSE CHANNEL
REINDEER / MOOSE CHANNEL
(TAGLU RESERVOIR)
25 km

Figure 4. Beaufort-Mackenzie basin regional cross-section.

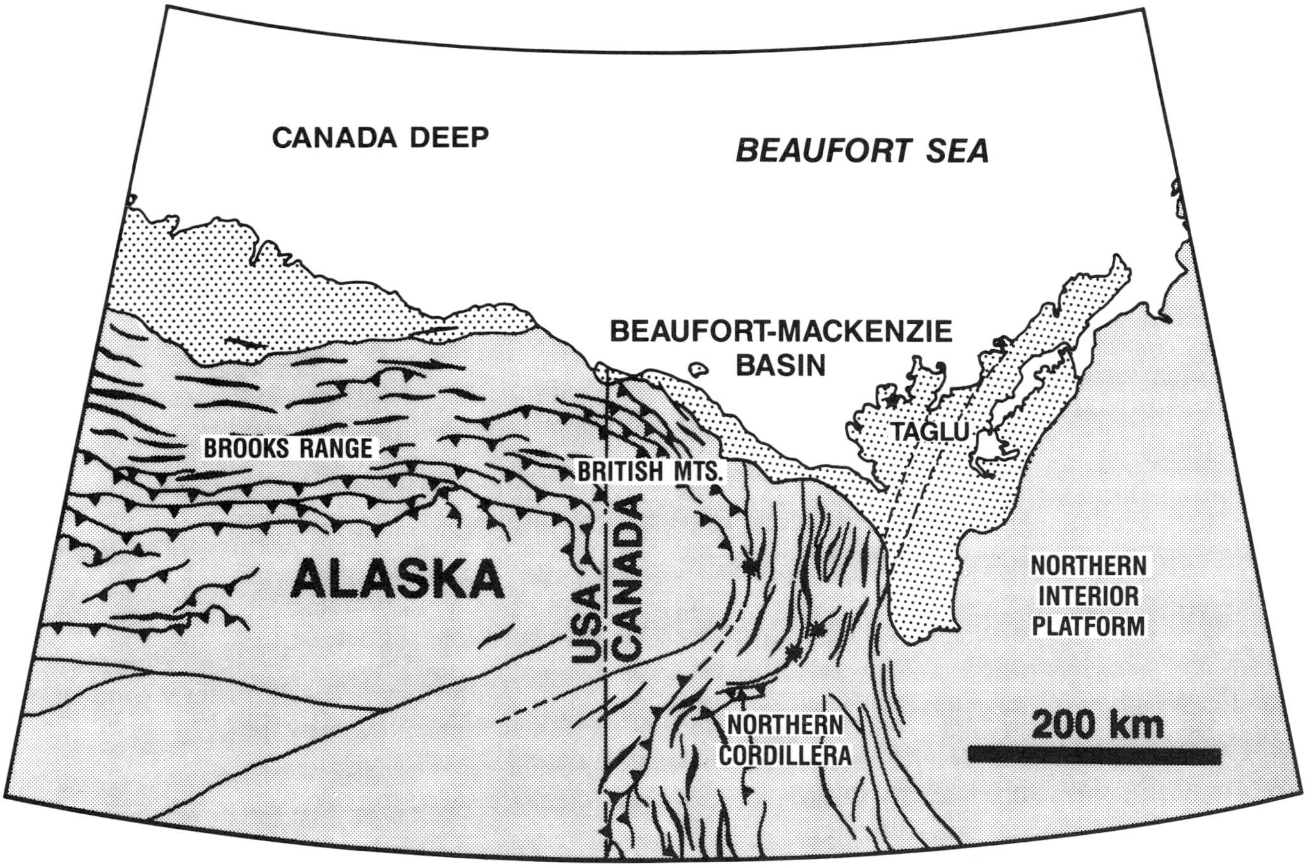
CANADA DEEP
BEAUFORT SEA
BEAUFORT-MACKENZIE
BASIN
TAGLU
BROOKS RANGE
BRITISH MTS.
ALASKA
USA
CANADA
NORTHERN
INTERIOR
PLATFORM
NORTHERN
CORDILLERA
200 km

Figure 5. Beaufort-Mackenzie basin, tectonic framework.

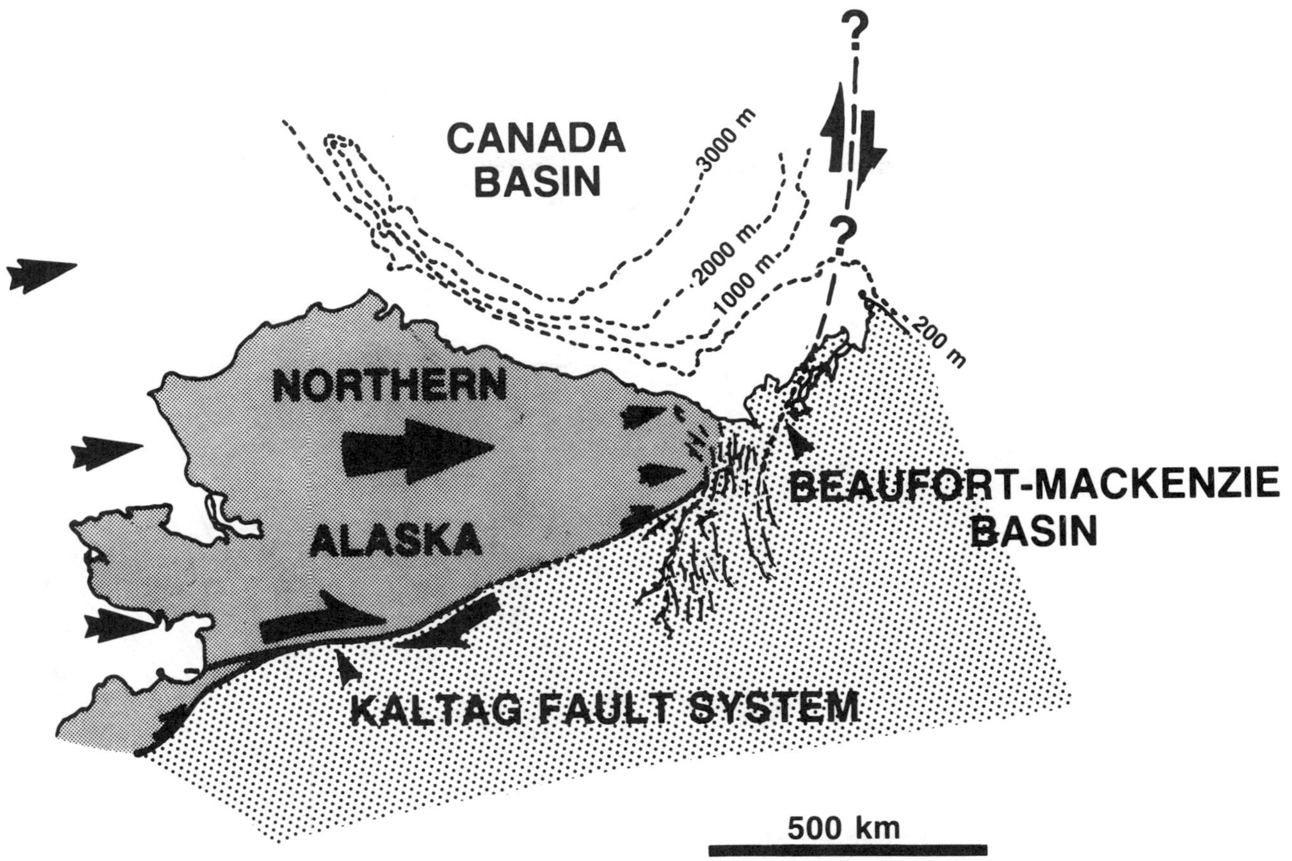
CANADA BASIN
3000 m
2000 m
1000 m
200 m
?
?
NORTHERN
ALASKA
BEAUFORT-MACKENZIE BASIN
KALTAG FAULT SYSTEM
500 km

Figure 6. Beaufort–Mackenzie basin, Late Cretaceous to mid-Tertiary tectonic events (Collot et al., 1984). Arrows show direction of plate movement.

The second stage of basin development occurred in Late Cretaceous to middle Tertiary time. During this period, the Mackenzie delta area underwent a strong compression as the northern part of Alaska was translated eastward along the Kaltag fault system of north-central Alaska (Figure 6). This compression led to further mountain building near the margins of the basin to the west and southwest and induced renewed sedimentation and progradation of the deltaic wedge in front of the mountain belt, and to strong deformation of this sedimentary wedge. As a result of the northeast-directed compression, numerous large faulted anticlinal ridges and smaller anticlines trending southeast-northwest developed throughout the western half of the Beaufort–Mackenzie basin (Figure 7). The intensity of this deformation increases from the North Point area, at the tip of Richards Island, to the western offshore area where large reverse faults are commonly associated with the anticlinal ridges. Wrench fault zones with minor lateral displacement have also been interpreted: northeast-trending, right-lateral fault zones in the Tununuk to Kugmallit Bay area, and east-trending, left-lateral fault zone below the Beaufort Sea, offshore from Richards Island (Simpson and Collot, 1986).

In addition to these compressional structures, extensional, gravity-related deformation occurred throughout the basin. This deformation includes down-to-basin normal faults trending dominantly east-west and diapiric anticlines in the deep offshore.

Taglu field is located on the upthrown side of a compressional anticline cut by an east-west, down-to-basin normal fault. Deep-seated shale movement may have modified the structural configuration.

STRATIGRAPHY *

A generalized time stratigraphic column of the Mackenzie delta area is shown in Figure 8. The Taglu reservoir is in the Reindeer sequence of Eocene age. Figure 9 provides a stratigraphic section of the Taglu field.

The stratigraphic column at Taglu consists of approximately 1000 ft (305 m) of Pliocene to Recent sediment, 2000 ft (610 m) of Neogene (probably

* Information from Bowerman and Coffman (1975).

BEAUFORT MACKENZIE BASIN STRUCTURAL FEATURES

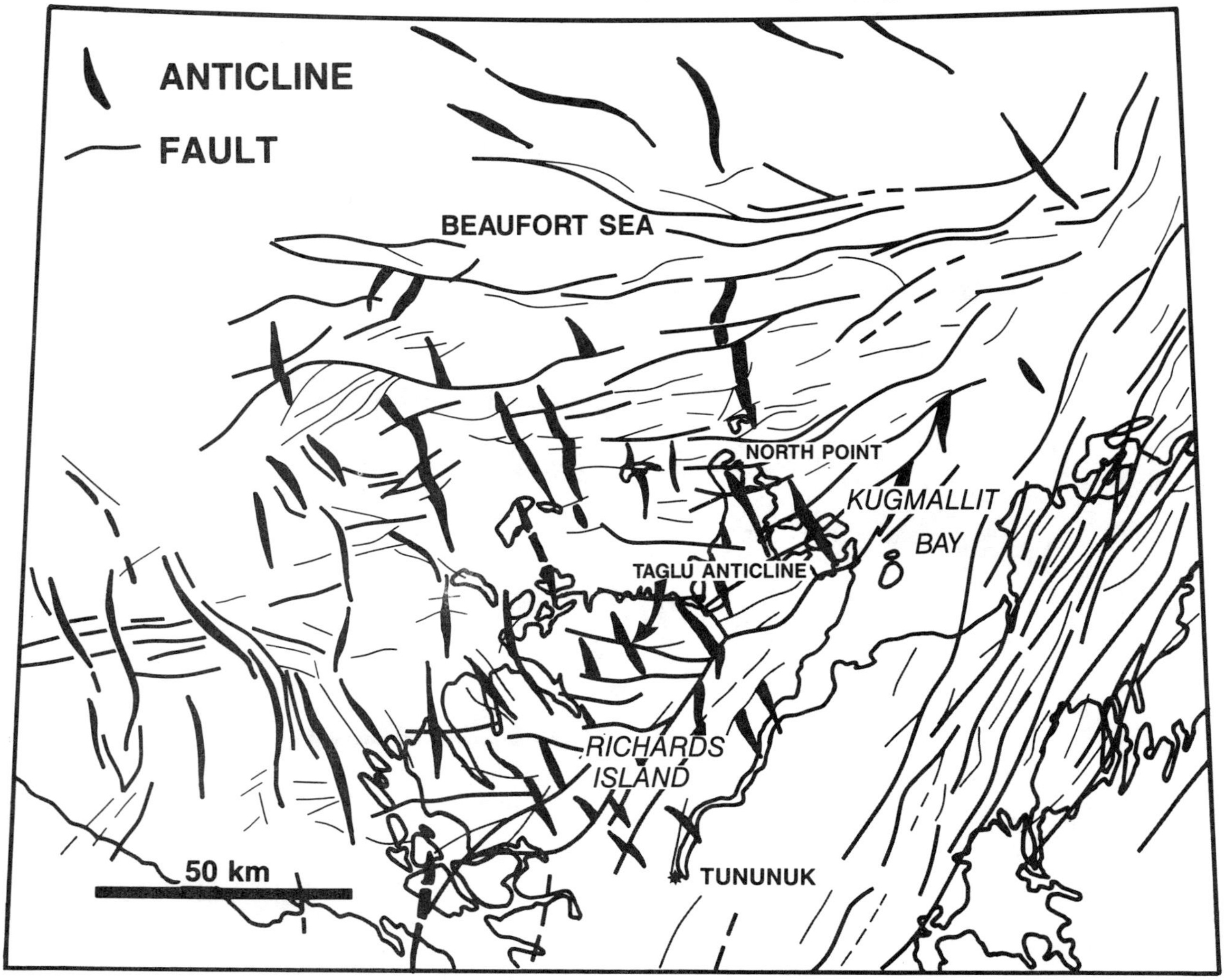
ANTICLINE
FAULT
BEAUFORT SEA
NORTH POINT
KUGMALLIT
BAY
TAGLU ANTICLINE
RICHARDS
ISLAND
TUNUNUK
50 km

Figure 7. Beaufort-Mackenzie basin structural features.

Miocene) strata, and more than 13,000 ft (3962 m) of Paleogene (probably Eocene) sediment. Seismic correlation indicates that total Cenozoic clastics in excess of 25,000 ft (7620 m) are present.

The entire stratigraphic column consists of interbedded sandstone and shale sequences. At the surface and down to the Neogene, the sediments are alluvial plain deposits. The Paleogene is composed of alluvial plain, delta front, and shallow prodelta sediment.

TRAP

Trap Type

The Taglu trap is structural, composed of an anticline on the upthrown side of a down-to-basin normal fault. Structural closure is about 2100 ft (640 m). Figure 10 is a structural map on the top of the first reservoir (A zone) showing its gas-water contact and the major east-west-trending field-bounding growth fault. The seismic line (Figures 11 and 12) show some of the minor faults and the field structural style in more detail.

Seal

Delta front and delta plain shales interbedded with the sandstone reservoirs provide inter-reservoir seals. The top seal for the reservoir is a thick section of overpressured, prodelta shales.

It is suggested that some of the gas in the Mackenzie delta is trapped in and at the base of permafrost and frozen into hydrate. This leads one to believe that the Mackenzie delta permafrost, developed during the Wisconsin glaciation period by

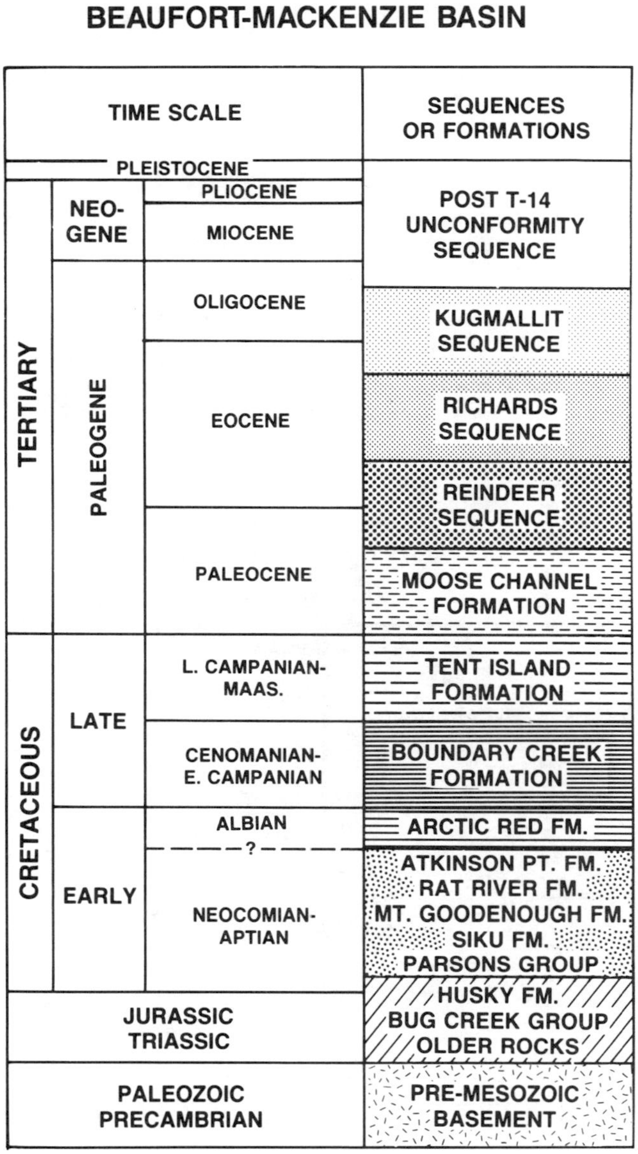
BEAUFORT-MACKENZIE BASIN
TIME SCALE
SEQUENCES OR FORMATIONS
PLEISTOCENE
TERTIARY
NEO-GENE
PLIOCENE
MIOCENE
POST T-14 UNCONFORMITY SEQUENCE
PALEOGENE
OLIGOCENE
KUGMALLIT SEQUENCE
EOCENE
RICHARDS SEQUENCE
REINDEER SEQUENCE
PALEOCENE
MOOSE CHANNEL FORMATION
CRETACEOUS
LATE
L. CAMPANIAN-MAAS.
TENT ISLAND FORMATION
CENOMANIAN-E. CAMPANIAN
BOUNDARY CREEK FORMATION
EARLY
ALBIAN
ARCTIC RED FM.
?
NEOCOMIAN-APTIAN
ATKINSON PT. FM.
RAT RIVER FM.
MT. GOODENOUGH FM.
SIKU FM.
PARSONS GROUP
JURASSIC TRIASSIC
HUSKY FM.
BUG CREEK GROUP
OLDER ROCKS
PALEOZOIC PRECAMBRIAN
PRE-MESOZOIC BASEMENT

Figure 8. Beaufort-Mackenzie basin stratigraphic column.

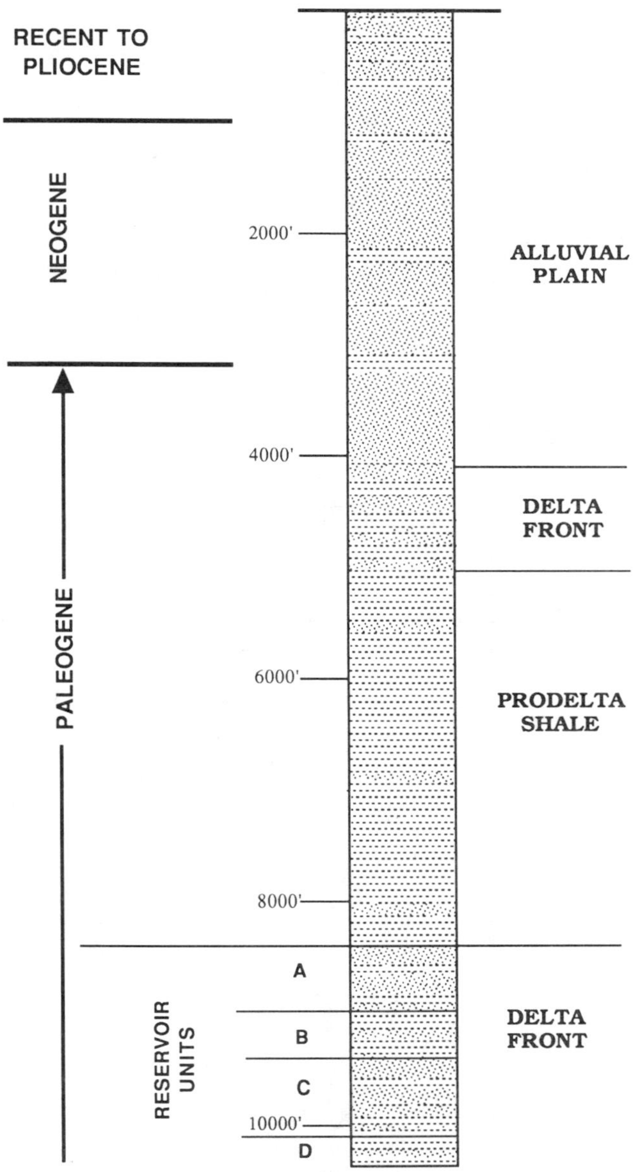
RECENT TO PLIOCENE
NEOGENE
PALEOGENE
2000'
4000'
6000'
8000'
10000'
ALLUVIAL PLAIN
DELTA FRONT
PRODELTA SHALE
DELTA FRONT
RESERVOIR UNITS
A
B
C
D

Figure 9. Stratigraphic column at Taglu.

freezing fresh and salt water together with the deltaic sediments shortly after deposition, probably forms a secondary regional seal for deeper hydrocarbons. The permafrost in the Taglu area extends from the surface down to approximately 1500 ft (457 m).

RESERVOIR *

The Taglu gas productive sequence covers a 1700 ft (518 m) stratigraphic section of deltaic sandstone bodies in the Reindeer sequence. The reservoir can be divided into three major depositional units (A, B, and C) based on the similarity of the depositional environments (Figure 13).

The upper unit, A zone (Figure 14), is about 500 ft (152 m) thick. It consists of beach and stream-mouth bar sands with some distributary channel sands making up the best reservoir deposits. In contrast, the B zone reservoirs (Figure 15) consist of thin sands deposited in the shoreface environment, and sand content decreases to the east. The C zone (Figure 15) consists of good beach and stream-mouth bar sands in the upper part and thinner distributary channel sands in the lower part. Tables 1 and 2 summarize the reservoir properties by depositional environment and unit.

Total net sandstones in the three zones vary from 450 ft to 600 ft (137 m to 183 m) in the wells.

* Information from Bowerman and Coffman (1975).

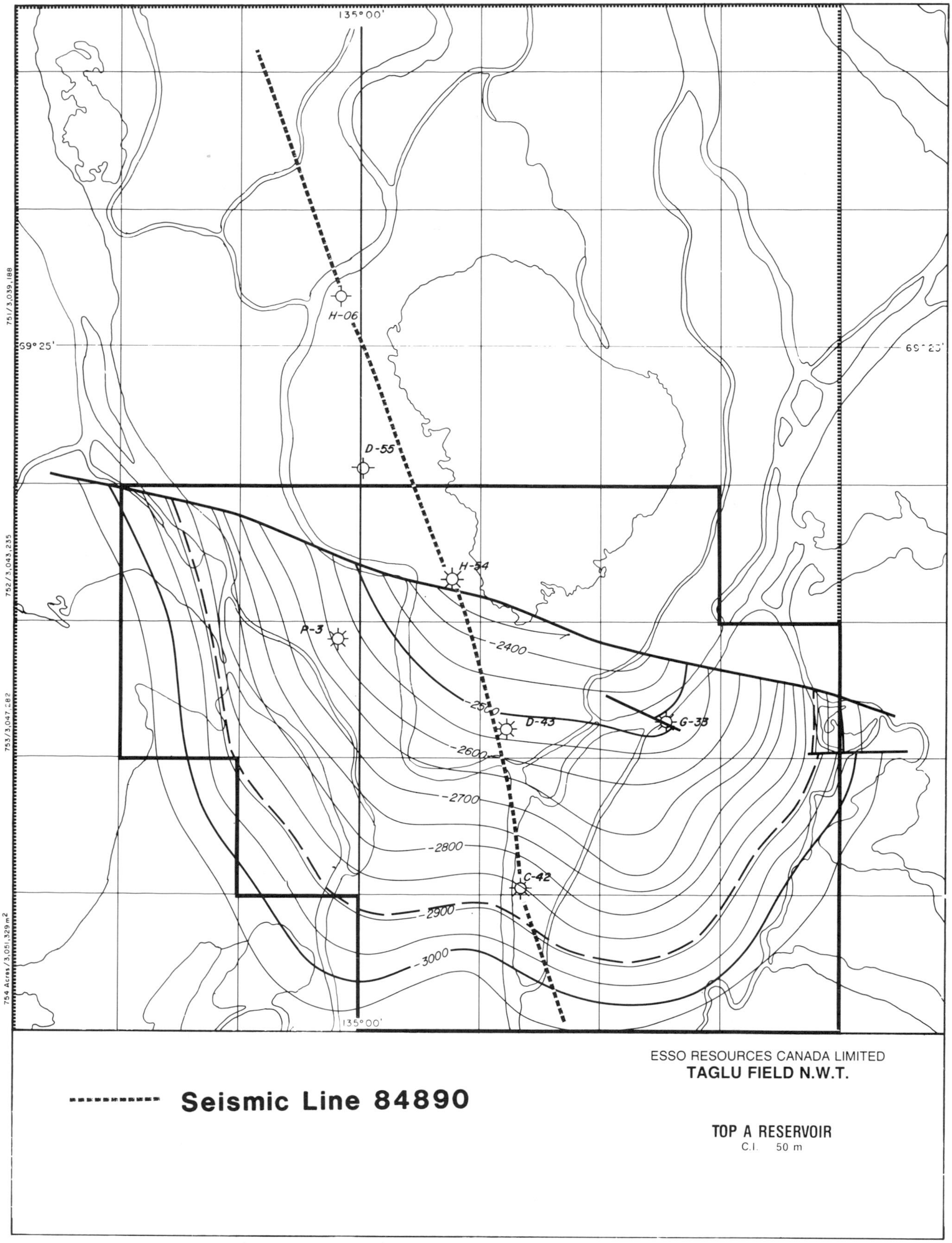
135°00'
69°25'
H-06
D-55
H-54
P-3
D-43
G-33
C-42
-2400
-2600
-2700
-2800
-2900
-3000
Seismic Line 84890
ESSO RESOURCES CANADA LIMITED
TAGLU FIELD N.W.T.
TOP A RESERVOIR
C.I. 50 m

Figure 10. Taglu structure map, top of A reservoir. Gas-water contact is shown by dashed line just above -2000 m. Growth fault is north boundary of the field.

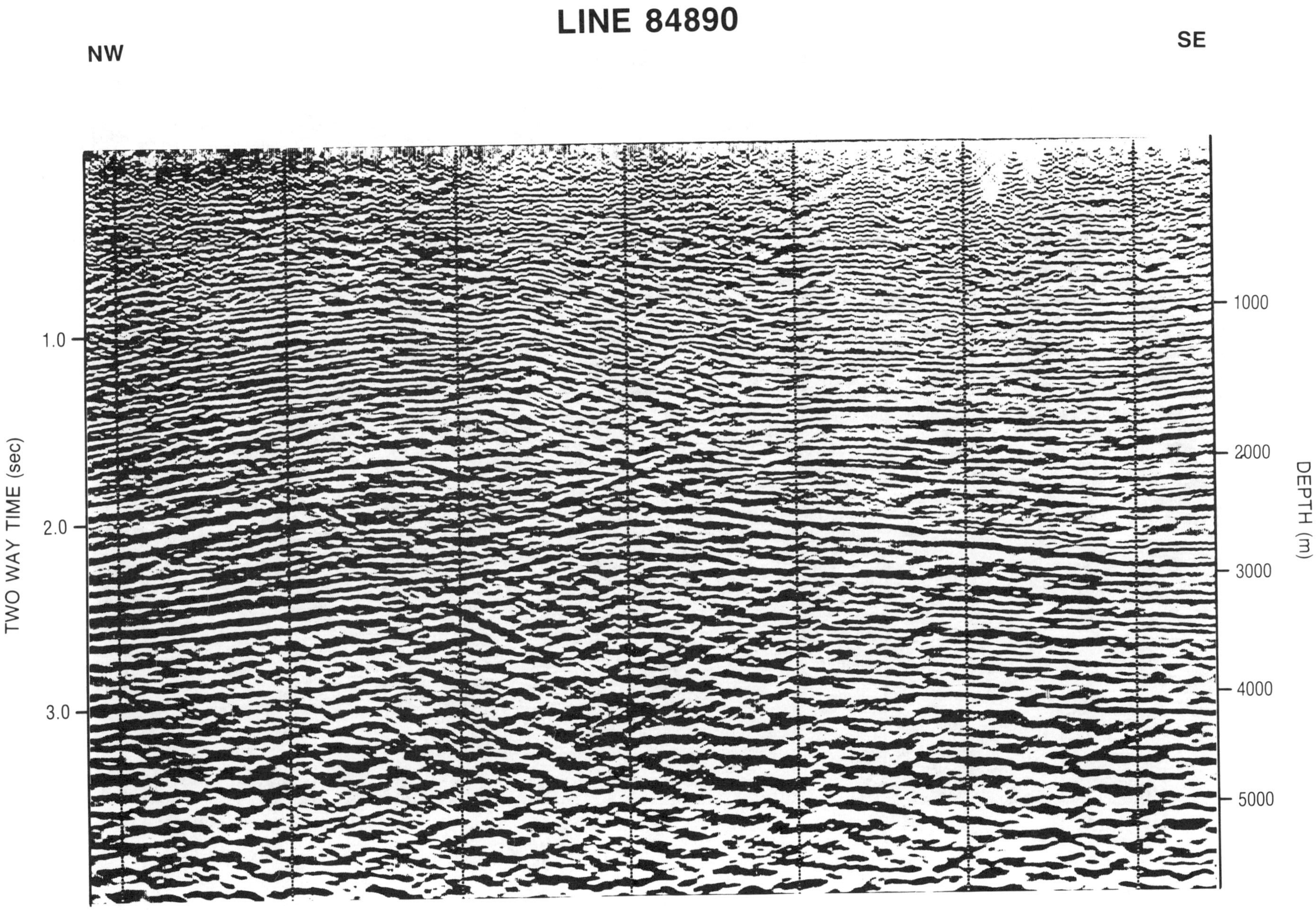
LINE 84890
NW
SE
1.0
2.0
3.0
TWO WAY TIME (sec)
1000
2000
3000
4000
5000
DEPTH (m)

Figure 11. Taglu seismic line, uninterpreted (15-fold CDP migrated line acquired by Esso in 1984).

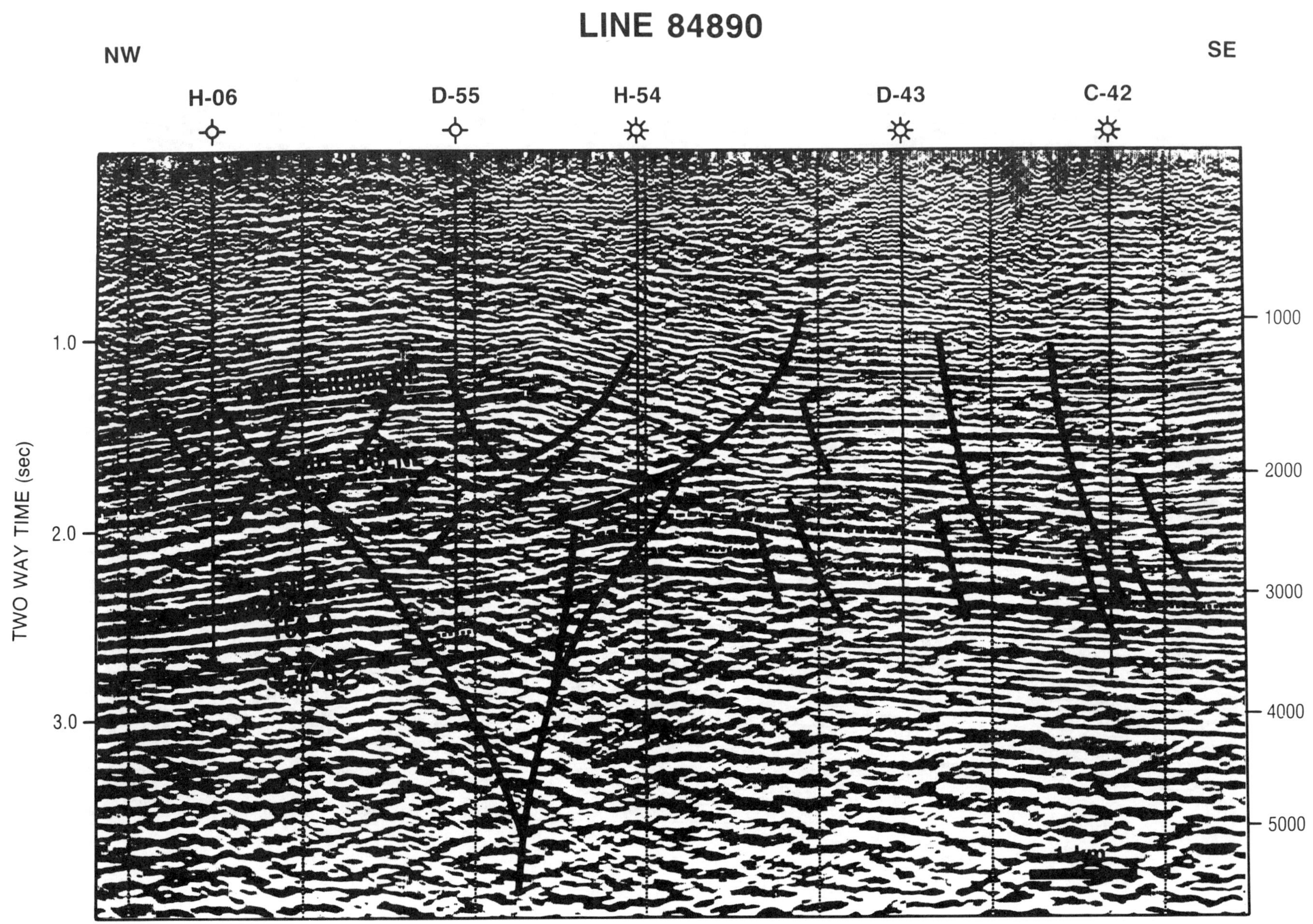
LINE 84890
NW
SE
H-06
D-55
H-54
D-43
C-42
TWO WAY TIME (sec)
1.0
2.0
3.0
DEPTH (m)
1000
2000
3000
4000
5000

Figure 12. Taglu seismic line with interpretation.

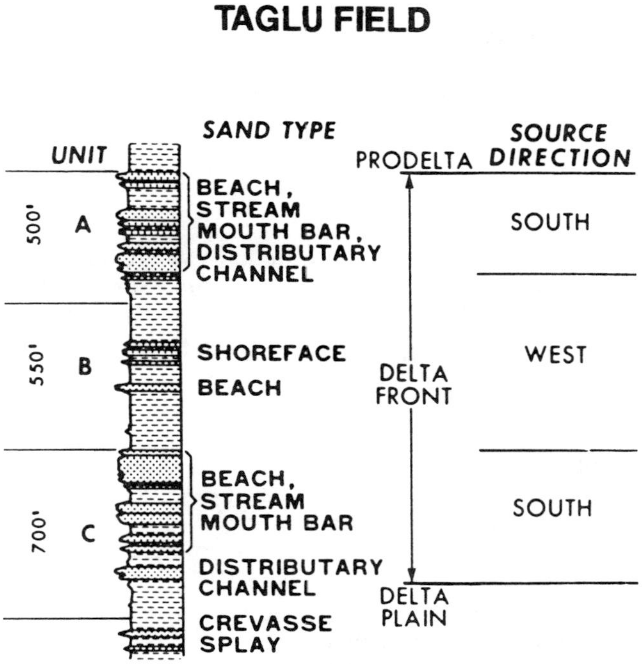
TAGLU FIELD
UNIT
SAND TYPE
SOURCE DIRECTION
PRODELTA
500'
A
BEACH, STREAM MOUTH BAR, DISTRIBUTARY CHANNEL
SOUTH
550'
B
SHOREFACE
BEACH
DELTA FRONT
WEST
700'
C
BEACH, STREAM MOUTH BAR
SOUTH
DISTRIBUTARY CHANNEL
DELTA PLAIN
CREVASSE SPLAY

Figure 13. Reservoir sequence at the Taglu gas field.

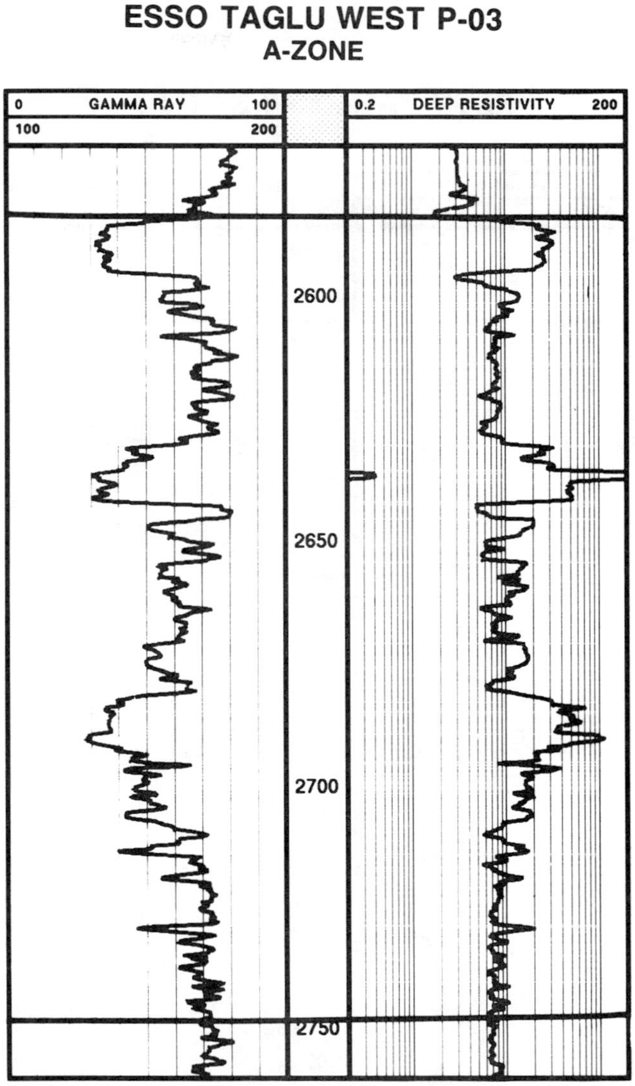
ESSO TAGLU WEST P-03
A-ZONE
0
GAMMA RAY
100
100
200
0.2
DEEP RESISTIVITY
200
2600
2650
2700
2750

Figure 14. Type log of zone A in IMP IOE Taglu West P-03 well.

Pressure data from tests run in the A zone sandstone of the four Taglu field wells indicate that they are in communication and have a common gas-water contact at -9470 ft (-2887 m) Figure 17. The B zone gas was tested only in the P-03 well. Gas was tested as low as 9200 ft (2804 m), and water was tested from sandstone approximately 100 ft (30 m) lower in the section. A gas-water contact is interpreted to be at -9507 ft (-2899 m). Similar to the A zone, the C zone is comprised of a series of sandstones with a common gas-water contact at -10,190 ft (-3106 m). A minor amount of oil was recovered from a test of one of the C sandstones in the IOE Taglu C42 well. The oil is considered to be trapped in a local pocket.

Porosity is greatest in the structurally highest and youngest A zone. Poorer reservoir quality in the B zone is due to a larger portion of shoreface sands than is present in the other units.

One of the features of Taglu sandstone is the large portion of nonquartz grains. Chert grains are common in the coarser-grained rocks. Mud matrix is lowest and porosity and permeability highest in the distributary channel sands.

Taglu gas consists of about 94% methane, 4% ethane, 1% propane, and 1% butane and heavier hydrocarbons. Its carbon dioxide content is about 0.2%. No hydrogen sulfide has been detected. The gas gravity is 0.6. Between 5 and 25 bbl of 47° API condensate have been recovered per million cubic feet of gas.

Low salinities in formation water are almost ubiquitous in the Tertiary of the Beaufort–Mackenzie basin. The phenomenon is due to freshwater invasion probably during each sea-level fall, at each sequence boundary. This fresh water, which contains oxygen and bacteria, has brought on the biodegradation of some of the oils in the Beaufort, where formation temperature is less than 60°C. The Taglu formation water recovered from the water sandstone is also quite fresh and mainly contains sodium bicarbonate. Table 3 shows the average composition of the reservoir water. The water in the gas sandstones may be saltier. If so, the calculation of water-saturation of the gas zones, which is based on water salinities in the water sandstones, may be too high (Wai, 1975).

Like many deltaic sequences, the Beaufort Tertiary basin is overpressured. This phenomenon, likely caused by undercompaction of fine-grained impermeable rocks, occurs when pore waters are trapped and

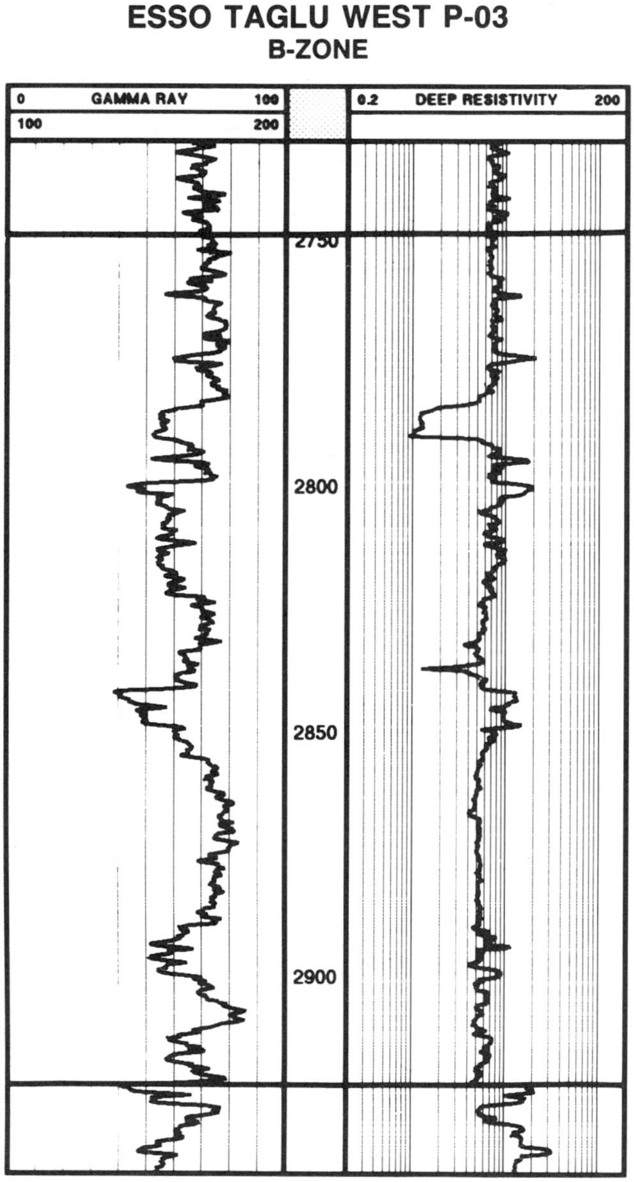
ESSO TAGLU WEST P-03
B-ZONE
0 GAMMA RAY 100
100 200
0.2 DEEP RESISTIVITY 200
2750
2800
2850
2900

Figure 15. Type log of zone B in IMP IOE Taglu West P-03 well.

support part of the overburden weight. It usually occurs in shales but can also occur in isolated sandstones. Figure 18 shows a pressure profile, in terms of mud weight equivalent, from the Taglu field. Overpressured zones are interpreted to be at depths of approximately 7000 ft to 8200 ft (2134 m to 2499 m) and from 10,000 ft to 14,000 ft (3048 m to 4267 m).

Overpressure seems to increase the sealing effect of the shales, and hydrocarbon reservoirs are commonly found beneath them. However, drilling is difficult and expensive in this area because of these abnormal pressure zones.

SOURCE AND MIGRATION*

Four different source rocks occur in the Beaufort-Mackenzie basin.

1. Tertiary: Probably Paleocene Moose Channel or Reindeer Formation. It is terrigenous delta plain/delta front associated source rock capable of generating significant amounts of gas and condensate with some oil.
2. Cretaceous (Albian-Cenomanian/Turonian): The so-called bituminous zone is a marine (clastic) often highly oil-prone source rock. It is immature over a large area of the Tuktoyaktuk Peninsula, but where mature can be correlated to Tuktoyaktuk Penninsula oils in the Devonian, Cretaceous, and Tertiary. Its extent beneath the North Delta is unknown. Total organic carbon (TOC) up to 14% has been recorded.
3. Lower Cretaceous Parsons Group: This is a second "terrigenous" type of source rock that has contributed gas and condensate to reservoirs in the South Delta. Total organic carbon (TOC) up to 20% with hydrogen indices up to 500 have been recorded.
4. Devonian: The Canol formation in the Paleozoic is a marine, carbonate type of source and is known from only one well in the basin where it was supermature. No hydrocarbons have been attributed to this source.

Migration in the Beaufort-Mackenzie basin is largely fault controlled with vertical migration in the onshore and dominantly cross-fault leakage in the offshore Tertiary zones. Biodegradation occurs throughout the Beaufort-Mackenzie basin, with most oils being affected. Marine-sourced oils ("bituminous zone" sourced) can potentially be biodegraded to "tar" whereas the terrigenous oils and condensates are only partly degraded with their physical properties (wax content, pour point, shrinkage, etc.), perhaps even enhanced.

The Taglu hydrocarbons are sourced from early Tertiary (Paleocene) Moose Channel/Reindeer Formation delta plain and delta front sediments. The gas is from a high maturity, off-structure position where vitrinite reflectance values exceed 1.5. The thin oil leg at Taglu C-42 is sourced from the same terrigenous organic matter as the gas but at a lower maturity. The condensates associated with the shallow reservoirs all show evidence of postaccumulation biodegradation.

Migration of the hydrocarbons into the reservoir is either along stratigraphically equivalent zones

*(S. Creaney, personal communication)

Table 1. Reservoir properties of Taglu by depositional environment.

DEPOSITIONAL ENVIRONMENT	DISTRIBUTARY CHANNEL	BEACH,STREAM MOUTH BAR	SHOREFACE
PORTION OF RESERVOIR	15%	75%	10%
GRAIN SIZE	MEDIUM - COARSE	MEDIUM - FINE	VERY FINE - FINE
DOMINANT GRAIN TYPE	CHERT	CHERT, QUARTZ	QUARTZ
MUD MATRIX	4%	7%	15%
POROSITY	14-23%	12-19%	12-17%
PERMEABILITY (md)	10 - 900	5 - 260	2 - 30

Table 2. Reservoir properties of Taglu by depositional units.

DEPOSITIONAL UNIT	A	B	C
POROSITY	13-22%	12-16%	12-18%
WATER SATURATION	25-55%	HIGH	25-55%
GAS/WATER CONTACT	-9,470'		-10,190'
GAS COLUMN	1700'		1400'
PRESSURE (PSIA)	4240	4370	4530
TEMPERATURE (F)	146	152	159

from synclinal deeps to the trap, vertically along fault planes, or by diffusion up through the section. However, overpressured shales can create barriers that hinder vertical migration (Evans et al., 1975).

EXPLORATION CONCEPTS

For hydrocarbons to accumulate, all the "play" elements have to be present. In the Beaufort area, the key play elements that distinguish the larger hydrocarbon discoveries and the smaller ones are seal integrity and source limitation. Discoveries with complex faulting in the structures are usually limited in hydrocarbon accumulation. The chance of leaking hydrocarbons from the trap increases as the number and complexity of the faults increase within the structure.

Structures that are relatively far from the mature source area that require long-range and complex migration paths are usually not full to the spill point and contain a less extensive hydrocarbon accumulation.

Taglu field, on the other hand, is located in an ideal hydrocarbon accumulation setting that satisfies four major "play" elements including seal integrity and source.

1. Thick deltaic deposits provide thick sand reservoirs (average net pay for the field is approximately 400 ft; 122 m).

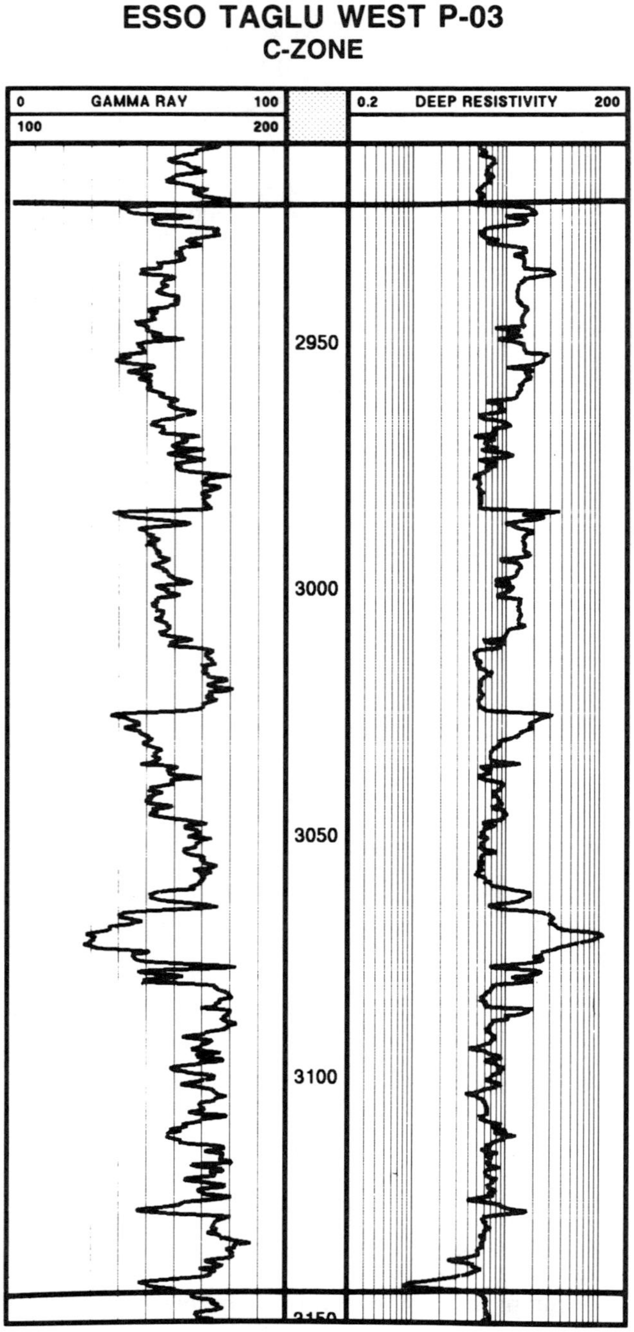
ESSO TAGLU WEST P-03
C-ZONE
0 GAMMA RAY 100
100 200
0.2 DEEP RESISTIVITY 200
2950
3000
3050
3100

Figure 16. Type log of zone C in IMP IOE Taglu West P-03 well.

2. Relatively simple large anticlinal structural trap (productive area up to 13 mi^2; 34 km^2) provides a large reservoir for hydrocarbon accumulation. Seal integrity was ensured by the small number of faults and their relatively minor throw within the structure.
3. Organic-rich sediments in the immediate offstructure lows provide mature source and easy migration into the trap.
4. Thick (70 ft; 21 m) prodelta shales provided a good top seal.

With these "play" elements in place, Taglu field is the largest gas field in the Beaufort area.

ACKNOWLEDGMENTS

Esso Resources Canada Limited (affiliate of Imperial Oil Limited) permitted publication of the paper. Personal communication with S. Creaney and B. L. Collot formed the basis for the geochemical and structural descriptions, respectively, of this paper. Reports and advice by other Esso personnel are the basis for the historical background leading to the discovery of the Taglu field.

REFERENCES CITED

Bowerman, J. N., and R. C. Coffman, 1975, The geology of the Taglu Gas Field in the Beaufort Basin, N.W.T., *in* Yorath et al., eds., Canada's continental margins: Canadian Society of Petroleum Geologists Memoir 4, p. 649-662.

Bruce, C. J., and E. R. Parker, 1975, Structural features and hydrocarbon deposits in the Mackenzie Delta: 9th World Petroleum Congress Proceedings, v. 2, p. 251-261.

Collot, B. L., A. P. Hemingson, G. L. Reed, and D. G. Simpson, 1984, Constraints on the Late Cretaceous to Middle Tertiary tectonics of the Beaufort-Mackenzie Basin of Arctic Canada: Esso Resources Canada Limited, Internal Report ERCL EX. 84.03.

Curtis, D. M., 1986, Comparative Tertiary petroleum geology of the Gulf Coast, Niger, and Beaufort-Mackenzie Delta areas: Geological Journal, v. 21, p. 225-255.

Dixon, J., G. R. Morrell, J. R. Dietrich, R. M. Procter, and G. C. Taylor, 1988, Petroleum resources of Mackenzie Delta-Beaufort Sea: Geological Survey of Canada Open File 1926.

Evans, C. R., D. K. McIvor, and K. Magara, 1975, Organic matter, compaction history and hydrocarbon occurrence—Mackenzie Delta, Canada: 9th World Petroleum Congress Proceedings, v. 2, p. 149-157.

Hawkings, T. J., and W. G. Hatelid, 1975, The regional setting of the Taglu Field, *in* Yorath et al., eds., Canada's continental margins: Canadian Society of Petroleum Geologists Memoir 4, p. 633-647.

Lerand, M., 1973, Beaufort Sea, *in* R. G. McCrossan, ed., The future petroleum provinces of Canada: Canadian Society of Petroleum Geologists Memoir 1, p. 315-386.

Simpson, D. G., and B. L. Collot, 1986, Structural framework of the Tarsiut area, offshore Mackenzie Delta: implication for exploration: Esso Resources Canada Limited, Internal Report ERCL EX. 86.02.

Wai, T., 1975, Reservoir properties from cores compared with well logs, Taglu Gas Field, Canada: MSc Thesis, Discipline of Earth Sciences, the Graduate School, The University of Tulsa.

SUGGESTED READINGS

Dixon, J., 1986, Cretaceous to Pleistocene stratigraphy and paleography, northern Yukon and northwestern District of Mackenzie: Bulletin of Canadian Petroleum Geology, v. 34, p. 49-70.

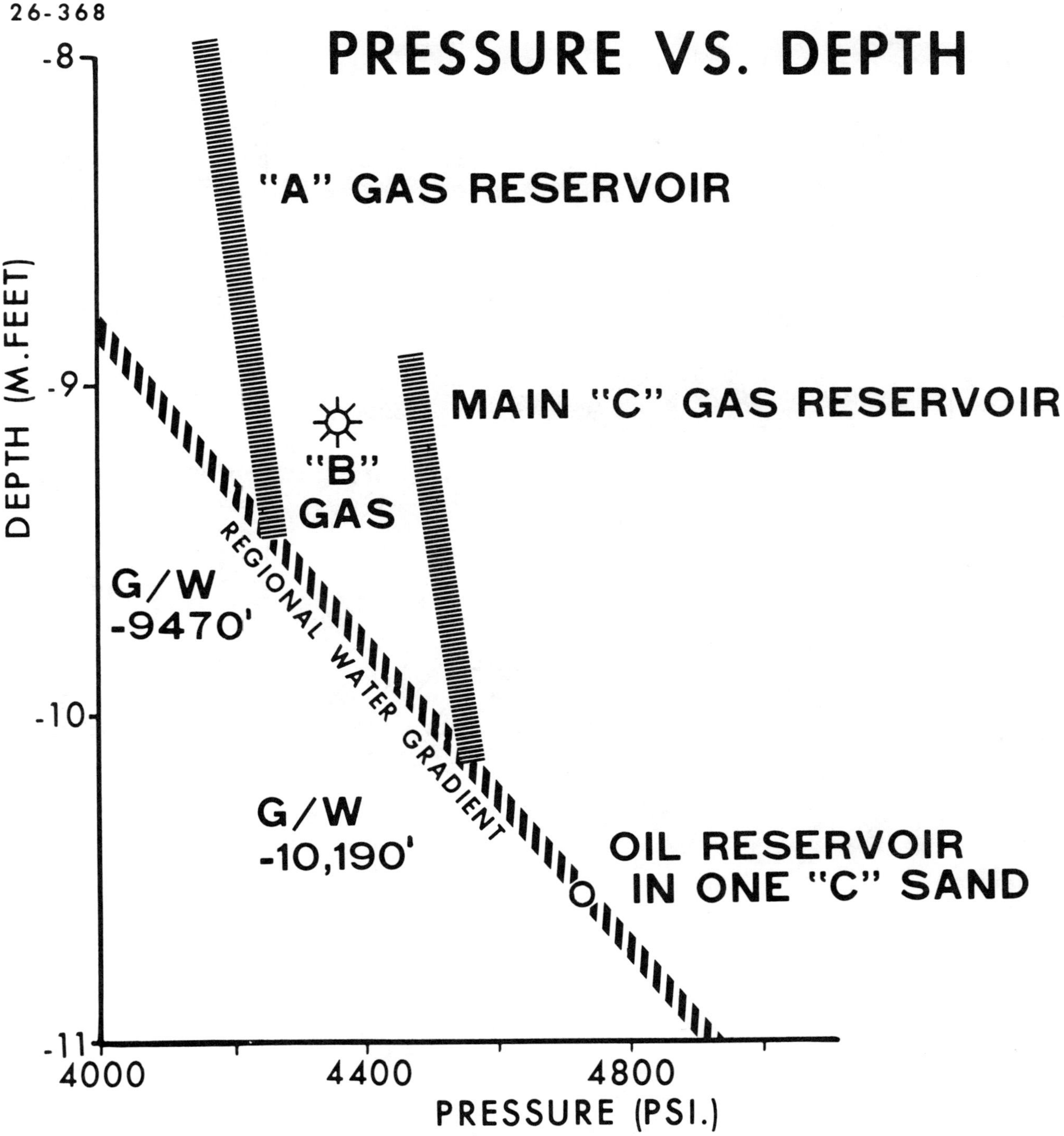
26-368
PRESSURE VS. DEPTH
"A" GAS RESERVOIR
MAIN "C" GAS RESERVOIR
"B"
GAS
G/W
-9470'
REGIONAL WATER GRADIENT
G/W
-10,190'
OIL RESERVOIR
IN ONE "C" SAND
DEPTH (M.FEET)
-8
-9
-10
-11
4000
4400
4800
PRESSURE (PSI.)

Figure 17. Taglu pressure vs. depth. (Depth in thousands of feet.)

Table 3. Composition of Taglu aquifer water.

	mg/L
Sodium	2600
Calcium	11
Chloride	1550
Bicarbonate	4170
Total solids	8400

R_w @ 70° F=1.1 Ohm-m.

Dixon, J., J. R. Dietrich, D. H. McNeil, D. J. McIntyre, L. R. Snowdon, and P. Brooks, 1985, Geology, biostratigraphy and organic geochemistry of Jurassic to Pleistocene strata, Beaufort Mackenzie area, northwest Canada: Canadian Society of Petroleum Geologists, Course Notes, 65 p.

Lane, F. H., and K. S. Jackson, 1980, Controls on occurrence of oil and gas in the Beaufort-Mackenzie Basin, Miall, ed., *in* Facts and principles of world petroleum occurrence, p. 489–507.

Morrell, G. R., and U. B. Schmidt, 1988, Sequence, stratigraphy, sedimentology: surface and subsurface: Canadian Society of Petroleum Geologists Memoir 15, p. 361–372.

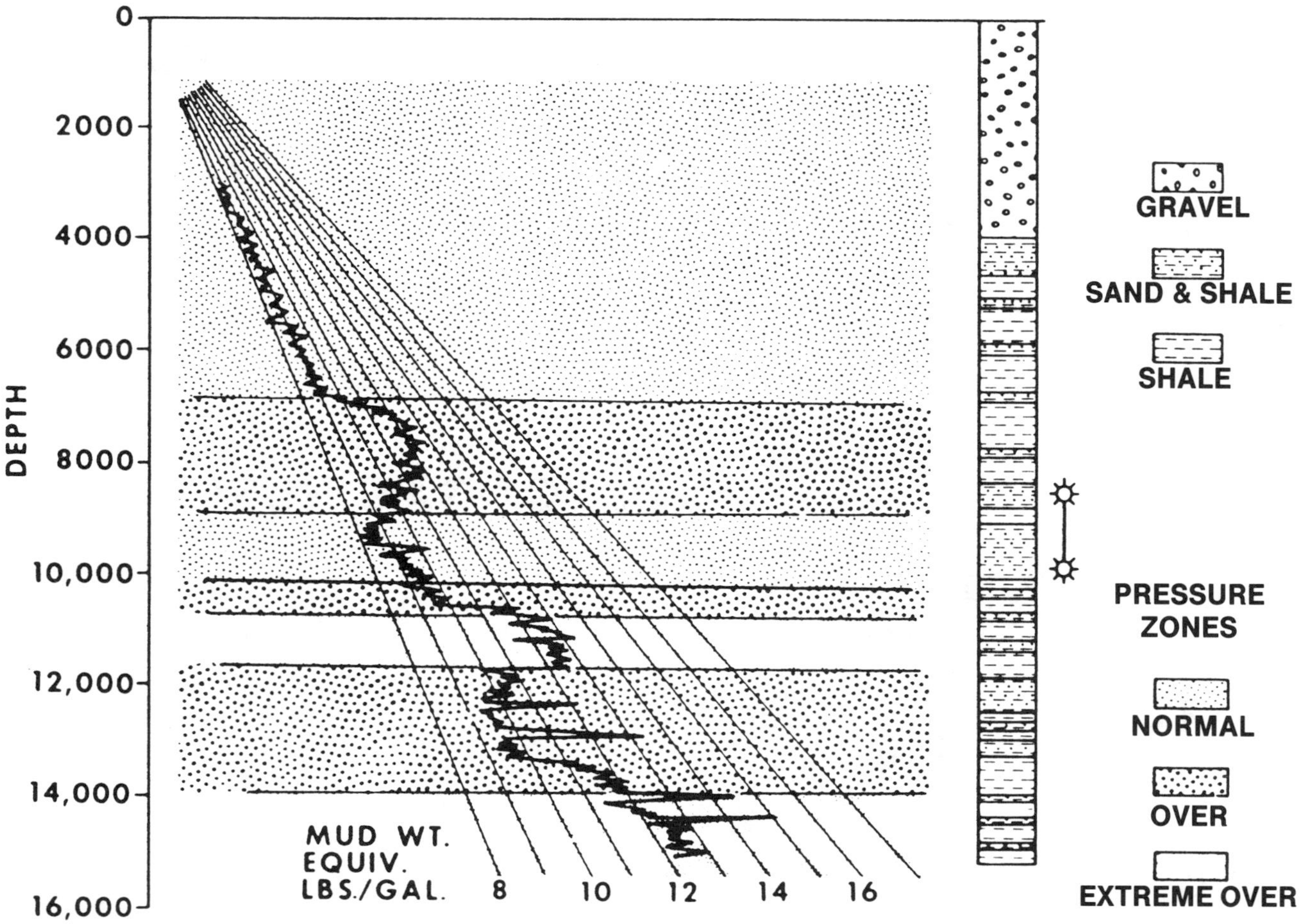
DEPTH
0
2000
4000
6000
8000
10,000
12,000
14,000
16,000
MUD WT.
EQUIV.
LBS./GAL.
8
10
12
14
16
GRAVEL
SAND & SHALE
SHALE
PRESSURE
ZONES
NORMAL
OVER
EXTREME OVER

Figure 18. Taglu, shale-pore pressure profile.

Appendix 1. Field Description

Field name .... *Taglu*

Ultimate recoverable reserves .... *3.2 tcf gas (90 × 10^9 m^3)*

Field location:

- **Country** .... *Canada*
- **State** .... *Northwest Territories*
- **Basin/Province** .... *Beaufort-Mackenzie basin*

Field discovery:

- **Year field discovered** *1971*
- **Year second pay discovered** .... *NA*
- **Third pay** .... *NA*

Discovery well name and general location
(i.e., Jones No. 1, Sec. 2T12NR5E; or Smith No. 1, 5 mi west of Sheridan, Wyoming):

- **First pay** .... *IOE Taglu G-33 69°30′N; 134°45′W*
- **Second pay** .... *NA*
- **Third pay** .... *NA*

Discovery well operator .... *Esso Resources Canada Ltd.*
(if more than one pay in field, list operators of discovery well in other pays)

- **Second pay** .... *NA*
- **Third pay** .... *NA*

IP in barrels per day and/or cubic feet or cubic meters per day:

- **First pay** .... *8.6–28.7 mmcfd (242,000–789,000 m^3/d)*
- **Second pay** .... *NA*
- **Third pay** .... *NA*

All other zones with shows of oil and gas in the field:

Age	Formation	Type of Show
None		

Geologic concept leading to discovery and method or methods used to delineate prospect, e.g., surface geology, subsurface geology, seeps, magnetic data, gravity data, seismic data, seismic refraction, nontechnical:

Geological play concept of deltaic environment with extrapolation of surface geology and seismic reflection survey.

Structure:

Province/basin type (see St. John, Bally, and Klemme, 1984) .... *Bally 1142; Klemme IV*

Tectonic history

The Beaufort-Mackenzie basin was formed in early Mesozoic by a major structural inversion of the Paleozoic land mass and the opening of the Beaufort Sea. Late Cretaceous to Tertiary compression from the west induced mountain building and deltaic sedimentation and the deformation of such formed structural traps such as the Taglu reservoir throughout the basin.

Regional structure

Located in the Mackenzie delta in the central part of the Beaufort-Mackenzie basin, which is bounded to the southeast by the Northern Interior platform and to the southwest by the mobile belt of the Northern cordillera and the British Mountains.

Local structure

Roll-over anticline formed on the upthrown side of a normal down-to-basin fault. Deep-seated shale movement may have modified the structural configuration.

Trap

Trap type(s)

Single structural trap; anticline on the upthrown side of a down-to-basin normal fault.

Basin stratigraphy (major stratigraphic intervals from surface to deepest penetration in field):

Age	Formation	Depth to Top in m
Recent		*Surface*
Pliocene		*300*
Neogene		*900*
Paleogene (probably Eocene)		*4900*

Location of well in field .... *NA*

Reservoir characteristics:

Number of reservoirs *4*

Formations .... *Taglu (Reindeer) Formation; depositional units A, B, C, and D*

Ages .... *Lower Tertiary, probably lower Eocene*

Depths to tops of reservoirs

A, 2484 m (G-33 well); B, 2763 m (G-33 well); C, 2789 m (G-33 well); D, 2989 m (G-33 well)

Gross thickness (top to bottom of producing interval)

A, 371 m (G-33 well); B, 185 m (P-03 well); C, 620 m (G-33 well); D, 71 m (P-03 well)

Net thickness—total thickness of producing zones

Average .... *A, 68 m; B, 13 m; C, 88 m; D, 8.5 m*

Maximum .... *A, 81 m (G-33 well); B, 14 m (P-03 well); C, 100 m (D-43 well); D, 11 m (P-03 well)*

Average ....

Maximum ....

Lithology

Medium- to fine-grained quartzarenite and sublitharinite, common chert occurrence

Porosity type .... *Intergranular porosity*

Average porosity .... *A, 18%; B, 16%; C, 16%; D, 16%*

Average permeability .... *Distributary channel ss, 14-23 md; beach, stream-mouth-bar ss, 5-260 md*

Seals:

Upper

Formation, fault, or other feature .... *Paleogene prodelta shale; possibly permafrost*

Lithology .... *Shale*

Lateral

Formation, fault, or other feature .... *Paleogene shale*

Lithology .... *Shale*

Source:

Formation and age .... *Reindeer/Moose Channel (Paleocene)*

Lithology .... *Coals and coaly shales*

Average total organic carbon (TOC) *1-50%*

Maximum TOC .... *Coal ~60%*

Kerogen type (I, II, or III) .... *III plus minor terrigenous II (spore, pollen resin)*

Vitrinite reflectance (maturation) .... $R_o = 1.50$

Time of hydrocarbon expulsion .... *NA*

Present depth to top of source .... *Unknown*

Thickness .... *Unknown*

Potential yield .... *Unknown*

Appendix 2. Production Data

Field name	*Taglu*

Field size:

Proved acres	*33.8 km^2 (3380 ha.)*
Number of wells all years	*5 on structure, 2 off structure*
Current number of wells (as of 1987)	*5 on structure, 2 off structure*
Well spacing	*NA*
Ultimate recoverable	*3.2 tcf gas (90.2 $\times 10^9$ m^3*
Cumulative production (1987)	*Not yet producing*
Annual production (1987)	*Not yet producing*
Present decline rate	*NA*
Initial decline rate	*NA*
Overall decline rate	*NA*
Annual water production	*NA*
In place, total reserves	*3.9 tcf gas (110 $\times 10^9$ m^3*
In place, per acre-foot	*NA*
Primary recovery	*83% of gas in place*
Secondary recovery	*83% of gas in place*
Enhanced recovery	*83% of gas in place*
Cumulative water production	*NA*

Drilling and casing practices:

Amount of surface casing set	*No production well yet*
Casing program	*NA*
Drilling mud	*NA*
Bit program	*NA*
High pressure zones	*2130–2500 m; 3050–3660 m; 4270 m*

Completion practices:

Interval(s) perforated	*NA*
Well treatment	*NA*

Formation evaluation:

Logging suites	*Full suite of open-hole logs*
Testing practices	*DST and production tests*
Mud logging techniques	*Mud logging from surface to TD*

Oil characteristics:

Type	*Aromatic-rich condensate; paraffinic-naphthenic oil*
(Tissot and Welte Classification in "Petroleum Formation and Occurrence," 1984, Springer-Verlag, p. 419)	
API gravity	*37°–47° (condensate); 29° (15 bbl tested)*
Base	*Condensate, aromatic; Gas, 94% methane; 4% ethane; 1% propane; 1% butane+; nil, H_2S; 0.2% CO_2*
Initial GOR	*Condensate/gas ratio 5–25 bbl/mmcfg*
Sulfur, wt%	*0.14%*
Viscosity, SUS	*40.5 at 100°F*
Pour point	*-54°C*
Gas-oil distillate	*NA*

Field characteristics:

Average elevation	*0.9–1.5 m above sea level*
Initial pressure	*A, 29,050 kPag; B, 30,405 kPag; C, 30,960–33,165 kPag*

Present pressure .... *As above*
Pressure gradient .... *Water, 9.73 kPa/m (0.43 psi/ft); gas, 1.81 kPa/m (0.08 psi/ft)*
Temperature .... *A, 60°C; B, 68°C; C, 71°C; D, 74°C*
Geothermal gradient .... *0.03°C/m)*
Drive .... *Bottom water drive*
Oil column thickness .... *NA*
Gas-water contact .... *A, -2887 m; B, -2833 m; C, -3107 m; D, -3167 m*
Connate water .... *A, 31%; B, 39%; C, 39%; D, 39%*
Water salinity, TDS .... *8400 mg/L (mainly sodium bicarbonate)*
Resistivity of water .... *0.3 at 150°F*
Bulk volume water (%) .... *NA*

Transportation method and market for oil and gas:

Not producing yet.

Bravo Dome Carbon Dioxide Gas Field

RONALD F. BROADHEAD
New Mexico Bureau of Mines and Mineral Resources
Socorro, New Mexico

FIELD CLASSIFICATION

BASIN: Bravo Dome
BASIN TYPE: Regional Arch
RESERVOIR ROCK TYPE: Sandstone
RESERVOIR ENVIRONMENT OF DEPOSITION: Beach; Tidal Flat; Fluvial
RESERVOIR AGE: Permian/Triassic
PETROLEUM TYPE: None (CO_2)
TRAP TYPE: Combination Structural-Stratigraphic-Hydrodynamic

LOCATION

The Bravo dome carbon dioxide gas field occupies southern Union County and northeastern Harding County, New Mexico (Figure 1). It straddles the boundary between the Raton and High Plains sections of the Great Plains physiographic province of Fenneman (1946). The nearest city is Clayton in Union County; Clayton is located approximately 24 km (15 mi) northeast of the presently known northernmost extent of the Bravo dome field. Surface exposures consist of Mesozoic and Cenozoic sedimentary units (Figure 2).

The Bravo dome field, as discussed in this report, comprises the Bravo Dome Carbon Dioxide Gas Unit (BDCDGU) east of the town of Bueyeros, and the old Bueyeros field, which lies mainly northwest and southwest of Bueyeros. The BDCDGU and the old Bueyeros field belong to the same areally continuous accumulation of carbon dioxide.

For the most part, the exact boundaries of the field have not been defined by drilling. As the demand for CO_2 continues or increases, and as existing reserves of CO_2 are depleted by production, outpost wells will be drilled in the field and exact field boundaries eventually will be established.

Most production in the field is from the Permian Tubb sandstone (Figure 3). A relatively minor amount of CO_2 gas is produced from the Santa Rosa Sandstone (Triassic). Shows of CO_2 gas have been encountered in the upper Clearfork member of the Yeso Formation, the Glorieta Sandstone (Permian), and in sandstones of the Chinle Formation (Triassic).

HISTORY

Discovery

The Bravo dome carbon dioxide gas field was discovered in 1916 by a petroleum exploration well, the American Production Corporation No. 1 Bueyeros, located in Section 32, T20N, R31E, Harding County, approximately 6.4 km (4 mi) south of the town of Bueyeros. That well was drilled in search of oil and is located on the northwest flank of the northeast-trending Sierra Negra anticline, a surface structure. It was one of the first petroleum exploration wells drilled in this remote area of northeastern New Mexico. The No. 1 Bueyeros well drilled to a total depth of 763 m (2506 ft); CO_2 gas was encountered in the Permian Tubb sandstone (Figure 3) from 3218 to 3251 m (2000 to 2020 ft) and flowed at a rate of 25 MMCFGD. There was no market for the CO_2 gas, and the well was allowed to blow open for one year before it was plugged in 1917 (Anderson, 1959).

Post-Discovery

Carbon dioxide in the Bravo dome area remained unproduced and undeveloped until 1931. In that year, the Southern Dry Ice Co. and Kummbaca Oil & Gas Co. drilled the No. 1 Kerlin, located in Section 34, T21N, R30E, Harding County, in search of CO_2 on the Baca anticline, a northwest-trending surface structure. The No. 1 Kerlin drilled to a total depth of 299 m (981 ft). Carbon dioxide gas flowed from the Triassic Santa Rosa Sandstone (Figure 3) at depths of 287 to 297 m (940 to 973 ft); initial potential was 3656 MCFGPD. Some CO_2 also was encountered in Chinle (Triassic) sandstones at depths of 87 to 91 m (287 to 300 ft) and 250 to 251.5 m (820 to 825 ft). The well was completed as the first producing CO_2 well in northeast New Mexico. It also was the discovery well for CO_2 in the Santa Rosa Sandstone. A small processing plant owned by the Timmons Dry Ice Company was constructed near the well; the plant converted CO_2 gas from the well into dry ice and liquid CO_2. Most of the dry ice was marketed in Denver, Colorado, and Amarillo, Texas. The liquid CO_2 was bottled and sold to carbonated-beverage bottlers in west Texas, Oklahoma, and Kansas (Anderson, 1959). The first processed CO_2 was shipped from the Timmons plant on June 4, 1938. Two additional wells were drilled to supply CO_2 to the Timmons plant. Operations at the Timmons

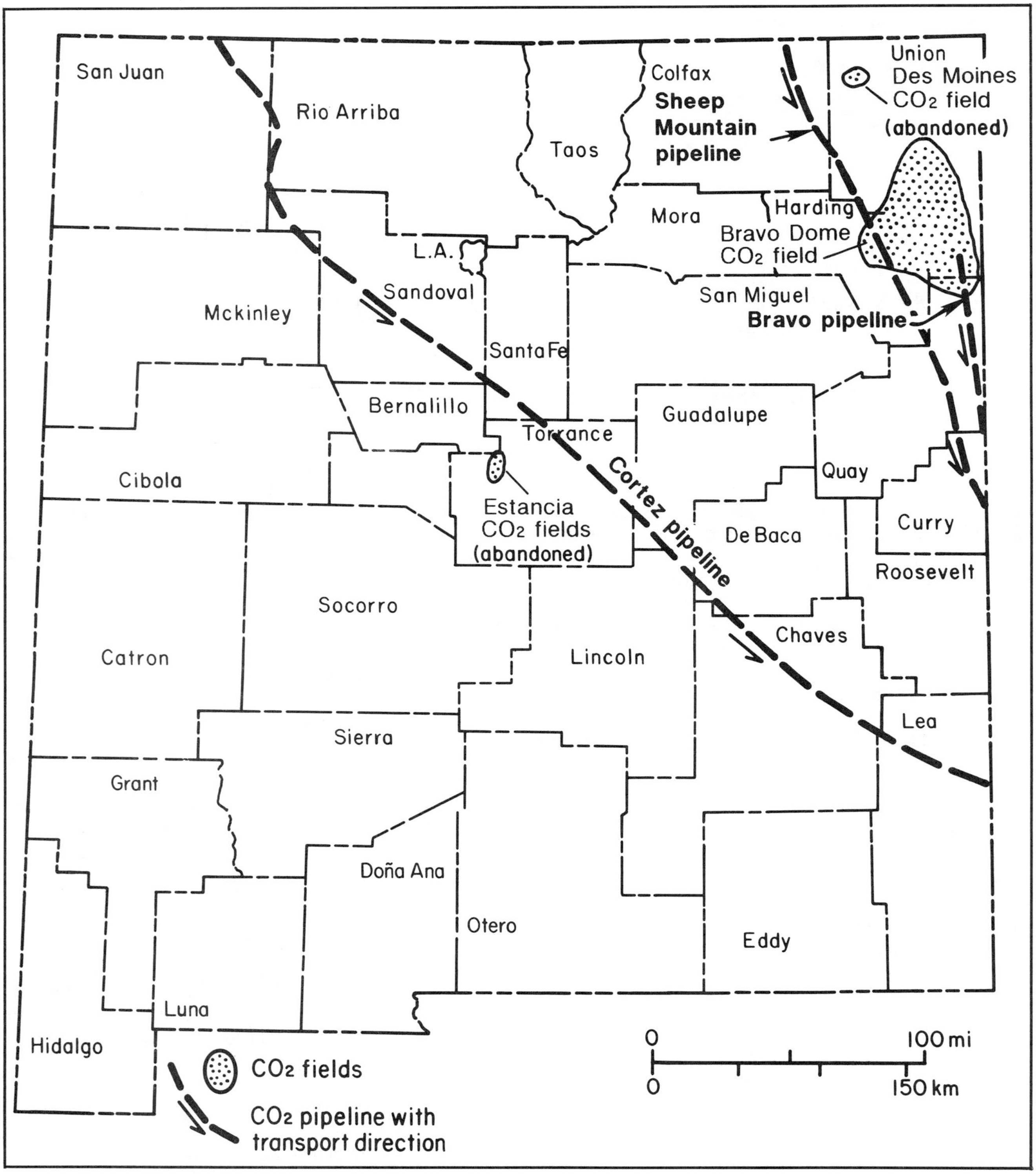
San Juan
Rio Arriba
Taos
Colfax
Sheep Mountain pipeline
Union
Des Moines CO2 field (abandoned)
Mora
Harding
Bravo Dome CO2 field
L.A.
Sandoval
Mckinley
San Miguel
Bravo pipeline
Santa Fe
Bernalillo
Guadalupe
Torrance
Cortez pipeline
Quay
Cibola
Estancia CO2 fields (abandoned)
De Baca
Curry
Roosevelt
Socorro
Catron
Lincoln
Chaves
Lea
Sierra
Grant
Doña Ana
Otero
Eddy
Luna
Hidalgo
0
100 mi
0
150 km
CO2 fields
CO2 pipeline with transport direction

Figure 1. CO_2 accumulations in New Mexico and locations of major CO_2 pipelines in New Mexico.

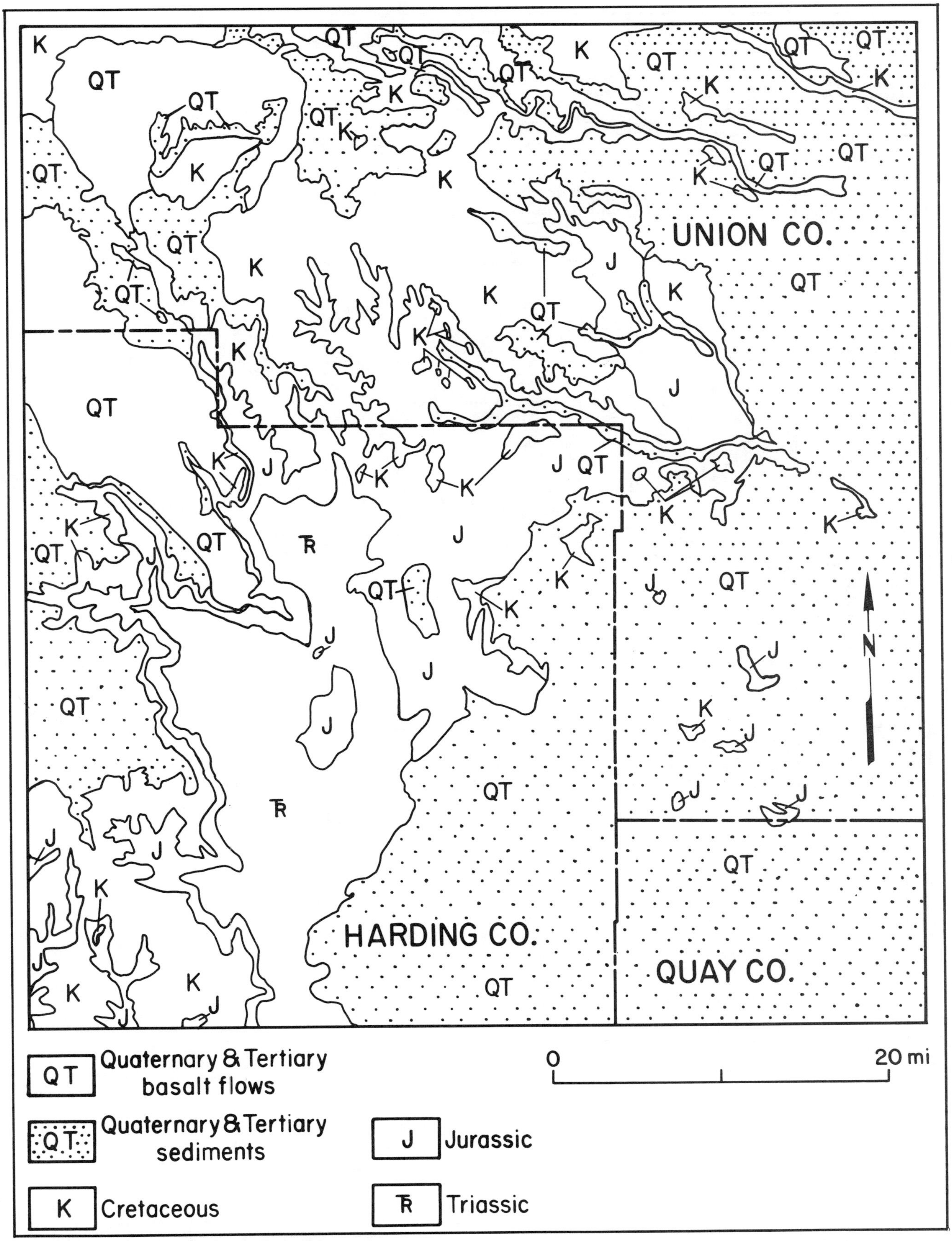
QT
K
J
TR
UNION CO.
HARDING CO.
QUAY CO.
N
0
20 mi
Quaternary & Tertiary basalt flows
Quaternary & Tertiary sediments
Cretaceous
Jurassic
Triassic

Figure 2. Surface geology map of the Bravo dome area in Union and Harding counties. Simplified from Dane and Bachman (1965).

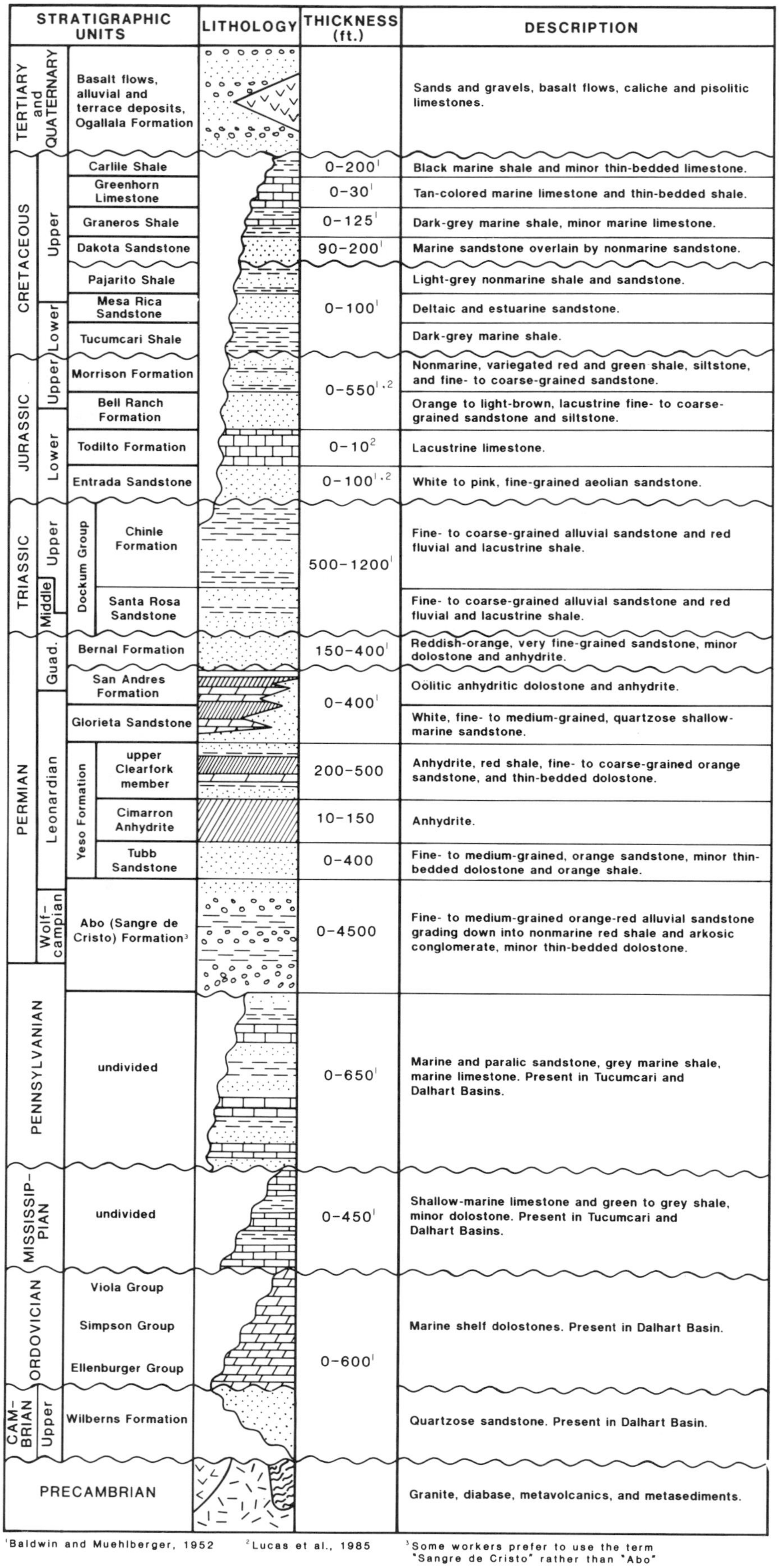

STRATIGRAPHIC UNITS
LITHOLOGY
THICKNESS (ft.)
DESCRIPTION
TERTIARY and QUATERNARY
Basalt flows, alluvial and terrace deposits, Ogallala Formation
Sands and gravels, basalt flows, caliche and pisolitic limestones.
CRETACEOUS
Upper
Carlile Shale
0-200[1]
Black marine shale and minor thin-bedded limestone.
Greenhorn Limestone
0-30[1]
Tan-colored marine limestone and thin-bedded shale.
Graneros Shale
0-125[1]
Dark-grey marine shale, minor marine limestone.
Dakota Sandstone
90-200[1]
Marine sandstone overlain by nonmarine sandstone.
Lower
Pajarito Shale
Light-grey nonmarine shale and sandstone.
Mesa Rica Sandstone
0-100[1]
Deltaic and estuarine sandstone.
Tucumcari Shale
Dark-grey marine shale.
JURASSIC
Upper
Morrison Formation
0-550[1,2]
Nonmarine, variegated red and green shale, siltstone, and fine- to coarse-grained sandstone.
Bell Ranch Formation
Orange to light-brown, lacustrine fine- to coarse-grained sandstone and siltstone.
Lower
Todilto Formation
0-10[2]
Lacustrine limestone.
Entrada Sandstone
0-100[1,2]
White to pink, fine-grained aeolian sandstone.
TRIASSIC
Upper
Middle
Dockum Group
Chinle Formation
500-1200[1]
Fine- to coarse-grained alluvial sandstone and red fluvial and lacustrine shale.
Santa Rosa Sandstone
Fine- to coarse-grained alluvial sandstone and red fluvial and lacustrine shale.
PERMIAN
Guad.
Bernal Formation
150-400[1]
Reddish-orange, very fine-grained sandstone, minor dolostone and anhydrite.
San Andres Formation
0-400[1]
Oölitic anhydritic dolostone and anhydrite.
Leonardian
Glorieta Sandstone
White, fine- to medium-grained, quartzose shallow-marine sandstone.
Yeso Formation
upper Clearfork member
200-500
Anhydrite, red shale, fine- to coarse-grained orange sandstone, and thin-bedded dolostone.
Cimarron Anhydrite
10-150
Anhydrite.
Tubb Sandstone
0-400
Fine- to medium-grained, orange sandstone, minor thin-bedded dolostone and orange shale.
Wolf-campian
Abo (Sangre de Cristo) Formation[3]
0-4500
Fine- to medium-grained orange-red alluvial sandstone grading down into nonmarine red shale and arkosic conglomerate, minor thin-bedded dolostone.
PENNSYLVANIAN
undivided
0-650[1]
Marine and paralic sandstone, grey marine shale, marine limestone. Present in Tucumcari and Dalhart Basins.
MISSISSIP-PIAN
undivided
0-450[1]
Shallow-marine limestone and green to grey shale, minor dolostone. Present in Tucumcari and Dalhart Basins.
ORDOVICIAN
Viola Group
Simpson Group
Ellenburger Group
0-600[1]
Marine shelf dolostones. Present in Dalhart Basin.
CAM-BRIAN
Upper
Wilberns Formation
Quartzose sandstone. Present in Dalhart Basin.
PRECAMBRIAN
Granite, diabase, metavolcanics, and metasediments.
[1]Baldwin and Muehlberger, 1952
[2]Lucas et al., 1985
[3]Some workers prefer to use the term "Sangre de Cristo" rather than "Abo"

Figure 3. Columnar stratigraphic section of Bravo dome area.

plant ceased during the 1950s and the plant subsequently was dismantled.

During the late 1930s, markets opened up for dry ice and liquid CO_2 produced in New Mexico. Additional wells were drilled in the Bueyeros area, and two additional processing plants were erected (Talmage and Andreas, 1942; Anderson, 1959). The Carbonic Chemicals Corporation plant was built near the village of Solano, approximately 14.5 km (9 mi) northwest of the town of Mosquero; the plant was supplied by 13 wells on the Mitchell Ranch (Anderson, 1959), which is located approximately 29 km (18 mi) east of the processing plant. Those wells produced from the Tubb sandstone (Permian).

The Iceco plant was built approximately 5 km (3 mi) south of Bueyeros. It was supplied with CO_2 by four wells located adjacent to the plant. Those wells produced from the Tubb sandstone.

Demand for bottled liquid CO_2 and dry ice has continued to the present. This has resulted in relatively slow but continued development of CO_2 reservoirs in the Santa Rosa and Tubb formations at the old Bueyeros field. Four plants presently process dry ice and CO_2 liquid from CO_2 produced in the Bueyeros field: (1) the Amerigas Inc. Valley plant located northeast of Mosquero; (2) the Amerigas Inc. Schwarz plant located south of Bueyeros; (3) the CO_2-in-Action Bueyeros plant located northwest of Bueyeros; and (4) the Ross Carbonics plant located northeast of Mosquero.

Main development of the Bravo dome field occurred in the early to middle 1980s, when several oil fields in the West Texas part of the Permian basin were designated to be flooded with CO_2 as enhanced oil-recovery projects. The Bravo dome field provides much of the CO_2 needed for those projects. The Bravo dome field was unitized (Bravo Dome Carbon Dioxide Gas Unit) under Amoco operation and aggressive development in the area east of Bueyeros followed. The number of CO_2 wells increased from 16 at the end of 1982 to 258 at the end of 1985. Annual CO_2 production at Bravo dome and Bueyeros increased from approximately 1 bcf in 1982 to 101 bcf in 1985. The CO_2 gas is transported by pipeline (Figure 1) from the Bravo dome field to the Permian basin.

DISCOVERY METHOD

Carbon dioxide was first discovered on the Bravo dome in 1916 by the American Production Corporation No. 1 Bueyeros well, located in Section 32, T20N, R31E, Harding County. That well was drilled as an oil test on a surface structure, the northeast-trending Sierra Negra anticline. Although historical records are sketchy, it appears that the discovery well was drilled solely to test the surface structure.

The areal extent of the CO_2 accumulation is so large that discovery by rank wildcat petroleum exploration wells was inevitable. The surface structure targeted by the discovery well is small compared with the Bravo dome structure, and does not appear to exert any measurable influence on trapping CO_2. The existence of the late Paleozoic-age Bravo dome and Sierra Grande uplift was unknown in 1916 because of a paucity of deep tests in northeast New Mexico. Indeed, the Sierra Grande uplift was thought to be a Tertiary-age feature as late as 1940 (Harley, 1940). The Bravo dome has almost no surface manifestations; it is buried by Permian and Mesozoic sedimentary units.

STRUCTURE

The Bravo dome CO_2 field is situated on the Bravo dome, a southeast-trending projection of the Sierra Grande uplift (Figure 4). The Bravo dome and Sierra Grande uplift initially developed as large fault blocks during Pennsylvanian time, and are just two of the many fault-bound basins and uplifts that formed in New Mexico during Pennsylvanian time (Figure 4). The Bravo dome is bordered on the east by the Dalhart basin. To the northwest, the Bravo dome merges into the Sierra Grande uplift. The Bravo dome is bordered on the south and southwest by the Tucumcari basin. On the southeast, it is bordered by the Palo Duro basin of the Texas panhandle. The Bravo dome is a subsurface feature that is not reflected by the outcrop patterns of rocks exposed at the surface (Figure 2).

The Bravo dome initially formed during Pennsylvanian time, when it rose out of the shallow epicontinental Paleozoic seas. By Strawn (Middle Pennsylvanian) time, Mississippian and Lower Pennsylvanian (Morrowan and Atokan) strata had been stripped off the Bravo dome, and the Precambrian granitic rocks that form the core of the dome were being eroded. Large volumes of Pennsylvanian-age and Early Permian-age arkosic sands and muds, derived from the erosion of those granitic rocks, were shed into the surrounding basins and deposited as large, basinward-thinning clastic wedges (Figure 3). The oldest Paleozoic sediments on top of the Bravo dome are Early Permian-age arkosic sands, the Abo Formation (Wolfcampian-Leonardian) and the Tubb sandstone (Leonardian). By the end of Abo time, major uplift ceased. The Bravo dome was buried and covered by its own detritus, the Abo and Tubb sands and muds. Marine to marginal marine evaporitic sediments of the upper Clearfork member of the Yeso Formation subsequently were deposited over the buried uplift.

A period of tectonic quiescence followed until the late Triassic or early Jurassic, when the redbeds of the Dockum Group (Triassic) were folded in northernmost Union County and overlain by the Entrada Sandstone (Jurassic) with angular unconformity (Lee, 1902). This post-Dockum, pre-Entrada deformation is not recognized in the Bravo dome area of central and southern Union County (Baldwin and Muehlberger, 1959).

Cretaceous sedimentary units were folded gently after deposition, presumably as a result of Laramide (Late Cretaceous to Early Tertiary) compressive deformation (Baldwin and Muehlberger, 1959). The Sierra Grande uplift and Bravo dome were rejuvenated as a result of Laramide deformation, but the relief that developed as a result of Laramide tectonic movements

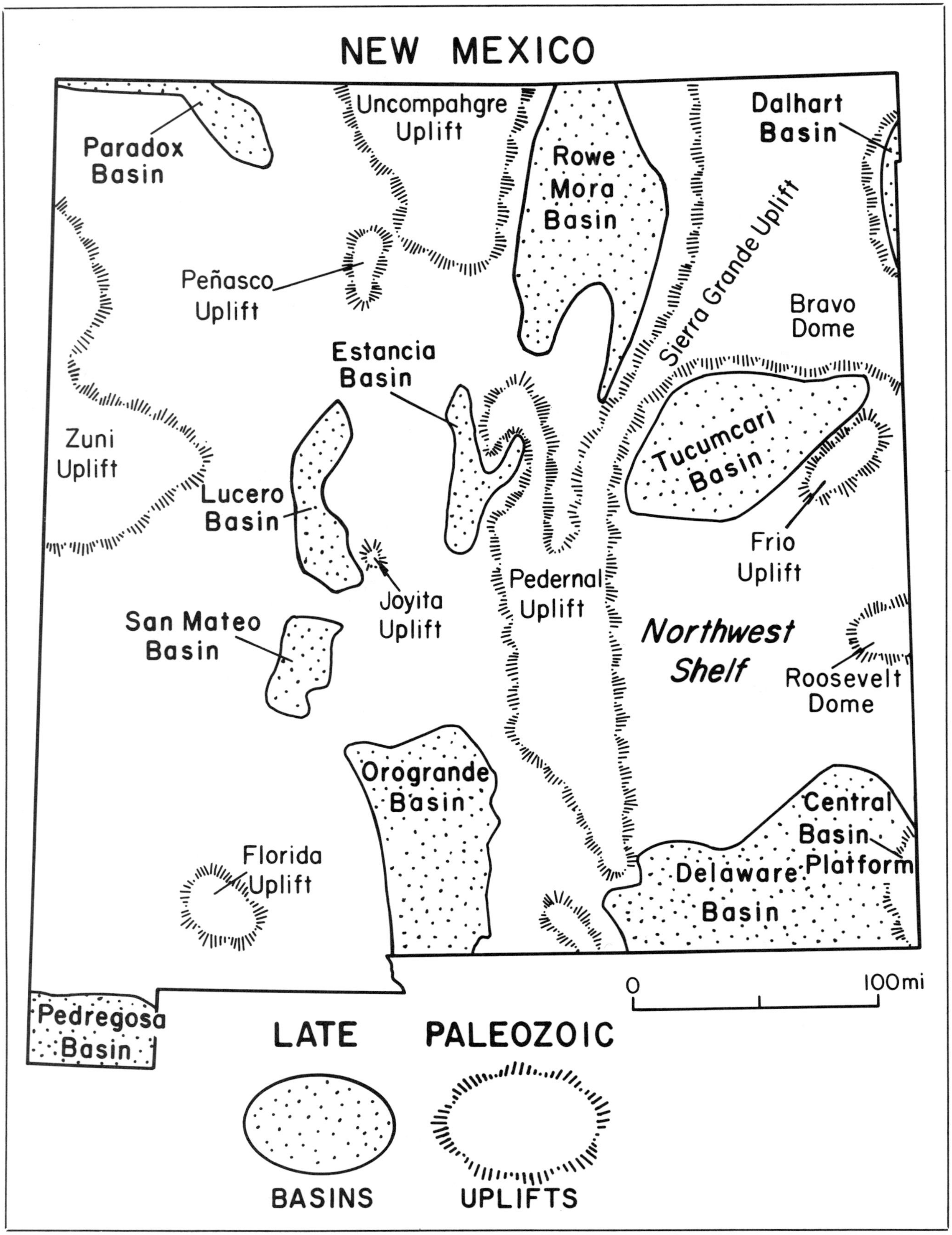
NEW MEXICO
Paradox Basin
Uncompahgre Uplift
Rowe Mora Basin
Dalhart Basin
Sierra Grande Uplift
Peñasco Uplift
Bravo Dome
Estancia Basin
Zuni Uplift
Lucero Basin
Tucumcari Basin
Frio Uplift
Joyita Uplift
Pedernal Uplift
San Mateo Basin
Northwest Shelf
Roosevelt Dome
Orogrande Basin
Central Basin Platform
Florida Uplift
Delaware Basin
0
100mi
Pedregosa Basin
LATE PALEOZOIC
BASINS
UPLIFTS

Figure 4. Late Paleozoic uplifts and basins in New Mexico. Modified from Kottlowski and Stewart (1970).

was much less than the relief that developed as a result of late Paleozoic tectonic movements, and was accompanied by only a relatively minor amount of faulting.

Laramide deformation was followed by Tertiary basaltic volcanism in northern and western Union County and in northernmost Harding County. The volcanic centers appear to be located primarily along northwest-trending fractures (Baldwin and Muehlberger, 1959). Some volcanic centers appear to be aligned along a subsidiary set of northeast-trending fractures (Baldwin and Muehlberger, 1959). Relatively minor Ogallala (Tertiary) age warping and uplift may have followed Laramide deformation (Baldwin and Muehlberger, 1959).

The Bravo dome is a large, faulted, southeast-plunging, basement-cored anticlinal nose (Figures 5 and 6). Large, high-angle faults form the east side of the dome and separate it from the Dalhart basin. Those faults cut Precambrian basement and the pre-Tubb sedimentary section, but do not appear to cut the Tubb sandstone, Cimarron Anhydrite, or younger sedimentary units (Figures 5 through 7). The presence of a great thickness of the Abo Formation in the Dalhart basin east of the Bravo dome (Figures 7 and 9) indicates that the faulting and major uplift were dominantly Late Pennsylvanian to Wolfcampian in age. Some of the relatively small folds superimposed on the Bravo dome (Winchester, 1933; Bates, 1942; Baldwin and Muehlberger, 1959) probably are Laramide features. Stratigraphic and structural studies in the Tucumcari basin (Broadhead and King, in press) indicate that a series of high-angle faults separate the Bravo dome and Sierra Grande uplift from the Tucumcari basin.

STRATIGRAPHY

The major reservoir unit in the Bravo dome CO_2 field is the Tubb sandstone (Figures 3, 10, and 11). The Tubb rests unconformably on Precambrian basement on the structurally highest parts of the Bravo dome (Figures 7 and 8). A wedge of the Abo Formation underlies the Tubb on structurally lower areas of the Bravo dome and on the flanks of the dome. The wedge of Abo thickens into the basins (Dalhart, Palo Duro, Tucumcari) that are adjacent to the Bravo dome (Figure 9) (Broadhead and King, in press). In the deeper parts of those basins, the Abo overlies Mississippian and Pennsylvanian strata (Baldwin and Muehlberger, 1959; Montgomery, 1986; Broadhead and King, in press). A westward-thinning wedge of Cambrian and Ordovician strata underlies the Mississippian section in the Dalhart basin.

The Cimarron Anhydrite and upper Clearfork member of the Yeso Formation conformably overlie the Tubb sandstone. The Cimarron consists of bedded anhydrite, and the upper Clearfork consists of interbedded red shale, sandstone, dolostone, and anhydrite. The Glorieta Sandstone overlies the upper Clearfork and intertongues southward with the dolostones and anhydrites of the San Andres Formation. In the Bravo dome area, the uppermost Permian unit consists of red beds of the Bernal Formation. Stratigraphy of Mesozoic units that overlie the Bernal Formation is illustrated in Figure 3.

TRAP

Trap Type

The nature of the trapping mechanism in the Tubb sandstone at Bravo dome is not entirely clear, but it definitely has structural and stratigraphic aspects. Hydrodynamics, combined with stratigraphic variations, also may play a role. Structural closure along the central, southeast-trending axis of the Bravo dome has controlled and localized the CO_2 accumulation on its northeast, southeast, and southwest sides (Figure 6). The impermeable Cimarron Anhydrite is a vertical seal for CO_2 trapped in the Tubb sandstone. Northwest thinning and localized pinchout of the Tubb (Figures 8 and 10) provide a lateral updip seal and overlying units apparently have acted as a barrier to updip migration of the CO_2 onto the structurally higher Sierra Grande uplift. The northwestern, updip limit of the CO_2 accumulation does not appear to be controlled by regional pinchout of the Tubb, but only by regional thinning. The northwest limit of the Bravo dome field may be controlled by internal permeability barriers (associated with regional thinning of the Tubb); alternatively, the northwestward thinning of the Tubb may create a sufficient constriction to groundwater flow so that a hydrodynamic barrier is present on the northwest side of the Bravo dome field. If the latter hypothesis is true, then the Bravo dome CO_2 pool in the Tubb is a combination structural-stratigraphic-hydrodynamic trap.

A paucity of geophysical logs through the Santa Rosa Sandstone makes determination of the trapping mechanism in that unit difficult. However, the areal extent of Santa Rosa production coincides approximately with the location of the southeast-trending Baca anticline (see Bates, 1942, for location of the Baca anticline). Therefore, structural closure created by the anticline probably is the principal factor controlling CO_2 accumulation. The Santa Rosa is sealed vertically by red shales of the Chinle Formation. Red shales within the Santa Rosa probably act as limited lateral seals.

RESERVOIRS

The major reservoir in the Bravo dome CO_2 field is the Tubb sandstone (Figures 3, 10, and 11). The Tubb consists of 0 to 122 m (0 to 400 ft) of fine- to medium-grained, well-sorted, orange, feldspathic sandstone and minor thin-bedded, orange-to-red shale and rare dolostone. The Tubb generally is thinnest on top of the Bravo dome and in the basins adjacent to the Bravo dome; it attains maximum thickness on the flanks of the Bravo dome (Figures 6 and 10). Depth to the top of the Tubb ranges from 580 to 900 m (1900 to 2950 ft). Gross thickness of productive Tubb ranges from 12 to 155 m (40 to 510 ft). Net thickness of perforated productive intervals averages 37 m (120 ft) and attains a maximum of 91 m (300 ft).

The orange color of the Tubb sandstone is derived from minor amounts of intergranular clays and iron oxides. Tubb sands generally are better sorted, more

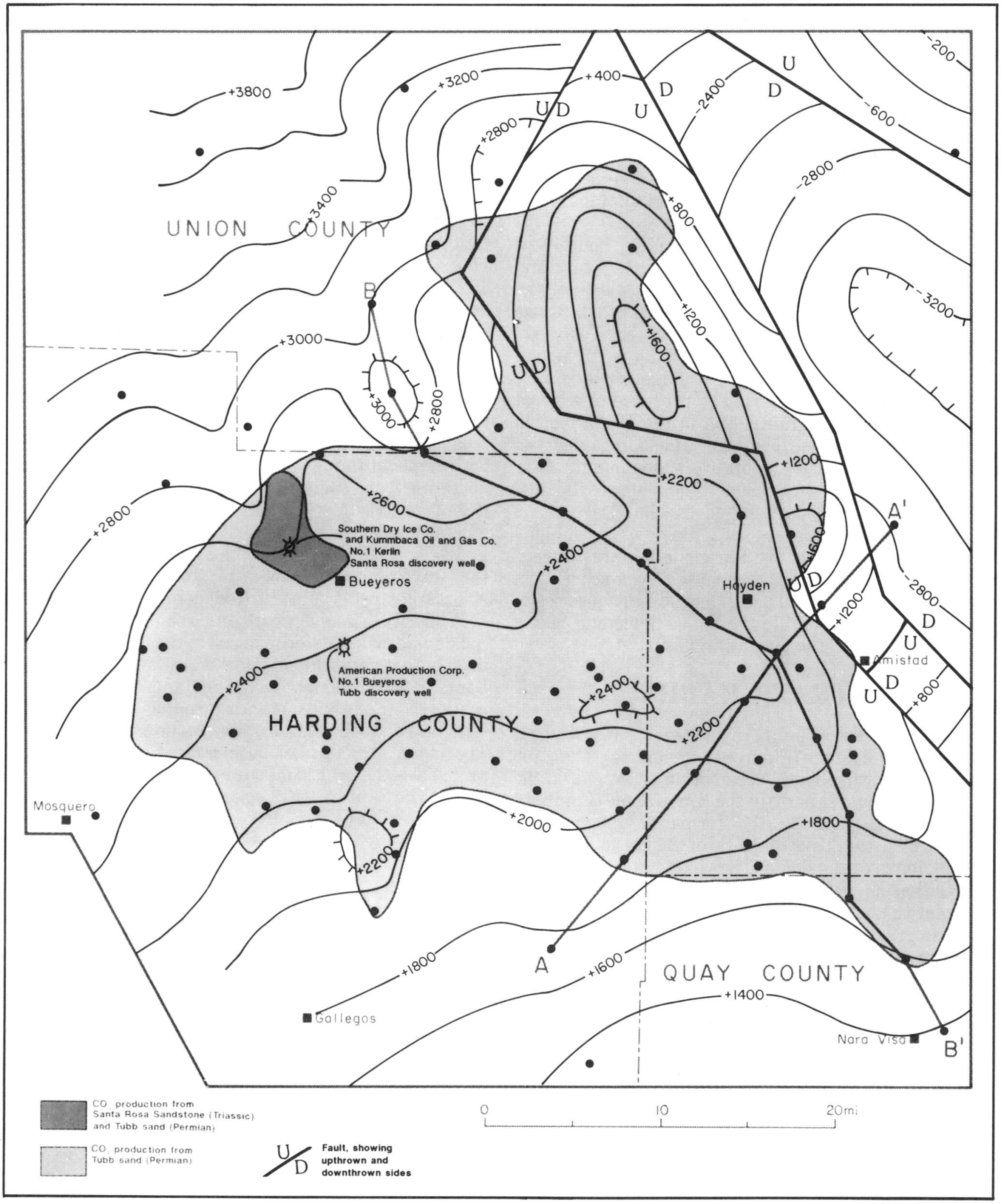
UNION COUNTY
HARDING COUNTY
QUAY COUNTY
Southern Dry Ice Co.
and Kummbaca Oil and Gas Co.
No.1 Kerlin
Santa Rosa discovery well
Bueyeros
American Production Corp.
No.1 Bueyeros
Tubb discovery well
Hayden
Amistad
Mosquero
Gallegos
Nara Visa
A
A'
B
B'
+3800
+3400
+3200
+3000
+2800
+2600
+2400
+2200
+2000
+1800
+1600
+1400
+1200
+800
+400
-200
-600
-2400
-2800
-3200
-2800
0
10
20mi
CO production from
Santa Rosa Sandstone (Triassic)
and Tubb sand (Permian)
CO production from
Tubb sand (Permian)
Fault, showing
upthrown and
downthrown sides

Figure 5. Structure contours on Precambrian surface. Contour interval equals 61 m (200 ft).

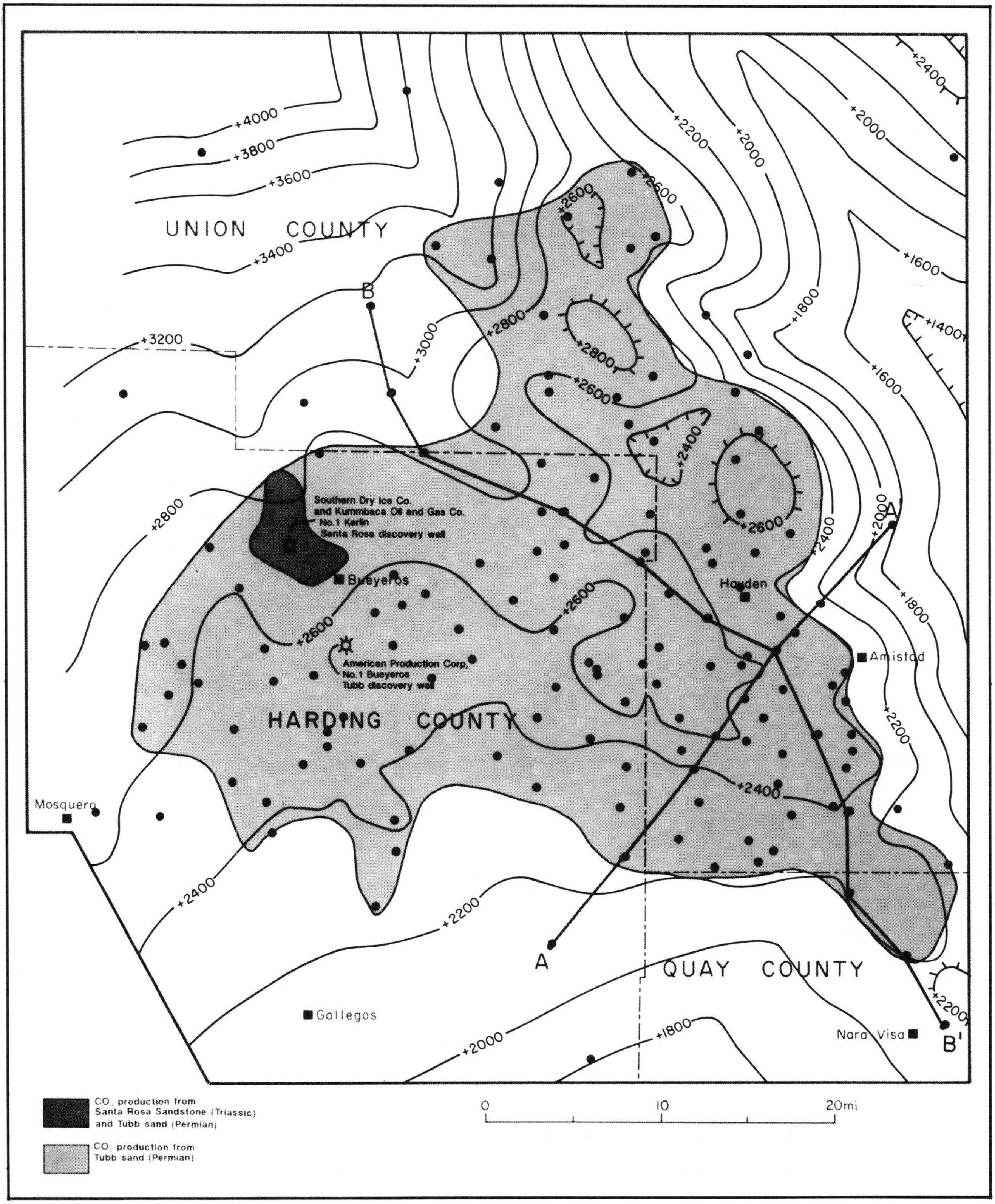
UNION COUNTY
HARDING COUNTY
QUAY COUNTY
Southern Dry Ice Co. and Kummbaca Oil and Gas Co. No.1 Kerlin Santa Rosa discovery well
American Production Corp, No.1 Bueyeros Tubb discovery well
Bueyeros
Hayden
Amistad
Mosquero
Gallegos
Nara Visa
CO_2 production from Santa Rosa Sandstone (Triassic) and Tubb sand (Permian)
CO_2 production from Tubb sand (Permian)
0
10
20mi

Figure 6. Structure contours on top of Tubb sandstone (Permian). Contour interval equals 61 m (200 ft).

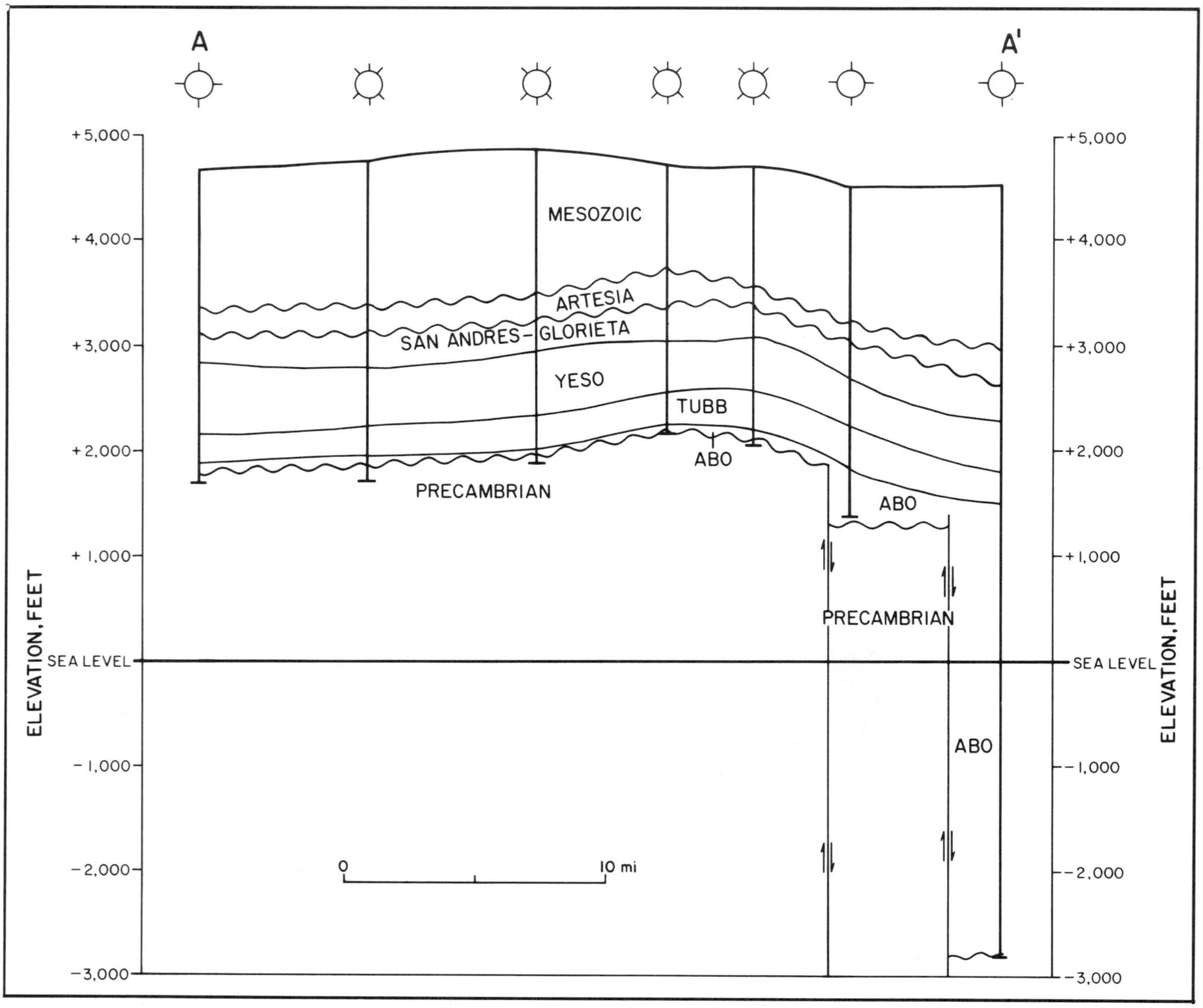
A
A'
+5,000
+4,000
+3,000
+2,000
+1,000
SEA LEVEL
-1,000
-2,000
-3,000
ELEVATION, FEET
MESOZOIC
ARTESIA
SAN ANDRES-GLORIETA
YESO
TUBB
ABO
PRECAMBRIAN
0
10 mi

Figure 7. Structural cross section A-A′. See Figures 5, 6, 9, and 10 for location.

rounded, and more porous than sands in the underlying Abo Formation. Johnson (1983) reported that Tubb sandstones have an average porosity of 20% and an average permeability of 42 md in the Bravo dome field.

Although the Tubb sandstone is a laterally continuous lithostratigraphic unit consisting predominantly of sand, examination of geophysical logs reveals internal reservoir inhomogeneities. Clean sands and shaly sands are interlayered within the Tubb. They are lenticular on the scale of the 640-ac well spacing within the Bravo dome field.

Determination of the depositional environment of the Tubb sandstone is problematic because of lack of access to cores or outcrops in the Bravo dome area. However, a beach and tidal-flat environment is indicated by four factors: (1) the wide lateral extent of the Tubb sandstone; (2) the generally well-sorted nature of Tubb sands; (3) the thin dolostones sparsely interbedded with Tubb sands; and (4) the association with underlying fluvial sediments of the Abo Formation and overlying shallow-marine evaporitic sediments of the Cimarron Anhydrite and upper Clearfork member of the Yeso Formation.

The secondary reservoir within the Bravo dome CO_2 field is the Santa Rosa Sandstone (Figure 3). Where it produces (Figure 5), the Santa Rosa consists of 76 to 113 m (250 to 370 ft) of fine- to coarse-grained quartzose sandstones interbedded with minor red shales. The sandstones bear minor amounts of mica and varied rock fragments that give them a "dirty" appearance. The sandstones are lenticular on a regional scale (Broadhead, 1984; McKallip, 1984) but generally appear laterally continuous in outcrop. The Santa Rosa Sandstone is a fluvial deposit laid down by braided streams (McGowen et al., 1979; McKallip, 1984; Broadhead, 1984).

Shows of CO_2 gas have been encountered in the upper Clearfork member of the Yeso Formation, in the Glorieta Sandstone, and in the Chinle Formation. Res-

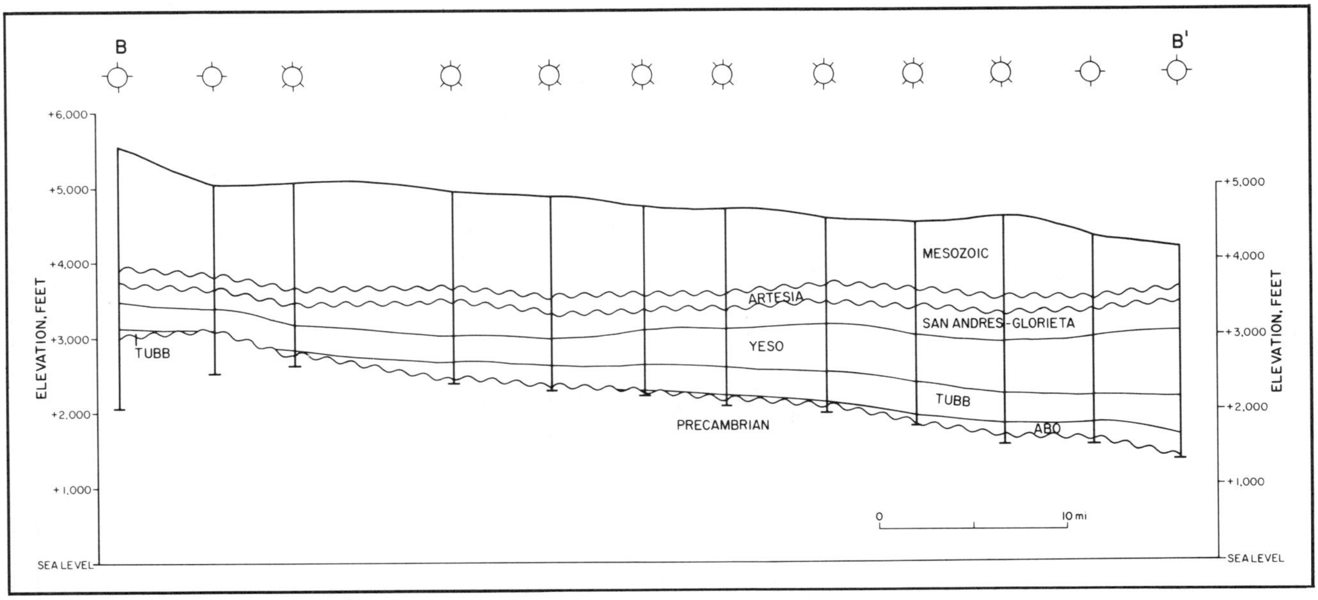
B
B'
+6,000
+5,000
+4,000
+3,000
+2,000
+1,000
SEA LEVEL
ELEVATION, FEET
MESOZOIC
ARTESIA
SAN ANDRES-GLORIETA
YESO
TUBB
ABO
PRECAMBRIAN
0
10 mi

Figure 8. Structural cross section B-B′. See Figures 5, 6, 9, and 10 for location.

ervoirs in the upper Clearfork appear to be fine-grained, orange, marine sandstones. A minor amount of production from some of the early (pre-1960) wells drilled in the Bravo dome field appears to come from the upper Clearfork. The Glorieta consists of fine- to medium-grained, rounded, white quartzose sandstone. The Glorieta was deposited in a marginal marine to shallow marine environment.

Shows in the Chinle Formation occur in fine- to medium-grained quartzose sandstones that are similar in appearance to Santa Rosa sandstones. Red Chinle sandstones were deposited by braided streams (McGowen et al., 1979; Broadhead, 1984). Interbedded Chinle shales are lacustrine deposits that seal the sandstones.

Faults

Faults were recognized in the subsurface by contouring of log tops. The faults mapped in this study (Figures 5 and 7) have significant vertical displacements and can be contoured with a good degree of confidence from log tops. A Bouguer gravity anomaly map (Keller and Cordell, 1983) and a residual aeromagnetic map (Cordell, 1983) were used to aid mapping. Seismic data were not used for mapping because they remain proprietary. Faults undoubtedly exist on the Bravo dome that do not appear on the Precambrian structure map (Figure 5). Those faults have not been mapped because they have relatively small displacements and because of a paucity of well control at the margins of the Bravo dome CO_2 field.

The faults appear to act as vertical migration paths for the CO_2. Most faults do not appear to penetrate as far upward as the Cimarron Anhydrite; therefore, the Cimarron acts as a seal for CO_2 in the Tubb sandstone. Small faults may cut the Cimarron locally in areas where CO_2 occurs in overlying units (e.g., Santa Rosa Sandstone, Glorieta Sandstone, and upper Clearfork member of Yeso Formation).

CO_2 SOURCE

The source of the CO_2 produced from the Bravo dome field has not been positively identified. Four hypotheses have been suggested: (1) juvenile or magmatic CO_2; (2) chemical breakdown of carbonate rocks by adjacent igneous intrusions; (3) dissolution of carbonate rocks by groundwater; and (4) bacterial or thermal decomposition of organic matter (Foster and Jensen, 1972). Geologic and geochemical aspects of the four hypotheses are reviewed below.

Juvenile or magmatic CO_2 could have been introduced into Bravo dome reservoirs as volcanic emanations, or it could have migrated into its present traps from a source in the mantle or lower crust through the deep-seated basement faults that form the Bravo dome and Sierra Grande uplift. Phinney et al. (1978) found support for a juvenile magmatic origin by analyzing isotopes of noble gases (Xe and He) that occur with the CO_2 in the Bravo dome area. Although Phinney's analyses indicate that the noble gases may have a juvenile magmatic source, it is not certain that the CO_2 has the same origin as the noble gases. Johnson (1983) pointed out that geochemists have tended to support a juvenile magmatic origin because of the isotopic analyses of the noble gases.

A problem with the volcanic origin theory is that the areas with the most volcanic rocks and the areas with the largest known CO_2 occurrences in northeast New Mexico are mutually exclusive. Very little volcanic activity has occurred in and around the Bravo dome field;

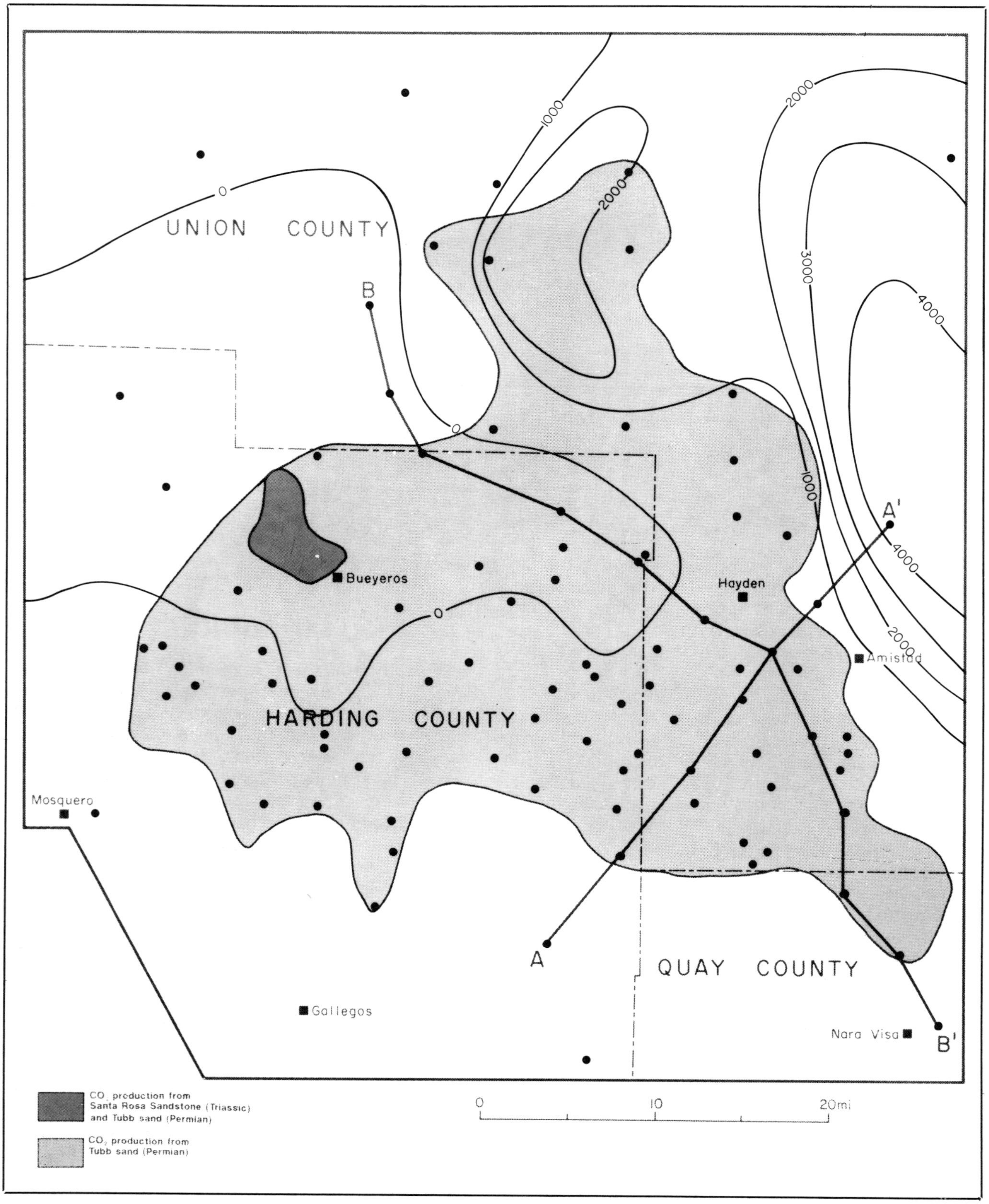
UNION COUNTY
HARDING COUNTY
QUAY COUNTY
Bueyeros
Hayden
Amistad
Mosquero
Gallegos
Nara Visa
A
A'
B
B'
0
1000
2000
3000
4000
0
10
20mi
CO₂ production from Santa Rosa Sandstone (Triassic) and Tubb sand (Permian)
CO₂ production from Tubb sand (Permian)

Figure 9. Isopach map of the Abo Formation (Permian). Contour interval equals 305 m (1000 ft).

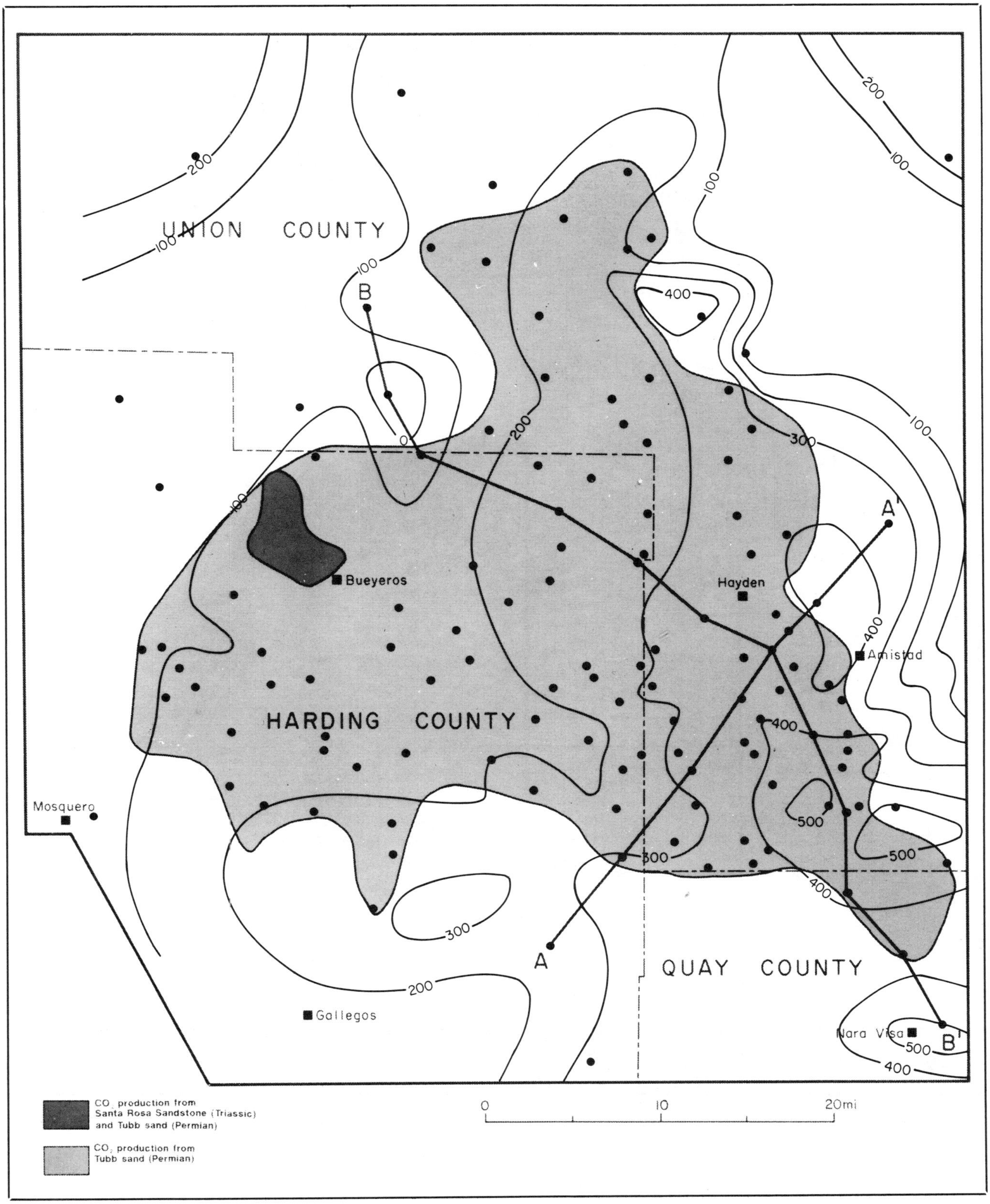
UNION COUNTY
HARDING COUNTY
QUAY COUNTY
Bueyeros
Hayden
Amistad
Mosquero
Gallegos
Nara Visa
A
A'
B
B'
100
200
300
400
500
0
10
20mi
CO_2 production from Santa Rosa Sandstone (Triassic) and Tubb sand (Permian)
CO_2 production from Tubb sand (Permian)

Figure 10. Isopach map of the Tubb sandstone (Permian). Contour interval equals 30.5 m (100 ft).

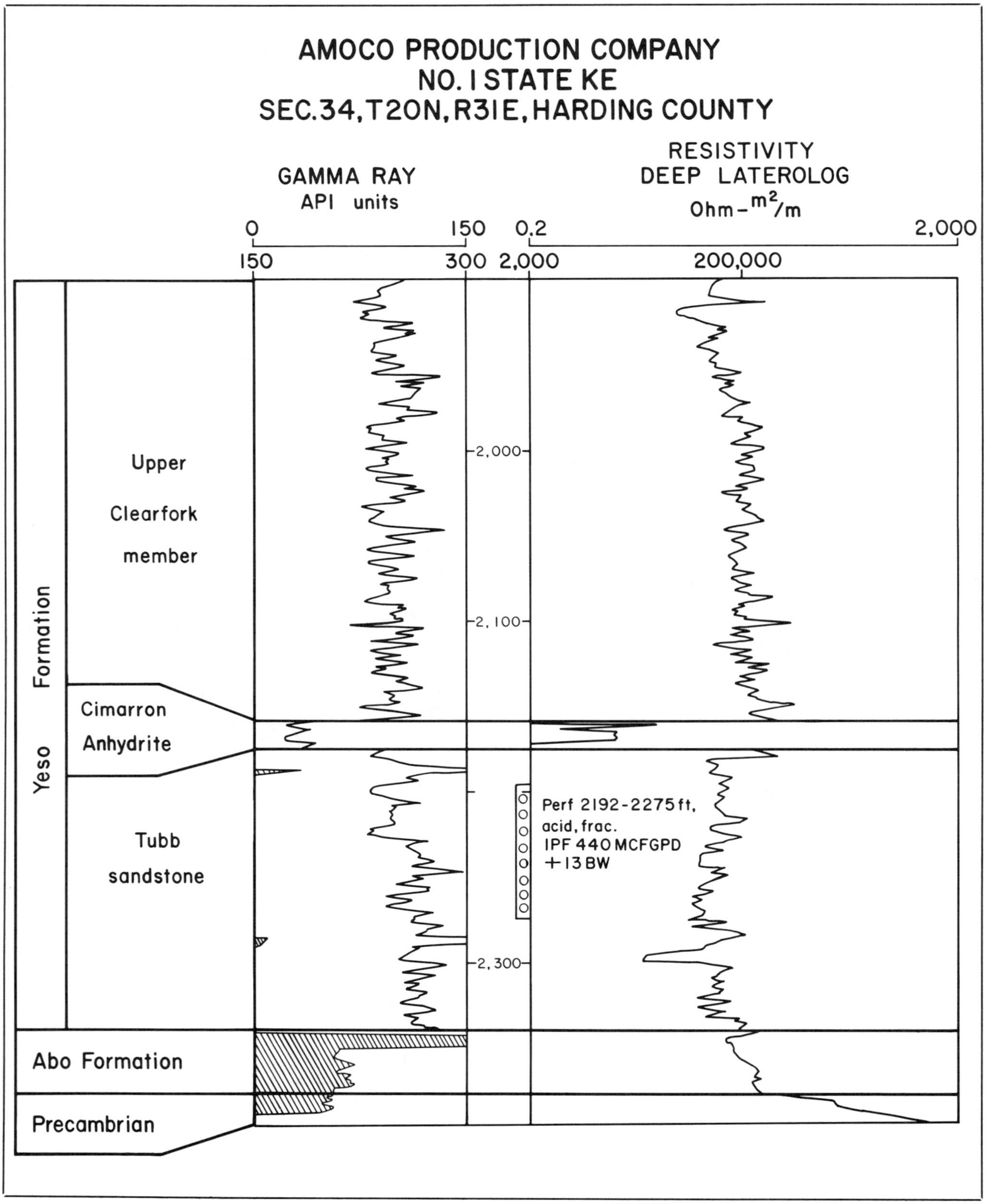
AMOCO PRODUCTION COMPANY
NO. I STATE KE
SEC.34, T20N, R31E, HARDING COUNTY
GAMMA RAY
API units
0
150
150
300
RESISTIVITY
DEEP LATEROLOG
Ohm-m²/m
0.2
2,000
2,000
200,000
Yeso Formation
Upper
Clearfork
member
Cimarron
Anhydrite
Tubb
sandstone
Abo Formation
Precambrian
2,000
2,100
2,300
Perf 2192-2275 ft,
acid, frac.
IPF 440 MCFGPD
+13 BW

Figure 11. Typical log through the Tubb sandstone, the major reservoir in the Bravo dome CO_2 field, showing the productive interval.

most of the volcanic centers are located in northwestern Harding and northern Union counties (Dane and Bachman, 1965), and are structurally updip from the Bravo dome field. However, the deep-seated basement faults that form the Bravo dome could have acted as migration paths from a CO_2 source or reservoir in the mantle or lower crust.

The petroleum industry generally has supported the hypothesis that the CO_2 derived from the chemical breakdown of carbonate rocks by igneous intrusions. Lang (1959) studied the $^{12}C/^{13}C$ isotope ratios of the CO_2 gases at Bravo dome. He concluded that the carbon in the CO_2 gases is isotopically heavy and derived from marine limestones. Galimov (1968), however, concluded that CO_2 derived from the mantle would have a similar isotopic composition to CO_2 derived from limestones. There are three principal geologic arguments against the hypothesis of an igneous-carbonate rock origin. First, there has been limited volcanic activity in the Bravo dome area. Second, very few carbonates occur within the principal CO_2 reservoir, the Tubb sandstone. Third, the Tubb rests on Precambrian basement over much of the Bravo dome; where it does not rest on the Precambrian, it overlies the Abo clastic red beds. Therefore, there is no apparent source of CO_2-bearing carbonate rock.

The hypothesis of groundwater solution of carbonate rocks requires that carbonates be present in the area of CO_2 accumulation or along a migration path leading to the CO_2 accumulation. As mentioned above, there is a paucity of carbonates at appropriate stratigraphic intervals in the Bravo dome area. It is possible, however, that carbonates in the Ordovician, Mississippian, or Pennsylvanian strata of the Dalhart basin are the sources of the CO_2. The CO_2 could have migrated onto the Bravo dome via the large faults that form the eastern margin of the Bravo dome. Carbon isotope ratios determined by Lang (1959) can be used to support the hypothesis of dissolution of carbonate rocks, but as Galimov (1968) pointed out, Lang's measurements are also consistent with a juvenile origin.

The hypothesis that the CO_2 originated by bacterial or thermal decomposition of organic matter largely has been discounted. Paleozoic strata on the Bravo dome and Sierra Grande uplift are mostly red beds, evaporites, or evaporitic carbonates, and contain little sedimentary organic matter. Paleozoic strata in the Tucumcari, Dalhart, and Palo Duro basins have fairly high contents of organic matter, but those strata appear to have generated mostly hydrocarbon gases and only minimal amounts of CO_2.

EXPLORATION CONCEPTS

Recent exploration for CO_2 in northeast New Mexico has concentrated on the Bravo dome and Sierra Grande uplifts. There are three geologic factors that justify this concentration of exploratory effort. First, the structural closure of the Tubb sandstone on the Bravo dome is a major factor that controls the areal location of the Bravo dome field. The CO_2 accumulations at the McElmo dome field (8 tcf recoverable CO_2) in southwest Colorado (Gerling, 1983), and the Sheep Mountain field (2.5 tcf recoverable CO_2) in southeast Colorado (Roth, 1983) also are structurally controlled. Second, the Bravo dome CO_2 field can only exist because there are no significant amounts of hydrocarbon gases trapped on the Bravo dome. Hydrocarbon gases, if present, would dilute any CO_2 that accumulated in traps. Hydrocarbon gases appear to be absent because there are no hydrocarbon source rocks in stratigraphic proximity to the CO_2 reservoirs. Third, if the CO_2 has migrated into traps from a source or reservoir in the mantle or lower crust, then exploration for CO_2 should concentrate on areas that have deep-seated migration pathways; those pathways may be either volcanic intrusions or deep-seated faults. Both types of potential migration pathways are present on the Bravo dome and Sierra Grande structures. The McElmo dome and Sheep Mountain fields also are associated spacially with volcanic intrusions (Fallin, 1984; Roth, 1983).

Any large CO_2 accumulation like the Bravo dome field requires the presence of a widespread sealing mechanism. The Cimarron Anhydrite overlies and seals the Tubb sandstone throughout the Bravo dome field. Although a widespread evaporite seal such as the Cimarron may not be a prerequisite for formation of a large CO_2 field like Bravo dome, the presence of a seal like the Cimarron enhances the possibility of a large CO_2 accumulation and should be considered a favorable factor in exploration.

ACKNOWLEDGMENTS

This report was published with the permission of Frank E. Kottlowski, Director of the New Mexico Bureau of Mines and Mineral Resources. Ben Donegan suggested this report be prepared for AAPG. Roy Johnson provided the author with much information on the Bravo dome area. The author's understanding of the Bravo dome CO_2 field and the regional geology of northeast New Mexico was enhanced by discussions with Dr. Robert J. Weimer and Dan Hartmann. Lynne McNeil typed the manuscript. Monte Brown, Rebecca Titus, Irean Rae, and Jean Moody drafted the illustrations.

REFERENCES CITED

Anderson, E. C., 1959, Carbon dioxide in New Mexico (1959): New Mexico Bureau of Mines and Mineral Resources Circular 43, 13 p.

Baldwin, B., and W. R. Muehlberger, 1959, Geologic studies of Union County, New Mexico: New Mexico Bureau of Mines and Mineral Resources Bulletin 63, 171 p.

Bates, R. L., 1942, The oil and gas resources of New Mexico (2nd ed.): New Mexico Bureau of Mines and Mineral Resources Bulletin 18, 318 p.

Broadhead, R. F., 1984, Subsurface petroleum geology of Santa Rosa Sandstone (Triassic), northeast New Mexico: New Mexico Bureau of Mines and Mineral Resources Circular 193, 22 p.

Broadhead, R. F., and W. E. King, in press, Petroleum geology of Pennsylvanian and Lower Permian strata, Tucumcari basin, east-central New Mexico: New Mexico Bureau of Mines and Mineral Resources Bulletin 119.

Cordell, L., 1983, Composite residual total intensity aeromagnetic map of New Mexico: New Mexico State University Energy Institute, Geothermal resources of New Mexico, Scientific map series, scale 1:500,000.

Dane, C. H., and G. O. Bachman, 1965, Geologic map of New Mexico: U.S. Geological Survey, scale 1:500,000, two sheets.

Fallin, J. A., ed., 1984, Carbon dioxide and its applications to enhanced oil recovery: Petroleum Frontiers, v. 2, no. 1, 63 p.

Fenneman, N. M., 1946, Physical divisions of the United States: USGS map.

Foster, R. W., and J. G. Jensen, 1972, Carbon dioxide in northeastern New Mexico: New Mexico Geological Society Guidebook to 23rd field conference, p. 192-200.

Galimov, E. M., 1968, Isotopic composition of carbon in gases of the crust: International Geology Review, v. 11, p. 1092-1104.

Gerling, C. R., 1983, McElmo dome Leadville carbon dioxide field, Colorado, *in* J. E. Fassett, ed., Oil and gas fields of the Four Corners area, v. III: Four Corners Geological Society, p. 735-739.

Harley, G. T., 1940, The geology and ore deposits of northeastern New Mexico (exclusive of Colfax County): New Mexico Bureau of Mines and Mineral Resources Bulletin 15, 102 p.

Johnson, R. E., 1983, Bravo dome carbon dioxide area, northeast New Mexico, *in* J. E. Fassett, ed., Oil and gas fields of the Four Corners area, v. III: Four Corners Geological Society, p. 745-748.

Keller, G. R., and L. Cordell, 1983, Bouguer gravity anomaly map of New Mexico: New Mexico State University Energy Institute, Geothermal resources of New Mexico, Scientific map series, scale 1:500,000.

Kottlowski, F. E., and W. J. Stewart, 1970, The Wolfcampian Joyita uplift in central New Mexico: New Mexico Bureau of Mines and Mineral Resources Memoir 23, part I, 31 p.

Lang, W. B., 1959, The origin of some natural carbon dioxide gases: Journal of Geophysical Research, v. 64, p. 127-131.

Lee, W. T., 1902, The Morrison shales of southern Colorado and northern New Mexico: Journal of Geology, v. 10, p. 36-58.

Lucas, S. G., K. K. Kietzke, and A. P. Hunt, 1985, The Jurassic System in east-central New Mexico: New Mexico Geological Society Guidebook to 36th field conference, p. 213-242.

McGowen, J. H., G. E. Granata, and S. J. Seni, 1979, Depositional framework of the lower Dockum Group (Triassic), Texas panhandle: Texas Bureau of Economic Geology, Report of Investigations 97, 60 p.

McKallip, C., Jr., 1984, Newkirk field: the geology of a shallow steamflood project in Guadalupe County, New Mexico: M.S. thesis, New Mexico Institute of Mining and Technology, 89 p.

Montgomery, S. L., ed., 1986, The Dalhart basin: patterns of success in the western panhandle: Petroleum Frontiers, v. 3, no. 1, 86 p.

Phinney, D., J. Tennyson, and V. Frick, 1978, Xenon in CO_2 well gas revisited: Journal of Geophysical Research, v. 83, p. 2313-2319.

Roth, G., 1983, Sheep Mountain and Dike Mountain fields, Huerfano County, Colorado; a source of CO_2 for enhanced oil recovery, *in* J. E. Fassett, ed., Oil and gas fields of the Four Corners area, v. III: Four Corners Geological Society, p. 740-744.

Talmage, S. B., and A. Andreas, 1942, Carbon dioxide in New Mexico, *in* R. L. Bates, compiler, The oil and gas resources of New Mexico (2nd ed.): New Mexico Bureau of Mines and Mineral Resources Bulletin 18, p. 301-307.

Winchester, D. E., 1933, The oil and gas resources of New Mexico: New Mexico Bureau of Mines and Mineral Resources Bulletin 9, 223 p.

Appendix 1. Field Description

Field name .... *Bravo dome carbon dioxide gas field*

Ultimate recoverable reserves .... *NA*

Field location:

Country .... *U.S.A.*

State .... *New Mexico*

Basin/Province .... *Bravo dome*

Field discovery:

Year field discovered .... *1916 (Tubb sandstone)*

Year second pay discovered .... *1931 (Santa Rosa Sandstone)*

Third pay

Discovery well name and general location
(i.e., Jones No. 1, Sec. 2T12NR5E; or Smith No. 1, 5 mi west of Sheridan, Wyoming):

First pay .... *American Production Corp. No. 1 Bueyeros, Section 32, T20N, R31E, Harding County, NM*

Second pay .... *Southern Dry Ice Co. and Kummbaca Oil & Gas Co. No. 2 Kerlin, Section 34, T21N, R30E, Harding County, NM*

Third pay

Discovery well operator .... *American Production Corp.*
(if more than one pay in field, list operators of discovery well in other pays)

Second pay .... *Southern Dry Ice Co. and Kummbaca Oil & Gas Co.*

Third pay

IP in barrels per day and/or cubic feet or cubic meters per day:

First pay .... *25,000 MCFGD*

Second pay .... *3656 MCFGD*

Third pay

All other zones with shows of oil and gas in the field:

Age	Formation	Type of Show
Permian	*San Andres*	*Oil*
Permian	*Upper Clearfork member of Yeso Formation*	*CO_2 gas*
Permian	*Glorieta*	*CO_2 gas*

Geologic concept leading to discovery and method or methods used to delineate prospect, e.g., surface geology, subsurface geology, seeps, magnetic data, gravity data, seismic data, seismic refraction, nontechnical:

American Production Corp. No. 1 Bueyeros was drilled on a surface structure, the Sierra Negra anticline. The Southern Dry Ice Co. and Kummbaca Oil & Gas Co. No. 1 Kerlin was drilled on a surface structure, the Baca anticline.

Structure:

Province/basin type (see St. John, Bally, and Klemme, 1984)

Bally, cratonic uplift; Klemme, IIA-Arch

Tectonic history

Broad cratonic fualt-block uplift developed during Pennsylvanian time. Pre-Pennsylvanian strata eroded from uplift. Bravo dome covered by Upper Pennsylvanian and Lower Permian nonmarine red beds. Uplift ceased during Early Permian. Middle Permian Triassic, Jurassic, and Cretaceous strata deposited over uplift. Late Cretaceous-Early Tertiary compression formed minor anticlines, which are exposed surface structures.

Regional structure

Bravo dome is a southeast-trending projection of the late Paleozoic Sierra Grande uplift.

Local structure

Buried southeast-dipping structural nose bounded on east side by major high-angle faults with combined displacements in excess of 1464 m (4800 ft).

Trap

Trap type(s)

Tubb sandstone: stratigraphic/structural trap formed by updip reservoir/permeability pinchout
Santa Rosa sandstone: structural trap on anticline

Basin stratigraphy (major stratigraphic intervals from surface to deepest penetration in field):

Age	Formation	Depth to Top in ft
Triassic	*Dockum*	*0-300*
Permian	*Artesia*	*1000*
Permian	*San Andres*	*1300*
Permian	*Yeso*	*1670*
Permian	*Tubb*	*2120*
Permian-Pennsylvanian	*Abo*	*2490*
Precambrian	*Granite and metasediments*	*2520*

Location of well in field .... *NA*

Reservoir characteristics:

Number of reservoirs *2*

Formations .... *Tubb sandstone; Santa Rosa sandstone*

Ages .... *Tubb, Leonardian (Permian); Santa Rosa, Upper Triassic*

Depths to tops of reservoirs .... *Tubb, 1900-2950 ft; Santa Rosa, 780-940 ft*

Gross thickness (top to bottom of producing interval) .... *Tubb, 40-510 ft; Santa Rosa, 250-370 ft*

Net thickness—total thickness of producing zones

Average .... *Tubb, 120 ft; Santa Rosa, 40 ft*

Maximum .... *Tubb, 300 ft; Santa Rosa, 50 ft*

Average

Maximum

Lithology

Tubb: Orange, fine- to very fine grained, feldspathic sandstone
Santa Rosa: Fine- to coarse-grained quartzose sandstone

Porosity type .... *Tubb and Santa Rosa: intergranular porosity*

Average porosity .... *20% (Johnson, 1983)*

Average permeability .... *42 md (Johnson, 1983)*

Seals (for Tubb sandstone):

Upper

Formation, fault, or other feature .... *Cimarron Anhydrite*

Lithology .... *Anhydrite*

Lateral

Formation, fault, or other feature .... *Northward lateral thinning of Tubb sandstone*

Lithology

Source:

Formation and age .... *Undetermined (see text)*

Lithology

Average total organic carbon (TOC)

Maximum TOC
Kerogen type (I, II, or III)
Vitrinite reflectance (maturation)
Time of hydrocarbon expulsion
Present depth to top of source
Thickness
Potential yield

Appendix 2. Production Data

Field name .... *Bravo dome carbon dioxide gas field*

Field size:

- **Proved acres** .... *~800,000 ac (324,000 ha)*
- **Number of wells all years** *281*
- **Current number of wells** *271*
- **Well spacing** .... *640 ac*
- **Ultimate recoverable** .... *5300 to 9800 bcf*
- **Cumulative production** .... *244 bcf*
- **Annual production** .... *127 bcf*
- **Present decline rate** .... *+26%*
  - **Initial decline rate**
  - **Overall decline rate**
- **Annual water production** .... *91,105 bbl*
- **In place, total reserves** .... *10,000 bcf*
- **In place, per acre-foot**
- **Primary recovery** .... *5300 to 9800 bcf*
- **Secondary recovery**
- **Enhanced recovery**
- **Cumulative water production** .... *184,081 bbl*

Drilling and casing practices:

- **Amount of surface casing set** .... *700 ft*
- **Casing program** .... *Set 9⅝-in. surface casing at 700 ft; set 7-in. production casing at total depth*
- **Drilling mud** .... *Salt mud or chemical gel*
- **Bit program**
- **High pressure zones** .... *None*

Completion practices:

- **Interval(s) perforated** .... *Tubb sandstone*
- **Well treatment** .... *Acidize with HCl; some wells fractured with CO_2 or sand-water mixture*

Formation evaluation:

- **Logging suites** .... *Dual laterolog with gamma ray and caliper; compensated neutron and formation density log*
- **Testing practices** .... *Flow-test through separator for 7 days (Johnson, 1983)*
- **Mud logging techniques**

Oil characteristics:

- **Type** .... *Gas is 98–99% CO_2; no oil produced with gas*

(Tissot and Welte Classification in "Petroleum Formation and Occurrence," 1984, Springer-Verlag, p. 419)

API gravity
Base
Initial GOR
Sulfur, wt%
Viscosity, SUS
Pour point
Gas-oil distillate

Field characteristics:

Average elevation .... *4760 ft*
Initial pressure .... *336 to 1082 psi*
Present pressure .... *240 to 390 psi*
Pressure gradient .... *0.13 to 0.40 psi/ft*
Temperature .... *92°F at 2400 ft*
Geothermal gradient .... *0.017°F/ft*
Drive .... *Gas expansion*
Oil column thickness
Oil-water contact
Connate water .... *0 to 45% (Johnson, 1983)*
Water salinity, TDS .... *85,000 mg/L*
Resistivity of water .... *0.132 ohm/m at 77°F*
Bulk volume water (%)

Transportation method and market for oil and gas:

Most gas transported via Bravo pipeline and Sheep Mountain pipeline to enhanced oil recovery projects in Permian basin. Relatively minor amount of gas is liquefied and bottled or converted to dry ice and shipped by truck.